Of stars and men

Reminiscences of an astronomer

The author, as photographed by Mr Charles Lowe, in his office at the University of Manchester on his sixtieth birthday in April 1974. The lunar mosaic in the background (now in the Manchester Museum) is based on photographs taken at the Observatoire du Pic-du-Midi in the course of the Manchester Lunar Programme in the 1960s.

# *Of stars and men*

# *Reminiscences of an astronomer*

### Zdeněk Kopal

Emeritus Professor of Astronomy
University of Manchester

**Adam Hilger**
**Bristol and Boston**

© IOP Publishing Limited 1986

All rights reserved. No part of this publication may be reproduced, stored in a retrieval system or transmitted in any form or by any means, electronic, mechanical, photocopying, recording or otherwise, without the prior permission of the publisher.

*British Library Cataloguing in Publication Data*

Kopal, Zdeněk
   Of stars and men: reminiscences of an astronomer.
   1. Kopal, Zdeněk  2. Astronomers ——
   Biography
   I. Title
   520′.92′4    QB36.K6/

ISBN 0-85274-567-2


Consultant Editor: **Professor A E Roy,**
                       University of Glasgow




Published under the Adam Hilger Imprint by IOP Publishing Limited
Techno House, Redcliffe Way, Bristol BS1 6NX, England
PO Box 230, Accord, MA 02018, USA


Typeset by Mathematical Composition Setters Ltd, Salisbury, UK
Printed in Great Britain by J W Arrowsmith Ltd, Bristol

# Contents

**Preface** vii

**1 The roots** 1
References 30

**2 Awakening** 31
Rendezvous with destiny 55
References 74

**3 Lehr- und Wanderjahre** 75
Charles University 78
Student years in Prague 85
First travels abroad 99
Postgraduate years and first trip to the Far East 111
1936/38: the doctorate and the first postdoctoral year 136
References 149

**4 The American years** 151
Harvard College Observatory and eclipsing variables 172
The war years: broadcaster to occupied Europe 188
War years at the Massachusetts Institute of Technology 191
Post-war years in Cambridge 203
References 224

**5 The Manchester years** 226
Space science and astronomy of the Moon 263
International travels and adventures 304
Summing up 357
Publish or perish 371
References 392

| | | |
|---|---|---|
| 6 | *The binary stars* | 396 |
| | References | 435 |
| 7 | *Astronomy of the past and in the future* | 437 |
| | Will our civilisation survive? | 460 |
| | References | 472 |
| **Index** | | 473 |

# *Preface*

When I received an invitation from The Institute of Physics to write an autobiographical volume I understood it in the following terms: you have now been around for more than seventy years—it is more than fifty since you joined the astronomical community of the world with your first contributions to the subject with which you have remained closely associated ever since—and you have had an opportunity to practise your profession in several countries: what have you learned in the course of these years that could be of interest to contemporary astronomers as well as to a more general public interested in science? The answer was not easy; for my work in those parts of the vineyard which have been my professional habitat for many years is still unfinished, and perhaps not yet ripe for a final summing-up. A reluctance stemming from this source has, however, been gradually weakened by a realisation that none of us will be here for ever; and that undue delay might bring about a permanent loss of information—of possible interest also to others—which could thus be confined to undeserved oblivion; and the result of this surrender is the present book.

In some ways, perhaps, my qualification to undertake this task may go beyond my number of years; for I happened to be born in a country which could have (then) been regarded as something of an astronomical backwater; and so through my teachers in Prague (who, as was the custom in those days, were mostly gentlemen of more than canonical age) I was still able to establish a more intimate contact with astronomy of the nineteenth century—the time when my masters themselves were educated. Thus, in this sense, my more-than-literary acquaintance with my subject extends, in effect, to almost a hundred years; and this is about one-quarter of the time that has elapsed since the discovery of the telescope.

The contemporaries of my age cannot indeed complain of having had a dull time. More than once in the course of my life I could regard myself as a 'stormy petrel'. Just think how many times my fate hung in the balance! Less than four months after I was born World War I broke out; and although I was too young to be called upon to serve in it as a combatant, merely to survive was no mean accomplishment for a child suffering from undernourishment as a consequence of the continental blockade in the first four years of its life. As is proverbially the case, it is the beginning which is difficult; and (as will be detailed in Chapter 2) my struggle for survival commenced, in fact, before I could walk.

Years later, in the summer of 1936, during my first trip to the Far East (the first of several) I experienced the first black-out of Tokyo in anticipation of World War II, and watched with fascination (from the upper floors of the Mitsukoshi building at Nihonbashi) the first attempt to protect Tokyo from air attacks by dissemination of artificial fog. (Plenty of good did these exercises do for the Japanese when war actually reached Tokyo some years later—perhaps as much as gas-masks issued to the British population in September 1939.)

The 'China incident' of 1937—which really started the war in the Pacific—was only months in the future during my first call on Shanghai at the age of twenty-two; and I still had a chance to visit there the famous local country club where 'dogs and Chinese were not admitted'; and whose members (of Somerset Maugham vintage) criticised their (refugee) Russian fellow-members for 'talking with their servants'. To me these were no stories made up by novelists, but observed facts. In the years which have elapsed since that time, I have seen many wrongs rectified, but others have arisen in their place; and to this day I do not feel competent to draw a balance.

But—to return to the past—fortune had also some smiles in store for me. The reader of Chapter 2 of this book will find out by what a slender thread my fate guided us out of Europe, just before my native country of Czechoslovakia was raped (and for six years occupied) by Nazi Germany, and led us to the United States which became our second home and adopted country. Although I eventually returned to Europe, and spent the second half (certainly the best years) of my life in Britain, my transatlantic connections have never been severed. With the sudden advent of the Space Age in 1957, I became an almost regular transatlantic commuter for several years till 1969 when the first men landed on the Moon; the interested reader may find out more about this in Chapter 5. Jet transport (which came into commercial use ten years before) made it possible to be in different places at almost the same time; and my travel bureau figured out that, in those days, my average speed over the year was about 40 miles per hour.

## Preface

Now, many years later, this first heroic period of the Space Age is receding rapidly into the past, and its accomplishments are only too easily being taken for granted by the younger generation who missed the thrill experienced by those who saw it happen. Yet I, for one, am satisfied that the years between 1959 and 1972—since 13 September 1959 (when the first man-made missile of Russian origin crash-landed on the Moon) to 11 December 1972 when the last two American astronauts left the lunar surface to return to the Earth—will remain forever a glory of mankind and pride of our turbulent century; certainly to be remembered long after all the more ephemeral events of our contemporary political history are justifiably confined to oblivion.

But even for me—old as I was to take any active part in cosmic astronautics—my role as a 'stormy petrel' was not yet over. Thus on my first transatlantic crossing in post-war years by air in October 1950, our plane (a Boeing Stratocruiser) lost two engines out of four in flight between Gander (Newfoundland) and Shannon (in Ireland); had they both been on the same side, ditching would have been inevitable. Or—during my later years at Manchester—my students and I left the Kottamia Observatory in the Egyptian desert (where we were taking photographs of the Moon with the observatory's 74-inch telescope) in 1967, shortly before the outbreak of the Seven-Day War between Egypt and Israel (but near enough to see our quarters and much of the observatory riddled with bullets); and, again, some years later, when I visited Iran as Pahlavi Lecturer in late spring of 1977, the Islamic Republic of Ayatollah Khomeini was already around the corner. Indeed, we would have to go back at least to the seventeenth century (in the European theatre) to encounter similarly tumultuous times; and the end of the present century may yet exceed them all!

This is all bound to be reflected in the pages of my reminiscences. The first chapter ('The Roots') should introduce to the reader the stock I came from, and is based largely on information I have gathered about my ancestry. It is not until the second chapter ('The Awakening') that I myself shall take the first tentative steps on the scene: to recapture the way of life still prevalent during the years of my youth (now irrevocably gone with the wind); and to explain how I became a student of astronomy. My subsequent career in this vocation, at home and, later, abroad, will be followed up in the subsequent three chapters. As their topics unroll, this writer himself will gradually retreat from the limelight, and increasingly restrict his role to that of an observer, reporting on what he witnessed in the course of his life.

Needless to stress, this part of the book cannot claim any particular completeness; and its earlier chapters may appear to some to be perhaps disproportionately long in comparison. There is, however, a design in this layout. For Chapters 1 to 3 deal with times which are no longer remembered by many; while the subject matter of Chapters 4 and 5 ('The American Years' and 'Astronomy at Manchester') still lives in the memories of my contemporaries; and may be (partly has been) recorded in reminiscences of their own. And, last but not least, certain topics mentioned in them may not yet be ripe for fuller disclosure. The reader desirous to know more than what could only have been hinted at between the lines is referred to the writer's correspondence of his lifetime (preserved for those who wish to consult it in the archives of the Manchester University Library) and other sources.

The same is even more true of Chapter 6 ('The Binary Stars') and parts of Chapter 7, which deal with the field in which I made my principal contributions in my lifetime. These are no longer based so much on personal reminiscences, as on other sources which I collected to introduce these subjects to the reader. To place their history in proper perspective has indeed never been as necessary as it is now, when events race past us at a speed which makes it increasingly more difficult to do so from momentary snapshots of everyday experience, and offer little encouragement to pause in quest of the sources.

Mere reminiscences of the past are, however, not enough to this end; and would lose much of their justification if they could not also be used as a basis for a broader historical survey of the past of our subject—insofar as a knowledge of this past can enable us also to anticipate the future. The aim of these recollections should not only be to bring home to us how much the lives of astronomers have changed in a few generations, but also to try to forsee whither all this may lead us in the future, and how much the lives of our grandchildren (be these astronomers or not) are likely to differ from the way of life at the present time. This we shall attempt to do in the concluding parts of Chapter 7. If, in doing so, we can offer any significant contribution towards this end, the time spent in collecting these reminiscences may perhaps have been justified. Of this, however, the reader—and, ultimately, the future—must have the last word.

In raising the curtain on the first seventy years of my life, I should like to thank all my friends and acquaintances (mentioned as well as unmentioned in the text)—above all, my teachers, colleagues and students—for making the subject of this book a story in which I often dwell in my recollections; and parts of which may perhaps be of sufficient interest to share also with others. Particular thanks are due to Mrs Ellen B Carling who typed the text of this book (as she has done with so many others for me in the past) for the press; and to Mr Brian McMahon whose careful editing of the script helped materially to eliminate many obscurities and bibliographical slips.

# Chapter 1

# The Roots

Tradition has it that we all derive our origin from Adam and Eve, though no-one has been able to trace his ancestry all the way back to Paradise. The present writer is indeed no exception; and the best he can do is attempt to follow the roots of his family in that part of the world in which his ancestors lived for at least several hundred years. Such an inquiry will take us to the very centre of the European continent, to a country known to the world in the past thousand years as the Principality and (since 1085) the Kingdom of Bohemia. This name was inherited by that land from its first historically known inhabitants—the Celtic tribe of the Boii who (as we learn from Tacitus) inhabited that part of the world in the last few centuries before Christ. They left nothing behind them but a few local names of rivers or mountains, which they bequeathed to subsequent inhabitants of their former homeland; and what happened to them since their disastrous defeat by the Dacians around 60 BC no-one knows for sure—they were probably dissolved in the great melting-pot of the lands traversed by the Danube without visible trace.

After the departure of the Celtic tribes the land of my ancestors did not remain empty for long; and in the first centuries after Christ it is known to have been inhabited by different Germanic tribes which had already come in direct contact with the Roman world at the time of Augustus. Their land had, however, never become part of the Roman Empire; and the Roman legions under the emperor Marcus Aurelius only temporarily set foot on the eastern province of the future territory of Czechoslovakia, far from the regions which, centuries later, became my ancestral home.

The Germanic tribes disappeared likewise from the future Bohemia during the decline and fall of the Roman Empire; and darkness descended once more on central Europe for a few hundred years— illuminated only by a few momentary glimpses recorded in chronicles of western Europe. When, however, the curtain rose again on the stage of central Europe in Carolingian times, the land had already changed hands once more; and the former Germanic tribes of Markomanni or Langobards disappeared once more without obvious trace to make room for Slavic tribes whom the last migration of nations brought over from the east to Europe in the fifth and sixth centuries; and which at the time of their maximum expansion managed to populate Europe almost to the river Rhine.

These Slavs did not, to be sure, manage to hold on to all their conquests for very long; and were eventually pushed eastward to the river Elbe. The Bohemian basin is, however, encircled by mountains which were then covered by impenetrable forests, and which offered its population a sufficient protection even from the armies of Charlemagne or from the nomadic Avars who invaded central Europe from the Asiatic steppes in the sixth century; so that when the first missionaries entered the land in the ninth century with the Cross in their hands to claim it for Christ and recorded what they saw, they found only Slavic tribes populating its fertile lands and valleys, rich with hunting game in its endless forests; and somewhere among them were already my distant ancestors.

Does this make me a Czech? Purists may doubt it. As is well known, the Czech tribes settled mainly in the western part of Bohemia, along the river Ohře (west of the Elbe), and only gradually gained ascendancy over the rest of the land, to which they gave also the first national Přemyslide dynasty. The ancestral cradle of our tribe stood, however, in north-eastern Bohemia, which at least till the end of the tenth century had its own tribal chieftains, with the family name of Slavník; according to the late Professor Dvorník of Harvard University, these were not Czechs proper, but Croatians, allied racially with the Persian Ants (the remnants of which still survive under their old name in southern Europe). At any rate, the migration of nations was a great melting-pot. Different tribes which settled in the Bohemian basin intermingled thoroughly in the course of time; and if the Přemyslide dynasty gave the new-born nation its first national saints in St Wenceslas and St Ludmila, the Croatian Slavníks gave the world a scholar-priest in St Vojtěch (Adalbert by his confirmation name), the second bishop of Prague, friend of the Emperor Otto III and a worthy member of the Ottonian renaissance,† around the turn of the second millennium.

†In one of the hagiographic writings extolling the virtues of St Vojtěch, the legendist stressed the fact that this saint 'did not waste his time playing chess'. Now the game of chess—invented in Iran some time in the seventh century of our era—was not much known in Central Europe in those days; but if it was known in St Vojtěch's entourage as the legendist seems to imply, Professor Dvorník's views on the Iranian (non-Slavic) origin of the Croatians settled in north-eastern Bohemia would receive further indirect support.

But to return from saints and scholars to my own ancestors: according to what we know about them (or can infer from scanty records), their cradle can be located within a relatively small corner of north-east Bohemia (well within the Slavník territory). This land, to be sure, was inhabited by man long before the advent of the Czechs or Croatians in their present homes; or even of the Markomans or Boii in the centuries before. A mound exists in the fields ('Na Tobolici') which were (until quite recently) owned by our family, showing evidence of human settlement which (according to a distinguished archaeologist, Academician Filip, whose cradle also stood in the same part of the country) goes back at least to 1500 BC. Unfortunately, the inhabitants who lived there so long ago left behind no clues as to their identity (at least, none that have so far been properly excavated); and certainly had nothing to do with the population waves that came and went in subsequent centuries.

In more recent times, that part of the country which eventually proved to be our ancestral cradle has been a small corner of north-east Bohemia, roughly between the (present) towns of Jičín and Turnov, referred to in tourist guidebooks as the 'Bohemian Paradise' (for reasons better known to tourists than to its inhabitants). It is too far from the river Elbe for the latter to share much of the prosperity of the fertile lands towards the south; while to the north it is bordered by the Krkonoše mountains whose chain can be clearly seen on clear days on the horizon some 40 km distant, with its highest point, 'Sněžka' (i.e. 'the snow-capped one'), true to its name for the better half of each year.

For many centuries this land has been populated by yeoman farmers tilling land unsuitable for large latifundia, but providing subsistence enough for those who reclaimed land from the wilderness, probably in the latter Middle Ages—with my forefathers among them. The ancestral village from which my father was descended is called Hřmenín—i.e. one founded by someone called Heřman, possibly identical with Heřman of Ralsko, entrepreneur ('locator'), who founded the town of Turnov at the northern limits of the Bohemian Paradise in the second half of the twelfth century. Certain it is that this Heřman's son, Markvart of Březno, founded a large village, named after him Markvartice (only about one hour's walk from Hřmenín), with a church and parsonage to which Hřmenín belongs to this day. It is that church (several times re-built) in which my parental ancestors were received or confirmed in the Christian Church, married, and—in the fullness of time—found their final resting place in the graveyard around it. My grandmother Ann was the last one of our line to be laid there in 1909 to await the resurrection; and my cousin Josef, still living in our ancestral farm in Hřmenín (though no longer owning its land), was married there in 1934.

## 4   The roots

The reason why I dwell on these scanty facts in this place is that the farming people from whom I stem have traditionally been sedentary folk, who spent all their lives from cradle to grave in the same place where they were born; and (before the advent of mechanised travel) but seldom ventured away from home for more than a day. It is within this perimeter that they traditionally found also a spouse; and they were too closely bound to their land to seek adventures in a wider world. It is, in fact, more than likely that my ancestors never left their district for fourteen centuries since the 'long march' of the last migration of nations brought them from their prehistoric sites, far to the east, to their new homeland; which the ceaseless work of countless generations has converted to what the world knows today as Czechoslovakia. My father was probably the first member of his clan since then to cross the ethnic frontiers of his fatherland, at the beginning of this century, in quest of higher education; and I am the first to have made my permanent home abroad. In the past fifty years I have travelled in many countries, and visited lands never heard of by my ancestors. But as I close my eyes in the evening of my life, I still see my native country, and hear its distant rhythm reverberating from the past, as it was when I was young—not as it is today.

But to return to the beginning, the names of Hřmenín or Markvartice (like those of most other places in the region of the Bohemian Paradise) recall land converted from the wilderness to arable fields in the latter Middle Ages; and colonised in the course of the Germanic *Drang nach Osten* in the thirteenth and fourteenth centuries of our era. However, when we first encounter the names of these villages in written records, Hřmenín at least already belonged to someone other than its original founders: namely, to Lord Arnošt of Pardubice (1305–1364), Archbishop of Prague between 1346 and 1364, and Chancellor to the Emperor Charles IV (1316–1378). The latter—one of the greatest rulers Bohemia ever had—left his mark on every corner of his kingdom (with Prague as its capital and the Emperor's seat of residence); and his deeds still dominate the skyline of that city. Not the least of these had been, in 1348, the foundation there of a *studium generale*—the first university in central Europe, now in the seventh century of its existence, and carrying the name of its founder to this day.†

---

† Of the three most important 'status symbols' in medieval Christian Europe—the seat of the emperor, papacy, and the university—Prague acquired two in the second half of the fourteenth century; and when Pope Innocens VI died in 1362, only a lack of money prevented Charles IV from securing the third—papacy—for Archbishop Arnošt (*prelato di grande autorità, esperto delle cose del mondo*, as he was described by contemporary Italian authorities) and from making (albeit temporarily) Prague also the residence of a successor to St Peter.

Archbishop Arnošt, himself a graduate of the University of Bologna, became (by virtue of his ecclesiastical office) also the first Chancellor of the newly-founded university; and as such he gave the village from which the Kopal family sprang a first opportunity to contribute also to human civilisation. For (as we learn from the late Professor Václav Chaloupecký, a son of the same part of the land and a good friend of my father) in 1349 Arnošt decreed that all income from the villages of Hřmenín and Važice (a sister-village about one and a half miles northeast of Hřmenín) be assigned to support a chair of theology at Charles University; and placed (until 1356) both villages under the jurisdiction of the cathedral chapter of Prague (cf Chaloupecký 1941). The amount involved was not large (corresponding to about 20 000 present Czechoslovak crowns, to be payable in silver); but was seven times as large as that levied on Važice; from which one should conclude that, at that time, Hřmenín was already a village of some substance.

After Hřmenín was detached from the cathedral chapter of Prague, its name vanishes again from the records; and its inhabitants no doubt shared in the political and economic upheavals of the country as a whole during the Hussite wars. Its continued existence is, however, attested by picture-points representing it on several contemporary maps; but such points tell us, unfortunately, nothing about its inhabitants. During the Thirty Years' War all territories around it belonged (for a time) to the Duchy of Friedland, eked out from largely confiscated Protestant estates by the grasping hand of Albrecht, Duke of Wallenstein (1583–1634), who in a few years of his meteoric career advanced his position from that of a local provincial noble to the Emperor's *generalissimo* and ill-fated *shogun* behind the throne, and whose equally spectacular downfall (as, according to him, his prior meteoric rise) was caused by his astrological predilections and faith in the stars.

The capital of Wallenstein's duchy was the town of Jičín, some three hours north of Hřmenín on foot; and we shall soon make its closer acquaintance in quest of my ancestors on the distaff side. Let us mention only briefly here that when my (paternal) ancestors brought their produce for sale on the market of Jičín square, they could have chanced to see (or even meet) a small and unusual-looking man, Johannes Kepler by name, walking (with his mind, we suppose, on the stars) to the gate of Wallenstein's palace (see figure 1.1) to wait on his patron. Poor Kepler! In spite of his best intentions, he did not bring his patron much luck (and got even less from him in return); and whether his mind was on the stars or on his terrestrial tribulations, he would have scarcely had time to notice a humble farmer offering the gifts of nature on the square. But the square was still there, and so was Wallenstein's palace, now housing less exalted functionaries (such as tax-collectors and other officials of local administration), when a little boy was jumping over its cobblestones in his tender years—almost seventy years before he sat down to write these recollections; and where, on 28 October 1918, together with older members of his family, he watched with excitement the local celebrations of the return of Czechoslovak independence, lost almost three hundred years before as a result of an ill-fated struggle in which Wallenstein played to perfection the role of a quisling.

**Figure 1.1** The town centre of Jičín, where I spent the first four and a half years of my life. In the right-hand corner of the square can be seen the (well-preserved) palace of Albrecht of Wallenstein, Duke of Friedland, who during the Thirty Years' War was (at times) also the patron of Johannes Kepler.

After the downfall of this ambitious *condottiere*, curtains descended once more on the lands formerly ruled by him, and darkness remained pretty well complete on the goings-on for more than one hundred years. There are several reasons to account for this. First was the fact that, following the conclusion of the Thirty Years' War in 1648, the country (formerly predominantly Protestant) was subjected to forcible counter-reformation, intended to reconvert its subjects to the Catholic faith. This process could, however, proceed only slowly for a lack of native priests; and more than half a century elapsed before all parishes could be occupied by local incumbents who could carry on records of the *status animarum* in their respective spiritual bailiwicks. And, secondly, most parsonage buildings which housed the church records were not built from stone—as they were in England—but from wood. Unlike in England (which was largely deforested at an earlier stage of its history), timber was still plentiful in central Europe at that time, and continued to offer cheap material for building. Only the castles of the nobles or burgher town houses were built of stone, thus ensuring a safe haven for family records; while in the country the parsonage buildings were burnt down (together with the rest of the village almost in each generation) until the second half of the eighteenth century, when the central government in Vienna enforced the requirement that at least the vital statistics of the local population had to be safeguarded in churches built of stone—and thus saved them for posterity (largely, of course, for purposes of taxation rather than genealogy).

But whatever the motives may have been, the preservation of such records ensured that, from the second half of the 18th century, I can at least look some of my ancestors in the face, and introduce them to the reader by name. The Kopal name emerges from the records as that of farmers who lived in Hřmenín from generation to generation, holding the same land; as I previously mentioned, my cousin Josef and his wife Mary live there to this day. A few foreign names (apparently of Swedish origin) are encountered among the inhabitants, probably of deserters from the armies of the Swedish hero-king Gustaf Adolf, which repeatedly invaded Bohemia during the Thirty Years' War. This happened particularly in 1639, when that part of the country was occupied by Swedish troops under General Banér (his wife and children lived for a time at the neighbouring castle of Kost); and again in 1643 when Bohemia was invaded by the armies of Field-Marshal Torstenson. Their troops remained in occupation of major parts of the land for some time; and it is more than likely that, on both occasions, the settled life of a farmer appealed to at least some of their numbers better than the lot of a mercenary soldier. The land was to be had for the asking (during the Thirty Years' War the population of the country dropped by more than a half). It is probable (judging from certain local names still in our native village) that some of these stragglers settled in Hřmenín; and as people used frequently to intermarry in the same village, it is not impossible that some of their northern blood still circulates in my veins.

But, in spite of this injection of foreign blood, fiscal records accidentally discovered (cf Teplý 1930) disclose that, in 1719, the village of Hřmenín consisted of only sixteen full-size farms (*lány*)—one of which (as will transpire later) was held by my ancestors—with a smaller number of the subdivided ones, which would correspond to a population of only 100-150 souls, no doubt much less than the village had in the days of Archbishop Arnošt more than 250 years before, and an eloquent testimony to the depopulation of the countryside during and after the Thirty Years' War. But, by and large, the ethnic character of the inhabitants of Hřmenín emerging from the oldest extant records appears to be solidly Czech; and their names, like our own (the word Kopal signifies in Czech a man who toils on the soil), relate to rural life.

Ours is not a very common family name in that part of the country; but neither is it very scarce: another of its current bearers of astronomical interest is JUDr Vladimír Kopal of the Czechoslovak Academy of Sciences in Prague—an expert in space law and currently a prominent functionary of the International Astronautics Federation (for whom I was more than once mistaken in the past). We know each other, of course, and have established that Vladimír Kopal is by fifteen years my junior in age; and that his family comes from the vicinity of the town of Turnov some 30 km north of Hřmenín. Perhaps because of this distance, the relation of his branch of the family to ours can no longer be traced. However, the spread of our family name over a rural area of such dimensions strengthens the surmise that the Kopals have been settled in that part of the country for a very long time; and that our ancestors were essentially farming folk who had not been lured away from their land for other pursuits. By their station, they could be described as 'yeoman farmers—very small beer', as old Jolyon Forsyte once described his own progeny.

Very small they perhaps were not; for the first one historically attested by extant records—Jan Kopal—was one of the largest landowners of the village. His year of birth is uncertain (because the relevant records have since perished); but he died in 1832 and left his land to his son Jan, born on 2 September 1802, who died on 24 March 1891 at the ripe old age of almost 89 years. This second Jan Kopal, still remembered in his old age by a grandson who happened to become my father, married in 1852 (rather late in life) and produced two sons: Jan and Francis; of whom the latter, born on 23 August 1854, was my grandfather.

When the time came in 1877 for my great grandfather (then almost 70 years of age) to leave his holdings to posterity, he did so in a way most unusual at that time: he divided the land between his sons in equal parts, thereby reducing their economic and social positions from those of *kulaks* to *seredniaks*—middle-size farmers whose land could no longer be subdivided further without irreversible economic consequences. If more than one son was born to such farmers, the oldest succeeded to the land, and the younger ones were to receive entail which would enable them to be married off elsewhere; or, if they were bright enough, they were sent to schools and supported there until they could make their living, but lost any further claim to the family estate.

My grandfather Francis, on coming to his part of the inheritance, married on 30 January 1877 Anna Pažoutová, a farmer's daughter from the same village, who was born on 20 July 1851 and was thus three years older than her bridegroom. Her father was, however, not born in Hřmenín, but in the neighbouring village of Skuřina, and he probably married into Hřmenín as a younger son of a family of some distinction. For, in the second volume of his delightful *Kniha o Kosti* (*Book on the Kost Castle*), describing the history of that part of the country since the Thirty Years' War, the great Czech historian Josef Pekař (himself a son of the 'Bohemian Paradise') reproduced in facsimile some important legal documents dated 1752, and certified (among others) by Václav Pažout, chief political officer (*Haupt-Richter*) of Markvartice—who, I reckon, would by my great-great-grandfather on the distaff side. Unfortunately, I could never ask grandmother Ann what she knew about him; for she died on 25 August 1909, five years before I was born: the only one of my grandparents I never knew.

Grandfather Francis and grandmother Ann had several children. The oldest, called likewise Francis after his father, was born in 1878 (he died in 1950), and, in the fullness of years, was to take over the family farm. No other child came for five more years until, on 25 April 1883, another boy was born to them who, at his baptism in Markvartice, received the name of Josef (the names Francis and Josef then being as widespread in the land, and for the same reasons, as Edwards or Georges were in England); and it was he who eventually became my father.

Many years, were, of course, to elapse before this became the case. The fact that he was second-born ruled him out of succession to the farm; and, truth to tell, he would have had even less inclination to that calling than his brothers. For, many years later, when father became a university professor and acquired a villa with quite a large garden in a suburb of Prague, he never evinced much interest in gardening; and I am afraid that I have only followed in his footsteps. But several circumstances prevented him from embracing the ancestral calling; and the most important one was the precocity which he already evinced at primary-school age, and which did not escape the notice of his teacher and of the priest. It did not take much persuasion on their part for father Francis to agree to send his bright son in his eleventh year to study at Latin schools in the regional town of Jičín, some three hours (on foot) from the parental village of Hřmenín, as the commencement of a career which was to lead young Josef to the summits of academic life in his country.

The *Gymnasium* of Jičín, together with one in Mladá Boleslav, were both founded by Wallenstein in the first third of the seventeenth century during the short rise of his meteoric career, and remained elite schools of secondary education in north-east Bohemia ever since that time. The one in Jičín was originally destined to be an academy of university rank (such as Wallenstein founded also in Rostock); but his downfall in 1634 cut short his plans, and the schools founded by him were handed over to the Jesuit (and, later, Piarist) orders of the Catholic Church. And several generations of their spiritual rule left a mark on these institutions which survived their secularisation in the second half of the nineteenth century, when they were taken over by the state. Although Wallenstein, founder of the Jičín *Gymnasium*, admonished its rector in 1625 not to admit too many *tölpische böhmische Janků* ('stupid Czech Johnnies'), that directive of the arch-renegade rested very lightly on the recipient and his successors.

For the clerical heads of Catholic schools were only too well aware that brains should be recruited from all social strata to keep their orders intellectually alive; and the less claim their pupils might have to preferment by virtue of birth, the more devoted they might become to their Church regardless of origin or nationality. Indeed, one of the early professors at the Jičín *Gymnasium* was the Jesuit Bohuslav Balbín—a great Czech patriot—whose name was carried by the street in which I spent the early years of my childhood in Jičín (see the next chapter); and, in the nineteenth century, the schools at Jičín and Mladá Boleslav could list among their alumni such outstanding scholars as Jan Gebauer (1838–1907), grand master of Slavic philology; or Josef Pekař (1870–1937), the greatest historian of the Czech nation since František Palacký. In the days when my father entered the Jičín *Gymnasium*, its outstanding scholastic tradition was still very much alive; and it educated a number of pupils who eventually ascended to university chairs (and some of whom perished between 1939 and 1945 in German concentration camps) within my lifetime.

But before I follow, in my recollections, my father to the University and beyond, let us record a few words on the town in which he not only spent his teens and acquired his secondary education, but also his future wife who became my mother. As her whole family (and all my ancestors on the distaff side) came from the same area and had their roots there, their geographic cradle should be described in a few words.

The region around Jičín, an area bordered by the medieval castles of Brada to the north and Veliš to the south, appears to have been inhabited by men since the end of the last Ice Age; and archaeologists have identified within its confines not less than four groups of different populations in the past five thousand years before the advent of the Slavs in the sixth century AD. The name of the town (as Gitzin) is first documentarily attested in a donation which Queen Guta (daughter of Count Rudolf of Habsburg in Switzerland—the first of his line to be elected Holy Roman Emperor in 1273—and youthful wife of the Bohemian King Václav II) made to a church in Lysá (another one of her dominions) in a document dated 1 August 1293—one of the oldest extant documents of that town. Queen Guta ( = Jitka, as she was called by the Czechs) died in 1297 (she was betrothed to her royal fiancé—not much older than herself—at the age of only sixteen years), and her

*History of Jičín* 11

property reverted to the King. It is, however, at least possible that the name Gitzin = Jitčin Hrádek (i.e. castle of Jitka) derived from her name.

After her death, the property was acquired by a local lord Beneš of Vartenberk (of a noble family descended from Markvart of Březno, whom we have already encountered) and remained in the family, which played a great role in the nation's history at the time of the Hussite wars, off and on for almost 150 years; having been granted the status of a township around the year 1300. Between 1348 and 1358, at the time when the University was founded in Prague by the Emperor Charles IV, Jičín belonged also to that ruler; and in 1416 his son Václav IV granted the town a royal charter, rendering its burghers freemen subject only to the King. This privilege expired during the Hussite wars, when the town came for the first time under the rule of the Valdštejn family; but after 1452 it reverted again to the King—the well-beloved Jiří of Poděbrady. After his death in 1471 that king's son Hynek (Prince of Münsterberg) sold Jičín back to the local nobles; and through the hands of the Trčka and Smiřický magnates it passed (by inheritance) to Albrecht Eusebius of Valdštejn—Schiller's Wallenstein, of whom we have already heard.

When my father came to that town as a boy of eleven and entered its *Gymnasium* as a pupil of its lowest form (*parva*), this all was already in the distant past; but its monuments were still visible wherever one turned. To be sure, the Veliš castle to the south, once the seat of Count Henry Matthias of Thurn—that firebrand of the first part of the Thirty Years' War—was in ruins (having been destroyed by imperial military engineers shortly after that war, in order not to offer—as it did repeatedly before—a strong-point for the invading Swedes); but Wallenstein's palace on the town square or the Jesuit college of the seventeenth century stood there in very much the same form as they would have been seen by Johannes Kepler; in fact, within the walls of that college in Balbín Street (rebuilt, of course) I was destined to spend my early childhood during World War I.

But more of that in the next chapter. At this point in our narrative, my father is still in the first form of the Jičín *Gymnasium*; but he did not stay there any longer than school rules dictated, climbing rapidly up its academic ladder till *octava*, which he reached in 1901, and took with high honours his *Abitur* which entitled him to enter university. The climb was rapid, but not without incidents; for twice in its course father was threatened with expulsion from the school, which would have been a real catastrophe for his future. On the first occasion, he was reported to the headmaster as having taken a swim in a local pond, clad in nothing more substantial than bathing trunks! Such behaviour was, in those days, deemed both frivolous and demeaning for pupils of Latin schools; and as a punishment, he was required to memorise I do not know how many pages of Homer's *Illias* (*Odyssey* would have been more appropriate for my father's tastes; but it encouraged swimming!); and to promise solemnly to abstain from such frivolities in the future—a promise which my father did not keep any too well even when, as senior university professor, he had a right on academic occasions to be addressed as *spectabilis*; for until quite an advanced age he enjoyed aquatic sports.

The second incident which could have nipped my father's academic career in the bud was, however, of a more serious nature. The late 1890s—when father was in the upper forms of his school—were turbulent years in the literary and political fields all over Europe, and Bohemia was no exception: but instead of cultivating decadence, its young intellectuals were politically-minded, and could never forget the subjugation of the interests of their lands to those of the Habsburg dynasty which conquered Bohemia during the Thirty Years' War. 'Long are the memories of the people who suffered injustice', proclaimed the seventeenth-century Jesuit Bohuslav Balbín, who had been professor at the Jičín *Gymnasium* 250 years before; and his words were not forgotten throughout the centuries any more than the Irish people have forgotten William of Orange; and the battle of White Mountain near Prague, lost in 1620, was being remembered by the Czechs around 1900 in the same way as the battle of the Boyne in Ireland (or the massacre of Glencoe in Scotland). In the revolutionary year of 1848 the romantic element among the Czechs too set up their 'Repeals', which were suppressed (but not extirpated) by subsequent reaction. In the 1890s, grandsons of the revolutionaries of 1848 were once more picking up the flag of state rights and national independence, and for the last time, for its hour had almost come; and the first President of the Czechoslovak Republic was already a professor at Charles University, educating the youth for their future tasks.

These aspirations indeed materialised twenty years later; but none too soon for young students of my father's age. No wonder, therefore, that father must have been itching to join others in the melée; though the only weapon he could wield at that time was his pen, with which he soon began to contribute articles and essays to the local (and, later, national) press. Unfortunately, to do so was not then allowed to pupils of the *Gymnasia* without the prior approval of their headmaster. The holder of the post at that time (Adam Fleischman by name) was no doubt a good and solid citizen; but like probably many of his contemporaries of that age, he believed only in 'moderate progress within the limits of the law'—a feeling not always shared by his young charges. Anything beyond these limits (especially if it did not agree with his own opinions) he would regard as a challenge to his official as well as spiritual authority—a trespass which he was unwilling to tolerate. As his opinions were well known, those with different ideas about the rate of progress could make their opinions known only under a pen-name.

My father, in his teens, was only too keen to avail himself of such a loophole (suggested to him probably by others); but, unfortunately, the pen-name he chose ('—opal') was too transparent to conceal the name of the real author for long. The outcome could have been serious if the entire faculty had not supported father's cause (many of them may have sympathised with the political views of their pupils more than the head); and so the Damocles' sword of expulsion which hung for a time over my father's head was eventually changed to a mark on the school certificate, giving father *e moribus primam, cetera eminenter*.

I recall this story mainly to remind my young readers of the way in which 'freedom of the press' was treated 90 years ago, at least for school students (and, of course, nothing was heard by them about student unions or strikes). The first brush with authority (about the swimming) left no noticeable mark on my father's subsequent career; but the second episode may have taught him something. For as far as my own memories go, as years went by father developed an increasing degree of tolerance towards opinions expressed by others—at home, at the university or in public life—and any manifestations of zealotry were apt to produce on his part only a tolerant but sceptical smile. Like the father of Silvestre Bonnard—that delightful character created for us by the pen of Anatole France in the last century—my father too was very hesitant in putting his full faith in any particular view, and always qualified his beliefs by many reservations, ending with a sceptical: 'but what do we really know about it?'

A gradual transformation of the simple farmer's son into a scholar who would have been at home in the Age of Enlightenment took, of course, many years to mature; and many events occurred in the meantime. The most important one (for me) was the fact that, before father obtained his *Abitur* ('*matura*') from the *Gymnasium* to continue his studies at the faculty of philosophy of the Charles University in Prague, he became engaged to be married to a young lady at Jičín (one year younger than he) named Ludmila Lelková; who eventually became my mother. It is, therefore, time to introduce this branch of our family in our narrative—which is particularly true of my maternal grandfather, Josef Lelek (1860—1930); not only because my earliest years were spent under his roof and much influenced by his strong and remarkable personality, but also because it was through him that my young mind was bent towards nature and its wonders even before I went to school. It was indeed a combination of gifts passed on to me by this grandfather, together with those from my father, which determined very largely my life as described in subsequent chapters of these reminiscences.

As far as extant records reveal, the Lelek† family also stemmed all from the neighbourhood of Jičín, and its tree can be traced there to the middle of the eighteenth century. Unlike the Kopals, the Leleks were no farmers, but artisans (or shop-keepers on the distaff side) and town people. The first historically attested one is Josef Lelek, a tailor in Trotín (north of Jičín), born still in the time of Maria Theresa, who departed to his ancestors after the fall of Napoleon. His son Josef (born on 26 December 1802 in the same town) became a cabinet-maker—a calling in which he was followed by his son Francis Josef, born on 2 April 1830 in Nová Paka (a town north-east of Jičín), who moved to Jičín some time in the course of his life and died there on 12 September 1886. His oldest son, likewise Josef (really, how many Josefs have we encountered in our pedigree?), born in Jičín on 26 August 1860, eventually became my maternal grandfather.

The cabinet-making in that family ended with my great grandfather Francis Josef; for (perhaps because the family moved from Nová Paka to Jičín with its old cultural tradition) his son was sent to schools to become a teacher. This my grandfather did at the Jičín State Teachers College (where Karel Václav Rais, subsequently a well-known writer, became his colleague); and since graduation spent his lifetime in the state teaching systems, first at primary, and later at secondary level; but never at Latin schools, for which a university diploma was required, and this my grandfather never possessed.

However, grandfather made up for this lack in other ways, some of which make us hold our breath today. Thus, after graduation from college, he set out (together with a school-mate of his, I believe Hartl by name) to travel on foot to the Holy Land—not so much out of piety (throughout his life grandfather was known as a free thinker), but for adventure. Adventure and excitement the two boys must have had aplenty as they pushed their way through Hungary and the Balkans to the Bosporus. They slept mostly in the open (it was summertime); for such money as they had was needed for food. But they were free at least of another inconvenience which started to bedevil travel when the Earth's human population exceeded its first thousand million: namely, passports and visas. In those days one could travel without them even to Turkey; the only European country where they were needed, even at that time, was Czarist Russia.

At any rate, my itinerant grandfather and his friend entered Turkey

† The name Lelek is the Czech word for the bird known as the nightjar (*Caprimulgus europaeus*).

without any problem, and continued to plod on through Asia Minor towards Palestine. That they did not get there was not due to any weakening in their determination, but rather to the fact that the outbound journey took the young travellers longer than they needed for their return—for they had to be back for the commencement of the new school year! And so they turned around somewhere near Ankara to retrace their steps, which they did successfully and on time, with but one incident. When they set out on their long journey, the young pilgrims were cautioned that bears were still at large in the Balkan forests; and they armed themselves for a possible encounter with such beasts with a military pistol (of which there were many around from the 1866 war between Austria and Prussia that had passed through Jičín not long before).

I doubt if my grandfather or his friend had ever fired a pistol before in anger; nor did they have to do so on their long return journey. When, however, they were already within sight of their paternal roofs in Jičín, they thought it would be a shame if they did not at least try to hear what it sounded like—and so grandfather fired the pistol in the air. In no time they were, however, apprehended by a gendarme, who fined them 'for making unnecessary noise'!

After the vacation which ended with this unnecessary noise, grandfather embarked on his vocation of schoolmaster which started in the neighbouring small town of Železnice (since better known as Spa Železnice, because of mineral waters which were discovered in its proximity); and there on 13 September 1883 he married Anna Neumannová (born on 13 December 1869 in Železnice, died on 20 March 1920 in Jičín), whose father Ignác Neumann (1828–1878) was a baker; her mother Anna (1833–1905) was the daughter of a cooper. They moved, however, almost immediately from Železnice to Libáň (a small town south of Jičín, within one hour's walk from Hřmenín, founded likewise by Archbishop Arnošt in the middle of the fourteenth century, and raised to township status in 1574 by the Emperor Maximillian II); and there, on 7 June 1884, my mother, who reached the ripe old age of almost 89 years (she died on 25 March 1973 in Prague) first saw the light of day.

However, my grandparents did not stay in Libáň long either, but soon returned to Jičín where they remained for the rest of their lives. It is there that my mother was educated and got married; and there, too, that both grandparents departed from life. It is also there that I spent the earliest years of my own life; but about these more will be said in the next chapter, while here we must return to the grandparents and accompany them briefly on their life's journey.

When they were young—before the advent of films or radio (let alone television)—cultural life (literary, musical or theatrical) in the provincial cities of Bohemia was much more creative and intensive than it has been ever since. The teachers as well as students were expected to take full part in it—and so they did! In the course of his life, from his teacher's salary grandfather managed to accumulate a library of several thousand books; partly out of personal interest, and partly because it was considered an honourable though unwritten duty for intellectuals of that time to provide a market for worthwhile books (which their authors often published at their own expense).

Neither grandfather nor grandmother had a good enough ear (or skill) to contribute to the musical life of the town; but what they failed to give there, they more than made up for in another direction: namely, in amateur theatricals. Grandmother Marie was an outstanding actress—a real talent—and was acknowledged as such by some of the best in the land. In those days, and especially since 1868 when the Czechs built (by private subscription) their own National Theatre in Prague (still today one of the most beautiful buildings in that city), theatre meant more to the people than all sports taken together do today; and the two greatest actors of pre-war times—Eduard Vojan (1853–1920) and Marie Hübnerová (1866–1931) (the likes of whom were never seen on the Prague stage since)—meant much more to the nation than all the Beatles or teenage crooners of more recent times; and special trains full of spectators from provincial towns used to come to Prague for important premières on the stage—as, I suppose, they come today to see the ball games.

During the summer seasons, when the National Theatre was closed, these great actors considered it a part of their *noblesse oblige* to offer free guest performances in provincial cities, and Jičín was usually on their circuits. Whenever they came, my grandmother was usually asked to partner them in their plays, and with Marie Hübnerová (herself the widow of a teacher) she developed a long-lasting friendship; to which several precious photographs of the great actress in her famous roles, with flattering inscriptions to her modest self-sister *in Musis* (still in the family's hands) bear ample witness. In my student years in Prague, I had the good fortune to see Hübnerová twice on the stage of the National Theatre before death claimed her in 1931; and have never seen anything like her anywhere in the world.

Grandfather's talents did not lie in exactly the same direction. However, what he failed to do on the stage, he more than made up for by his services as producer and stage-manager of these performances. Sometimes, my mother told me years later, when he was called upon to stage a play which was also being given at the National Theatre in Prague, he was curious to see how it was staged there. The quickest solution would be to take the train and visit Prague to see the performance. Unfortunately, the round trip by rail would have been too

## Peregrinations of grandfather and grandson

much for the slender family budget of a young teacher. But grandfather would not allow such a triviality to stay in the way of his plans; and so after school-hours he took to the road—on foot! The distance from Jičín to Prague is approximately 60 miles; but what was that for a former pilgrim to the Holy Land? Bicycles were not yet invented; but grandfather was a tall man whose long steps through the night enabled him to reach Prague well before noon the next day. He then took forty winks in an inn, but no more than he could afford in order to be in the National Theatre for the beginning of the performance (where, unless he had a free ticket, he stood at the gallery). And the rest was simple: when the performance was over, grandfather took to the road again back to Jičín, where he arrived—let us hope—not too late to meet his class in the morning; for he was supposed to set his pupils a good example.

Such expeditions, repeated on occasions a few times a year, were not reasonable; they constituted, in fact, an abuse of health for which grandfather paid the price in later years. But although he was told so by his wife (who was as reasonable as she was gentle) grandfather would not pay much attention. Mother (who listened to it as a child, at an age when children were supposed to be seen and not heard) thought grandfather could be opinionated, if not obstinate on such occasions. Although, in most respects, he was the model of a solid and rather stern citizen (without much sense of humour—a quality which my mother inherited from him), it must have been a sense of adventure which sent him occasionally on these long jogs even in middle age—an inclination which, in turn, was perhaps inherited from him by his grandson.

For when, many years later, I retired from my Manchester Chair in October 1981, the grace passed by the Senate on that occasion described me too as an inveterate traveller. What my colleagues did not add (because they did not know) was that it was my grandfather's legs which carried me on my various exploits—from the top of Fujiyama in 1936 in the days of my youth, to the Matterhorn (not to the summit any more, but high enough) when I was well past seventy; and earned me membership of the Explorer's Club of New York. If, moreover, my colleagues (in that grace of 1981) mentioned that 'rumour has it that there is an observatory in Tibet which he' (i.e. myself) 'has not visited, though this is doubtful', I have since been able to settle that doubt as well (the observatory is not there yet, but will be!).

But let us return to the grandfather of eighty years ago. Even past his middle age he remained a man of wide interests; and not all of them were absorbed by artistic activities. Adult education was for him always in the forefront; and he was also one of the organisers of the 'extension cycles of lectures', traditionally held in provincial towns by

Prague university professors during their academic vacations. Like great actors, well-known professors considered it in those days also a part of their mission to lecture in the provinces on subjects of topical interest; and one of those who was most in demand was Professor Thomas G Masaryk—the future President of the Czechoslovak republic. My grandfather, together with Dr Jakub Všetečka (1850–1927), then Professor of Classics at the Jičín *Gymnasium* and once Masaryk's fellow-student at Vienna University, were the prime movers of this invitation. Not everyone in Jičín appreciated this move (for Masaryk in his day was no conservative!); but grandfather was on good terms with him, and his feelings were apparently reciprocated. At least the story came down to me (which I can no longer authenticate) that, during his brief stay in Jičín, Thomas Masaryk visited also my grandparents in their home; and got at least a glimpse of their teenage daughter Ludmila who, in the fullness of years, became my mother.

But, to come back to grandfather, although his field of specialisation at secondary school was natural sciences, he is best remembered today outside the family as a literary historian, for his contributions to the history of the life and times of Božena Němcová, the greatest writer of the Czech literary awakening in the nineteenth century, whose work *Grandmother*, published in 1862, has since been translated into more than ten languages and still remains one of the most precious Czech literary heritages of that age.

The life of Božena Němcová (1820–1862)—a writer who combined the pen of George Sand with the beauty of Madame Récamier—was as short as it was tragic; and her immediate family were very close about the last years of her martyrdom. Chance willed it, however, that her youngest daughter Dora was a teacher at the same school as my grandfather, who gradually acquired her confidence and thereby gained access to many documents which she allowed him to use (and some of which, on grandfather's death, passed to my father and eventually to myself). Spurred on by this new evidence of literary interest (which he published in a number of articles), grandfather set out systematically to consult other sources and to interview surviving witnesses of the tragedy. He eventually published all in a major book, entitled *Božena Němcová*, which appeared in Prague in 1920, on the hundredth anniversary of Němcová's birth, and was followed a year later by a smaller book on a more detailed part of the same subject— the latter dedicated to the memory of his wife Marie (my grandmother) who had departed from this world in the meantime. Her end was as tragic as it was unexpected—of cancer which was not detected in time. She was operated on by Professor Otakar Kukula of the Medical School of Prague University, himself born and educated in Jičín (where, in his youth, he may have seen my grandmother on the stage);

*Grandfather's literary works*

but in spite of his help and care the end came on 20 March 1920, when grandmother was only 58 years old.

Grandfather's 1920 book was his most important legacy to posterity, and remains of considerable interest to literary historians today. It did not, however, attempt any final synthesis of his material; and, some said, he may not have seen the wood for the trees. This may have been by design; for grandfather certainly lacked the professional training of the literary historian (after all, the subject he taught was natural sciences); and he may have been aware of his limitations in the literary world. Nevertheless, he accumulated a large amount of evidence (which otherwise could have been lost to posterity) to be used by others; and, in so doing, had an opportunity to correct also a number of mistakes committed by Prague university professors (in particular, Václav Tille)—a duty which (I was told) grandfather discharged with some satisfaction.

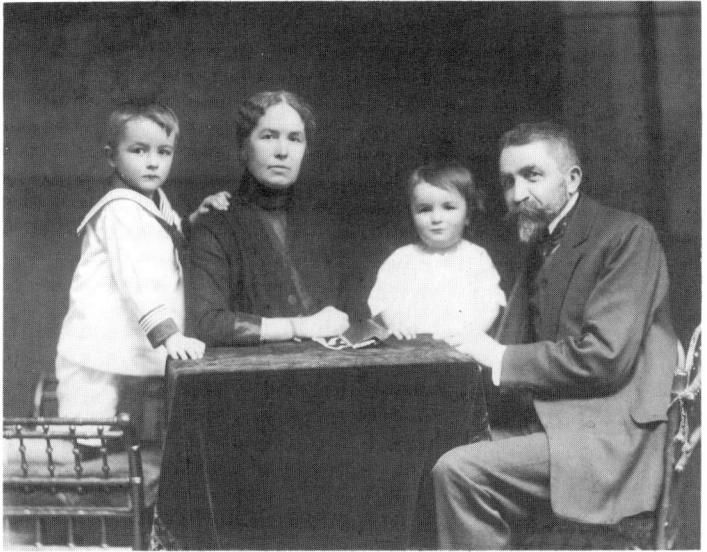

**Figure 1.2** Grandfather Lelek (1860–1930) and grandmother Marie (1869–1920), with grandchildren Miloš (1910–1948), left, and Zdeněk (1914–    ), taken in 1917 in the middle of World War I.

Grandfather Josef retired officially from public service after grandmother's death in 1920, but accepted an honorary appointment as the town's historiographer. In this new capacity he embarked on the task of preparing a history of Jičín from the fourteenth century to the present, and accumulated to this end a large amount of material, some of which I have used in writing this chapter. This work was, however, never completed (its torso being now in my hands); for by that time his working days were almost at an end. The first stroke which struck him in 1921 impaired only (to some extent) his ability to walk; but the second stroke confined him permanently to bed in 1924. Although since that time he could no longer write, his mind remained clear almost to the end, which came on 13 April 1930, at the age of seventy years. I was present at his funeral; and remember well the moving service in front of the school where grandfather taught for so many years; and where the main eulogy was delivered by the colleague of his (of the same age) with whom grandfather had set out for the Holy Land, in the prime of their lives, more than half a century before. They both rest now in a better land; and their ashes, as well as those of grandmother Marie (who predeceased her husband by 10 years), are buried at Jičín cemetery among their contemporaries only a few hundred yards from the school where grandfather taught.

But having taken leave of my grandparents, I wish to return now to their children, my father and mother. We have already remarked upon their engagement in 1902. From there it was, however, still very far to the actual wedding (which did not take place until 3 August 1909 in the Church of St Ignace in Jičín, within a few steps of the bride's home); for, in those days, a dutiful father would not have given away his daughter in marriage before the prospective bridegroom could properly support a wife; and to this end my father had first to finish his university studies as well as military service (the latter representing no mean constraint at a time when World War I was not much more than ten years in the future).

A gradually deepening shadow which its portents were casting on civilian life in those years did not, however, deter my father from following his chosen path. In the autumn of 1902 he left Jičín to enrol as a student of the Faculty of Philosophy of Charles University of Prague. His preceding academic record ensured his admission; and a Government scholarship (then by no means automatic, and given only to the best students) ensured his support. The field which father chose as the subject of his course was modern philology; with the immediate aim of qualifying to teach such languages (in particular, French and German) at secondary schools of *Gymnasium* rank. Such a position would, in turn, guarantee the possibility of offering a modest support for the family, and would provide also a springboard for a subsequent career.

From his *Index Lectionum* which is still in my hands, I gather that father attended lectures by Professors Jan Gebauer and Vlček in Slavonic philology, Jan Mourek and Oskar Kraus or Josef Janko in Germanic philology; Jaroslav Vrchlický or Jan U Jarník in the philology of romance languages; by Thomas Masaryk and František

Drtina in philosophy and pedagogy, and several others. Many of these are no longer remembered by the students, but a few are. Thus the name of Jan Gebauer still shines like a star of first magnitude in the field of Slavistics; while Jaroslav Vrchlický, a phenomenal linguist and a great poet, still occupies a high position on the Czech Parnassos; but his professorship was essentially honorific, and he did not lecture with any regularity at all. He is being mentioned in this narrative mainly for the fact that, many years later, my father was elected to the chair originally created for Vrchlický at that time (and held after him by Šalda; father himself held it between 1937 and 1958). Needless to say, father was no poet; nor could he equal his great predecessor in his monumental linguistic abilities (Vrchlický—apart from poetry in his native Czech language, which by volume remains unsurpassed to this day—also translated into Czech poetical masterpieces from twelve other languages, from Dante and *Chanson de Roland* to Shakespeare, Goethe or Pushkin—even the *Mahabharata* or *Bhagavad Gita* from Sanskrit). Nevertheless, when the time came for father to join his ancestors, he left behind something which Vrchlický could not equal: namely, a school of modern philologists of romance languages, manning, in the fullness of time, academic positions at all Czechoslovak universities, and still doing so today.

Father's most interesting contact with his professors during his student years was perhaps that with Thomas G Masaryk, the future President (between 1918 and 1936) of the Czechoslovak Republic; then Professor of Philosophy at Charles University. During father's student years Masaryk was at the peak of his intellectual powers, exerting a powerful influence on Czech public life—more, perhaps, outside the university than in the classroom (for, as a member of parliament and heavily engaged in political life, Masaryk did not lecture very regularly at that time).

Father told me once of his first encounter with Masaryk in his first year at the University, when he took Masaryk's course on sociology and needed from his professor a report (*colloquium*) on his progress (required under the conditions attached to his scholarship award). In order to catch Masaryk in Prague between his frequent trips to Vienna he had to call on his professor at his flat in the morning (before the latter left home); expecting that Masaryk would examine him on the history of the subject (as was then the custom, commencing with the staple statements 'Already the old Greeks ...' etc), or on what everyone else wrote about it before.

But he was mistaken! On the way to the professor's flat father bought some morning newspapers, which he read and afterwards stuffed into his pockets (all his life father was an avid newspaper reader). Masaryk noticed these and remarked: 'I see that you already know that the

Government has fallen' (quite a frequent occurrence in those days); and when father confirmed this, Masaryk asked: 'Comment upon it.' He did not have to ask twice; for father was likewise politically-minded, and a member of the party which Masaryk represented in parliament. What he said should, therefore, have pleased Masaryk; but it did not. Instead he questioned the student closely for some time on the reasons behind the events, to make sure that he understood the issues involved and could judge them critically. When father's answers satisfied him in this respect, he wrote the top grade on his report; and the examination was over!

In spite of this auspicious beginning, father never attached himself to Masaryk too closely, then or later; and Masaryk would probably have considered him too 'bookish'. Years later, in his capacity as President of the Republic, he signed *ex officio* all decrees appointing my father to his university chairs; and, for his part, father (together with the whole University Senate) accompanied Masaryk on his last journey through Prague, after that grand old man passed away on 14 September 1937, aged 87 years. Masaryk's funeral on 21 September of that year was a memorable and historical occasion—one I watched myself from the University Library (where I was then briefly employed before leaving for Cambridge in January 1938; and where Masaryk must have spent many an hour during his University years). Although it was not fully realised at that time, Masaryk's passing marked the golden sunset of Czechoslovak democracy; and people are awaiting the dawn of a new day to this time.

Only once more did a brief contact take place between the Kopal family and the Masaryks: not with Thomas Masaryk, to be sure, but with his son Jan. In 1947, when Jan Masaryk was still Foreign Minister of post-war Czechoslovakia, he came (during one of his trips to the United Nations) to address students of the Massachusetts Institute of Technology in Cambridge, Massachusetts (where I then served as associate professor); and MIT's President Karl T Compton asked me to show Masaryk around the laboratories with which I was associated. Encouraged by Masaryk's interest, I asked him to call up, on his return to Prague, my father and give him our greetings (he had already two grandchildren in Cambridge he had not yet seen). Masaryk not only promised, but actually did so, not many weeks before his untimely death in front of the Czernin Palace which housed the Foreign Office in Prague—an event whose official explanation still leaves much to be desired to this day.

Thomas Masaryk had four children—two daughters (Alice and Olga), and two sons (Herbert and Jan). Herbert Masaryk was killed in World War I, and his brother Jan in the spring of 1948; while both daughters attained a ripe old age, like their father. Alice (the older) never married, and died in the United States (her mother was an American from the East Coast) in the 1950s. Olga Masaryk, the youngest of the four, married in Switzerland and presented her father with two grandsons who were his pride and joy. Both escaped from Czechoslovakia in 1939 for England, joined the RAF as fighter pilots; and both were shot down, in the prime of their lives, during the Battle of Britain in helping to defend its capital, where their grandfather once served as visiting professor at King's College (and where the Chair of Slavonic Studies carries Masaryk's name to this day).

In the mid-1950s, the only surviving members of the Masaryk family were daughters of his son Herbert, who remained in Czechoslovakia and married there. According to post-war Czechoslovak law, at the time of the wedding the spouses may choose the name to be carried by their children, and in this particular case they opted for Masaryk, so that this name will not become extinct; and its holders may live long enough to see better times.

But let us return in our thoughts to *philosophiae studiosus* Josef Kopal at the beginning of this century; and to his plans for the future. As his university studies progressed, he became aware that his future was in romanistic studies; and in his last years he concentrated on this subject. Moreover, in order to gain added fluency in the French language, in 1905 he spent one semester at the University of Neuchâtel (in the French part of Switzerland); and, somewhat later, at the University of Grenoble in France. In the year of his return he passed his final examinations in Prague; and after his military service (for university graduates, this lasted one year) he embarked on his probationary year of grammar-school teaching, which was then the usual way of preparation for an eventual university career. He took this probation in Prague, at an elite secondary school, on the staff of which father became acquainted with several senior colleagues who played an important role in his career later on.

Such was Prokop Miroslav Haškovec (1876–1935), already at that time an outstanding romanist destined for a university career, whose successor at the Masaryk University in Brno father became 25 years later; such was František Žákavec (1878–1937), renowned historian of art; or Karel Skála (Rocher) (1863–1934), outstanding exponent of comparative linguistics, whose textbooks of the French, Italian and Spanish languages as derived from colloquial Latin enjoyed for a time a European reputation. Approaching loss of eyesight prevented Skála from following a regular academic career; but even when he became totally blind and could no longer leave home, his flat became a regular postgraduate seminar of comparative linguistics, to which particularly gifted students were referred by their university teachers to receive the final polish in their speciality and inspiration for future work.

Father received much inspiration from this brilliantly gifted linguist in the course of his probational year and later: it was Skála who paved his way in post-war years to his university career; and he was still alive when I myself entered university in 1933; a few words on our brief encounter at that time will be postponed till the next chapter. In the meantime, with the satisfactory conclusion of his probational year, father received a permanent appointment in the school system of the (then) Austro-Hungarian monarchy; and thus fulfilled a prerequisite for the conclusion of a suit for his bride; and they were married during the summer vacation in August 1909, to embark on the joint life's journey which was to last for almost sixty years.

**Figure 1.3** My father and mother, Josef and Ludmila Kopal (*née* Lelek) on their wedding day, 3 August 1909, in Jičín, Czechoslovakia.

The first posting father had in this new capacity was in the town of Kolín, some 50 km east of Prague, and my parents stayed there for three years (1909–1912), in the course of which my older brother Miloš (1910–1948) was born on 13 June 1910. At the end of the school-year 1911/12, father was transferred to a more important post in the historical town of Litomyšl, about 120 km east of the capital, where I was to see the light of day on 4 April 1914—four months before the outbreak of World War I. A fuller story of my early years is postponed to the next chapter; while in the remainder of the present one we shall briefly accompany my parents (whom we shall no longer meet much in the rest of these recollections) till the end of their lives.

## Father's academic career

*Inter arma silent Musae* was an old Roman proverb which has retained its validity up to the present. The years 1914–1918 father spent in military uniform—first Austrian, and later that of the new-born Republic of Czechoslovakia (as successor to the ancient Kingdom of Bohemia); and after demobilisation he returned in 1919 to his school teaching in Litomyšl. However, with a return to civilian life father returned also to his plans for an academic career, encouraged by the fact that whereas, before the war, the Czechoslovak nation had only one university (the ancient Charles University of Prague, founded in 1348) with Czech as the language of instruction, two new such universities—in Brno and Bratislava (capitals of Moravia and Slovakia)—were founded after 1920, thus offering young scholars better chances for promotion.

The first step in this career was father's return to Prague in 1921 (where the family could not follow him till two years later because of an acute housing shortage); and the preparation of a book which was to be the basis of his application for *veniam legendi* at the University. Such an application constituted in those days a much more formal process than it is in Britain (let alone America). The candidate was expected to submit to the faculty a major book based on original research (a collection of papers would not do) which had to appear in print, and became the basis of the process of *Habilitation*: if approved by an appropriate professorial committee (since the Middle Ages, the faculty had a right to elect its own members) the appointment had to be carried out by the Minister of Education, and countersigned by the Head of State. Such was the respect in which scholarship was held in those days!

For the subject of his *Habilitationsschrift* father chose 'The Literary Theory of Boileau'; and in submitting its text for publication by the faculty, he sought the support of Professor F X Šalda, the greatest Czech literary critic of the twentieth century, who had succeeded to the chair once Vrchlický's at Charles University some years before. His written reply to father is still in my hands; and is not without interest to this day. In it Šalda promised to recommend indeed to the faculty that father's work should be printed (at the University's expense) in book form—it did appear in 1927—and also to support his *Habilitation* proceedings at the faculty; but after some flattering comments on his work (surprising enough; for Šalda was a rather abrasive character) he added ' ... but do not promise yourself too much from my support; for in the faculty I do not carry much weight; politics there is conducted by others.'

In re-reading his message, one is at first surprised that F X Šalda— who together with the historian Josef Pekař was a star of the first magnitude of the Faculty of Philosophy at Charles University of that time—would have assessed his influence in this manner; but later I learned for myself that this was true—not only at Charles, but at other universities as well; namely, that one can be a good scholar, or a good committee man skillful in pulling strings behind the scene—but seldom both at the same time! For life is too short to develop a talent for both; and while committee men are only too glad to relinquish tacitly any pretence to scholarship for what must obviously be pursuits more satisfying to their nature, real scholars would be impatient with the trivialities discussed mainly in committees, while their minds are occupied with other issues.

But to return to my father's first steps on the academic stage, Šalda's pessimism was superfluous; indeed, the young man to whom he then offered a helpful hand would in ten years become his successor! In the meantime, in 1928 father was sent out to Bratislava, to start a department of romance languages at the newly-founded Comenius University, and received a personal (extra-ordinary) chair in his subject in 1930, which in 1935 was elevated to a statutory one. But only a year later his old teacher P M Haškovec died (on 20 December 1935) in Brno; and Masaryk University offered father his chair there. As there was no successor for him at the Comenius University of Bratislava, father temporarily agreed to serve on both of these schools for two days each week of the term; and when, in 1936, the health of Šalda (then 69 years old) began to fail, father was asked to lecture in Prague in his place as well, so that in 1937 he professed his subject, lectured and examined at all three Czechoslovak universities! No wonder that we seldom saw him at home in those days! But this, fortunately, did not go on for long: for when Šalda died in April 1937 after a protracted illness, father was elected to succeed him at the most prestigious chair in his subject in the country, which he was to hold for another 21 years.

But—alas—the years of the tenure of his new post were anything but conducive to scholarly preoccupations. Soon after the Munich betrayal of September 1938 and the Nazi occupation of what was left of Czechoslovakia in March 1939, Charles University (as well as all other universities and academies of higher education) was closed by the Nazis, not to reopen for almost six years till the summer of 1945; and when it did, more than one-sixth of the staff (more than 160 of its scholars, young or old) could not return from battlefields or concentration camps.

I was at that time already in the United States; and our home in Prague was visited more than once by the Gestapo in quest of incriminating material. In May 1942 (after the Czechoslovak parachutists sent out from England did away with 'protector' hangman Heydrich) father received warning from the underground that his name was on the list of hostages facing arrest, and was urgently advised to leave home. For more than six weeks, father told me many years later, mother and he did not dare to sleep more than once in the same place, until the main wave of terror subsided and they could return home once more to continue unfinished work.

**Figure 1.4** The Kopal family in 1935. From left to right: father, Professor Josef Kopal (1883–1966); JUDr Miloš Kopal (1910–1948); RNC Zdeněk Kopal (1914–    ); and mother, Ludmila Kopal (1884–1973).

The vacations enforced by the closure of Czechoslovak universities during World War II had at least one advantage: they provided father with enough time to bring to completion his life's main work, the monumental *History of French Literature*, on which he worked throughout the war years, and which he published in 1949, when he was 65 years old, as the largest work in this field which has so far appeared in the Czech language.

Between the wars, before university teaching and other activities began to claim an increasing fraction of father's time, the interest of his young years led him to serve in the 1920s as co-editor (with E Čapek) of an elite literary journal, *Nové Čechy* (*New Bohemia*); and, since 1936, as co-editor (for romance languages) of the *Czechoslovak Journal of Modern Philology*, of which he became Editor-in-Chief in 1947; he did not relinquish this function until 1953 at the age of 70 years. Chance willed it that, in the 1953 March jubilee issue planned for father, he had to share the honours with J V Stalin, who died the same month (and who, in the last year of his life, tried his hand also on philological matters); and with Klement Gottwald, a Czechoslovak Communist Party boss who (though no philologist himself) failed to return alive from Stalin's funeral in Moscow to Prague the same month.

I mention father's editorial activities in this place because (as the reader may find out in Chapter 7 of these reminiscences by his son) they proved to be contagious! But, in those years, father's own personal work was restricted largely to literary monographs. Unlike his predecessors Vrchlický and Šalda who liked to survey large portions of romance literature, be these French, Italian or Spanish, from bird's-eye perspectives (and whose main contributions to Czech culture were made outside the University), my father confined his attention predominantly to modern French literature by specific writers. Two of them stand, in particular, in my memory: Gustave Flaubert (1821–1880), to whom father devoted several monographs (the main one was published in 1932), and Romain Rolland (1866–1944) with whom he was personally acquainted and corresponded for several years, until Rolland's death in Vézelay on 30 December 1944, in liberated France. Born on 20 January 1866 in the nearby provincial town of Clamecy in the heart of Burgundy, Rolland is buried in a little cemetery at Brèves (some 10 km east of Clamecy), where his paternal ancestors farmed the land from time immemorial; and each time we pass through Burgundy on our way to summer vacations in the Alps, we bring some roses from our garden to lay on his grave.

Our ties with Rolland and his legacy are of old standing; for already in the years following World War I, father had translated into Czech his ten-volume cycle *Jean Christophe*, which earned its author the Nobel Prize for literature in 1916. I recall to this day the interest with which I listened, in my early teens, to father as he dictated its translation to my mother; and this book (which has since appeared in Czech in six editions) has remained very dear to me my whole life; as did Anatole France's *Le Crime de Sylvestre Bonnard Membre de L'Institut*, which father also introduced into the Czech language before I could read it in the original (or in an excellent English translation by Lafcadio Hearn; with whom I share also our joint affection for Japan).

This is perhaps not the place to recount all the scholarly contributions my father made to the history of French literature, recognised by the French, who enrolled him in the Legion d'Honneur. On 12 December 1940 father was elected a full member of the Czech Academy of Arts and Science; and, on 15 January 1947, Fellow of the Bohemian Royal Society of Sciences (founded in 1785; in the publications of which Christian Doppler (1803–1853) published in 1842 an article 'On the colours of the components of double stars' enunciating his well-known principle; and Ernst Mach (1838–1916) published his

# An academic career draws to its close

principles which are causing us headaches up to the present time). A bibliography of father's contributions, collected by his students and printed on the occasion of his seventieth and eightieth birthdays, runs to over 270 individual items.

Instead, let us accompany him from the end of the war to the end of his life. On account of heavy duties which devolved on his department, his retirement had to be postponed for several years. In his seventieth year he retired as Editor-in-Chief of the *Czechoslovak Journal of Modern Philology*, a function which he had carried out conscientiously for twelve years; but (in the absence of a suitable successor) he continued to serve as head of the department of romance philology at Charles University till 1958, when he was almost 76 years old; and after a teaching career which encompassed (with breaks caused by two World Wars) a full half century. His predecessors at the University (Jaroslav Vrchlický and F X Šalda) made their great contributions to the culture of the nation mainly outside the walls of academe; but on departure from his chair father left behind more than one generation of scholars who carry on the tradition of his subject in Czechoslovakia today.

After 1958 father still continued to write and give occasional lectures at the University and on the radio; but after his eightieth birthday his remaining strength began to wane (see Figure 1.5); and (after a brief

**Figure 1.5** (left) My father on his retirement in 1958, decorated by the Government after almost half a century of academic service. (Right) Father's last photograph, taken in summer 1966, when he was more than 83 years old.

illness) he passed away on 29 December 1966, aged 83½ years. During his long life he served his country under two Emperors and five Presidents; the first of whom—Masaryk—had been his teacher at the University of Prague, and the last—Husák—his student at Bratislava. His funeral on 5 January 1967 in Prague was like a congregation of a large family, in which a dwindling band of his contemporaries and fellow-students from the days when the century was young mixed with his students, and the students of his students. All came to pay him their last respects, together with official representatives of the three universities in which father taught during his lifetime. Decorations he received in the course of his life were assembled at the foot of his coffin. The presence of one person would have pleased him more than anything else as a messenger of fraternal greetings to those who had departed before—Madame Romain Rolland, who waged a long journey from Switzerland to Prague to pay her farewell to a modest interpreter of her husband's work in another language.

My mother survived father by almost seven years, attaining the ripe old age of almost 89 years when she passed away on 25 March 1973, as the last link with much of the past and witness of events gone by, as related in this chapter.

**References**

Chaloupecký V 1941 'Arnestus of Pardubice, the first Archbishop of Prague' *Spol. přátel starožitností*, Prague
Kopal J 1927 'Literary Theory of Boileau' *Práce z Vědeckých Ústavů* (Prague: Charles University)
—— 1930 *Romain Rolland* (Bratislava: Comenius University)
—— 1932 'Gustave Flaubert' *Spisy Filosofické Fakulty* (Bratislava: Comenius University)
—— 1949 *History of the French Literature* (Prague: Melantrich)
Lelek J 1920 'Božena Němcová' *Učitelské nakladatelství*, Prague
Teplý F 1930 in *Pekařův Sborník* (Prague: Historický Klub) vol **2** pp 130ff

# Chapter 2

# *Awakening*

It was in Litomyšl that the author of these reminiscences was born on 4 April 1914, and opened up his young eyes for the first time on this old Universe of ours, now some seventy years older. However, before we say anything here about him, let us introduce to the reader his birthplace (see figure 2.1) where our little subject spent several of his formative years; and in which he dwells fondly in his recollections whenever there is time to reminisce.

In the preceding chapter of this book we described to the reader another Czech town, Jičín in north-east Bohemia, and its surroundings from which my family stems both on the sword and distaff sides (and to which we shall return, albeit temporarily, once more in this chapter). However, it is in Litomyšl that I spent the major part of the first decade of my life; and although this was well before the stars entered my mind for the first time, at least the Moon did (through the instrumentality of Jules Verne); and the story perhaps deserves to be told in this place.

Litomyšl, now but a modest-size town in eastern Bohemia (some 150 km east of Prague) and smaller in size than Jičín, has, however, a longer and even more distinguished history. Its existence as a local castle (situated at an important commercial route connecting, since time immemorial, the Middle East with western Europe) as far back as 981 AD is attested already in the *Bohemian Chronicles*† written (in Latin) between 1117 and 1125 by Kosmas, Dean of the Prague Chapter; and tradition has it that it was along this trail that St

---

† We refer to these *Chronicles* in this place, not only because of their earliest mention of the name of Litomyšl (probably that of its first known possessor), but also because the first printed edition of them (which appeared in Hanover in 1607) was known to Johannes Kepler, and used by him as a source of the early Bohemian legends referred to in the opening parts of his (posthumous) *Somnium, sive Astronomia Lunaris*.

32   *Awakening*

Methodios, Archbishop of Panonia, travelled in the latter part of the ninth century to Prague (or, rather, Budeč in the vicinity of the newly founded Prague castle) to baptise St Ludmila, the wife of the first historically attested duke of the national Přemyslide dynasty which ruled the Duchy (and, since 1085, Kingdom) of Bohemia until 1306.

**Figure 2.1**   My birthplace, the town of Litomyšl. The thirteenth-century Gothic church in the background is where I received baptism and my first name on 22 April 1914. The large building below the seventeenth-century baroque church is the old *Gymnasium*, for more than 250 years the centre of higher learning in that part of the country, in which my father taught for several years before he was recalled to Prague in 1921.

With the advent of the second millennium AD the traffic on the trail passing through Litomyšl increased to such an extent that it became profitable for the local chieftains to provide armed escort for caravans (especially through the heavy forests along the frontiers between Bohemia and Moravia) against payments in cash (or kind), from which they built their first fortified places in those parts; the ruins of one (Hrutov by name) still stand to this day. With a gradual disappearance of frontier forests in the later Middle Ages (and a shift of commercial routes to south of the Alps) the 'protection racket' ceased to be sufficiently profitable; but the subsequent efflorescence of Litomyšl came from another (and more honourable) source, which left its imprint on the town's history for centuries to come.

In the year of 1141, the Bishop of Olomouc in Moravia (in whose diocese Litomyšl then belonged), Jindřich Zdík ( = Zdislav; or, in colloquial form, Zdeněk), founded in Litomyšl a monastery for the (then modern) Order of Premonstrates, who brought with them also the coat-of-arms of their mother house in Prémontré in France, which then became (and still is) that of the town itself. In keeping with the tradition of their Order, these Premonstrates proved to be good organisers; and thanks to their initiative the community growing around the monastery prospered to such an extent that King Přemysl Otakar II—that Richard the Lionheart of Czech medieval history who was extolled by Dante in Canto VII of his *Commedia Divina*—raised the town to the rank of royal borough by a privilege dated 27 July 1259 (the original of which can still be seen in the Museum of Litomyšl to this date).

The most famous hour of medieval Litomyšl struck in 1344, when King (and later Emperor) Charles IV induced Pope Clement VI of Avignon to create in Litomyšl (*locus nobilis et insignis* in the words of the respective papal bull) a new bishopric—which Litomyšl remained until 1425—and some of its incumbents became men of international renown. Such was the second bishop Jan II of Středa (Johann von Neumarkt), formerly Chancellor of the Emperor Charles IV (successor in this function to Archbishop Arnošt), who was one of the forerunners of early humanism in central Europe, a personal friend and life-long correspondent with Petrarca†. With advancing years (we suspect) the aging Chancellor (who, unlike Arnošt, did not die in harness) was 'pensioned off' to the see of Litomyšl, where he also died in 1380 and is buried in his cathedral church (still standing)—only a few steps from the font where on 22 April 1914 I was received into the Christian Church, in the arms of my paternal grandfather Francis (the maternal grandfather Lelek performed the same honorific function for my older brother Miloš at Kolín four years before).

But before the good bishop Jan II was laid there to await resurrection, he deserved well of the town's future by one more meritorious deed: namely, by bringing to Litomyšl (in 1356) a new religious order,

---

† Francesco Petrarca (1304–1374) is known to have visited the Emperor Charles IV in Prague in 1356. The two knew each other from previous meetings in Avignon and Rome; in their correspondence the Emperor styled the poet *amice* (an unusual courtesy in those days). The reason for Petrarca's visit was (to use today's language) to consider acceptance of a chair of classical languages at the newly-founded *studium generale* in Prague (since known as Charles University), offered to him by the Emperor. Petrarca (then over 50 years of age) did not accept; and remained in Prague only for some months. However, on his return to Italy he recorded that '... nowhere in Europe did I find an atmosphere less barbaric, and more touched with humanism than at the court of the Emperor and of his principal advisers—Arnestus of Pardubice or Jan of Středa ... These personalities were so educated, and of manners so courteous and pleasant as if they hailed from Attica itself'!

the Augustinians from Brno. He set up a cloister for them in the form of a new Gothic church, already the third in the town, which he attempted to glorify by an acquisition (from Paris through the intermediary of the French dauphin Charles) of a part of Christ's cross from Calvary. The bill showing how much the burghers of Litomyšl had to pay for this (allegedly) priceless relic has, unfortunately, not been preserved. But we are still in the Middle Ages in our story, when laymen used to believe what they were told by the clerics; and possibly both sides were satisfied with their bargain—which is the best way of doing business in any age!

And not only in Litomyšl itself, but also outside their city did its inhabitants make their existence felt. Not long after the University of Prague was founded in 1348, we read among its professors of the faculty of arts the name of Mikuláš of Litomyšl, who in 1386 was elected Rector of the University; and was among the first to discuss and defend from his chair the teachings of the Oxford reformer John Wycliffe. As such, Mikuláš became also the teacher of Jan Hus (1369–1415), the greatest religious reformer of central Europe before Martin Luther, who to the end of his life remembered Mikuláš as a 'man of holy memory'.

Perhaps the best known (or notorious) bishop of Litomyšl was Jan IV, who held that see for 30 years (between 1388 and 1418), and was one of the first Czechs on whom Rome conferred (in 1426) the cardinal's red hat. At the time when the majority of the population in Bohemia was inclining towards the need of church reform, symbolised by the name and life of Jan Hus, Jan IV of Litomyšl stood rigidly on the side of orthodoxy; and on his return from the Council of Constance (where Jan Hus was burned at the stake in 1415), bishop Jan IV, having taken an inimical position at Constance against Hus, was chased by the latter's followers away from Litomyšl; and ended his life in 1430 as Archbishop of Ostřihom in Hungary.

During the Hussite wars between 1419 and 1436, the see of Litomyšl languished for a lack of incumbents; and has not *de facto* been re-opened ever since. This fact did not, however, mark the end of the role of Litomyšl as a cultural centre of eastern Bohemia. By no means! For until almost the middle of the sixteenth century, Litomyšl became an Utraquist city and cultural centre of the Unity of Bohemian Brethren—perhaps the finest flower of the autumn of the Middle Ages in central Europe. The Brethren were literary men—the name Jan Amos Komenský (Comenius), their last bishop, is known all over the world as that of the founder of modern pedagogy—and Litomyšl became almost a national centre of their publishing activities.

And more: for in 1492, the Brethren in Litomyšl, dissatisfied with Christianity as they saw it practised in western and central Europe,

*History of Litomyšl* 35

sent a number of their trustees on expeditions to eastern Europe and the Near East, in quest of the original Evangelic Church from the days of Christ. They travelled together to Constantinople; and there their ways parted: one went to Muscovite Russia in search of Orthodox Christianity; another to Greece with the same goal; and the third to Palestine and Egypt. This last, the longest and most difficult part of the quest, was entrusted to Martin Kabátník, burgher of Litomyšl, who (unlike my grandfather several centuries later) *did* visit the holy places of Palestine and got as far as Cairo—no mean feat for a traveller knowing only the Czech language. But he (as well as his fellow travellers from other parts of the world) returned home disappointed; for wherever they went, of 'old Christianity' (as they imagined it) they found no trace. At least Martin Kabátník, a man who could neither read nor write, later dictated his experiences to a cleric; and his notes were published (for the last time, in 1948) to enable us to accompany him and the other expeditionaries on their long journey.

In following in their footsteps the modern reader must, however, be sadly disappointed: for neither the Acropolis of Athens (which they saw before it was irreparably damaged by the Venetians), nor the pyramids of Giza were of much interest to our pious fifteenth-century travellers; and if they did meet a community of early Christians who preserved the traditions of the evangelists, how would they have recognised them? Did they really think that any section of the human community can survive contact with others lasting over a thousand years in unadulterated form?

As Martin Kabátník and his fellow brethren were returning empty-handed from the east in 1492, the caravels of Columbus were already on the high seas to discover the New World; and his sponsors (the Most Catholic Majesties of Castile and Aragon) had other things on their minds than early Christianity. Gold was what they were after, and gold they received. Within little more than half a century after Kabátník's death in 1503 its tentacles were to reach even Litomyšl—in 1567, to be specific; when the town and lands belonging to it were acquired by the Lords of Pernštejn in 1567. The first holder of the fief, Vratislav of Pernštejn, was a true Renaissance oligarch. He was converted to the Catholic faith for political reasons (his wife, Maria Manrique de Lara, came from a family of Spanish Catholic grandees); but after his death in 1582 his Spanish widow (and, from 1608, her daughter Polyxena) unleashed a counter-reformation, soon to become (from 1627) the official policy of the Government. Before, however, the Pernštejn family became extinct (for lack of male issue) in 1631, they left behind two monuments which have adorned the town and endured up to the present time: namely, their Renaissance castle built by Lord Vratislav at the peak of his power; and the *Gymnasium* founded by Frebonia of

Pernštejn (the last childless scion of her famous family; she died in 1646) which was founded in 1640, and opened its doors to students in 1644.

This school—the most outstanding one in eastern Bohemia for three centuries so far—was under the spiritual jurisdiction of the bishops of Hradec Králové (Königgratz) and, at times, was a part of the University of Prague. The teaching in it was entrusted by Lady Frebonia, not to the Jesuits (though it was founded at the time of forcible counter-reformation), but to the Piarist order, which held on to it until the secularisation of the school by the Government in 1874. Between 1644 and 1874 the institution educated many outstanding men. Perhaps the one best known abroad was the famous nineteenth-century physiologist Jan E Purkyně (1787–1869), to whom the world owes the term *protoplasma*; and whom astronomers (at least, the visual observers of earlier generations) may remember as the father of the 'Purkyně ( = Purkinje) effect'.

Other, and equally famous, men were born in Litomyšl, who left their native nest too soon to receive all their education in their home town. The best known of these was, of course, Bedřich Smetana (1824–1884)—apart from Antonín Dvořák, the greatest composer of music whom the Czechs have given to the world. Born in Litomyšl on 2 March 1824, he quickly attracted local attention as a musical prodigy (he gave his first concert in Litomyšl at the age of barely six years!); and though his life's journey took him away from his native town soon thereafter, he returned often (for the last time, in 1880); and although buried in Prague (in the memorable cemetery at Vyšehrad) he still lives in the grateful memory of his fellow citizens of Litomyšl.

But to come back to the old *Gymnasium*, by 1874 the Piarist order was already languishing, and could no longer provide enough teaching brothers from their own ranks to provide adequate instruction in all requisite subjects. Therefore, the State stepped in and secularised the institution, still functioning for another fifty years in the old (280 years old!) building, built to outlast centuries, which was not replaced by a modern one until 1923. The old historical edifice has continued since then to house the Municipal Museum; and the Town Founding Charter of 1263 from King Přemysl Otakar II is one of its priceless exhibits. When father was transferred in 1912 from Kolín to Litomyšl, he was still called upon to teach in the building of the old *Gymnasium*; and did so until he left for Prague in 1921. I visited him there many times; and recall distinctly having seen, in the school's collection, a large colourful picture of the Japanese volcano Fujiyama—little knowing that in fifteen years' time I should stand on its summit!

However, a respect for chronology compels me to confess that, having been born in Litomyšl on 4 April 1914 (see figure 2.2) I did

*Early years in Jičín* 37

not linger there very long; for only three months after that event, at the beginning of the summer vacation at the end of June, the Kopal family decided to spend that summer with the maternal grandparents in Jičín, and it was not to return to Litomyšl for more than four years! The reason was, of course, the outbreak of the Great War. The dual monarchy of the Emperor Francis Josef (of which the Kingdom of Bohemia was then a part) precipitated its outbreak on 28 June 1914 by a declaration of war on Serbia, an event which quickly spread into a more general conflagration, and in its flames the dual monarchy itself was to perish in the autumn of 1918. This was, however, to happen only after more than four years of brutal fighting, in the course of which some twenty million people all over the world were to lose their lives. In the meantime, however, my father (a reserve officer) was promptly called to the colours; and remained in the army for more than four years. That my mother with the children would meanwhile stay in Jičín with her parents went without saying.

**Figure 2.2** (left) The house in Litomyšl where I was born. (Right) The first school I attended in Litomyšl between 1920 and 1923.

The first memories in my life which I can recall go back to the middle of the Great War, to the time of the battle of Verdun on the western front. I still remember quite vividly even today that one evening (it was already dark) my mother's cousin sneaked into the house and (quite oblivious of the presence of a child) informed my mother jubilantly of the German retreat from Verdun! I do not know from which source she could have received that news; but I recall that my mother (a rather unemotional person, and generally mindful of decorum) joined in the jubilation without restraint.

The reason was, of course, that the Czechs were hoping for the victory of the *Entente* to help them recover their national independence, lost since the Thirty Years' War. This expectation was not disappointed; and the twenty years between 1918 and 1938—the years of my youth—were the 'golden sunset' of modern Czechoslovak history, without parallel since the days of King George of Poděbrady (1448–1471). Between 1939 and 1945 the Czechs learnt what a German victory in World War I would have meant for them twenty years before; just as the events which followed 1948 (and especially after 1968) cured them of the faith in a benevolent 'big uncle' in the east.

The German setback at Verdun (when General Falkenhayn broke off the battle in which more than half a million Frenchmen as well as Germans lost their lives) would date my earliest recollections as June 1916, when I was only a little more than two years old. Subsequent memories, however, are emerging in ever-increasing clarity and numbers, mostly connected with my father's military life. By 1917, father was stationed as military commander of munition factories (making grenades and shrapnel)—what a suitable job for a modern philologist!—in the east Bohemian city of Pardubice, where his family could visit him from time to time. As my older brother Miloš was already of school age at that time, I accompanied mother on these visits mostly alone, and thus got an early opportunity to get mixed up in 'high society' at a tender age.

The local hotel (*Koubek*) in which my father (and his occasional guests) were accommodated offered an occasional haven also to Princess Windischgrätz, daughter of the ill-starred Crown Prince Rudolf (the only son of the Emperor Francis Josef I of Austro-Hungary, whose life ended so abruptly at Mayerling in 1889), and the Emperor's granddaughter. During the horse-racing season at Pardubice (which had a reputation somewhat equivalent to Ascot in England) the princely pair put in a frequent appearance there; and (in the days when she was in a good mood—which was not always!) the Princess was said to have paid the little Kopal tot (then three or four years of age) a rather unusual amount of attention. I remember none of it myself; but was told so by my parents (who described her husband as a very reasonable man, who often had his hands full with efforts to calm down the commotion caused by his sometimes unbalanced spouse).

I remember, however, other things from those days of greater portent for my future: for it was there that I first saw the domestic use of electrical light, and—even more important—was allowed to compose (under suitable guidance) my first letter to grandfather in Jičín on a typewriter. In grandfather's home, only kerosene lamps were in use up to the end of his life; and later in life, when illness confined him to

*Early years in Jičín*

bed for several years, he disdained to use radio, or even a fountain pen which was presented to him; and remained faithful to his ordinary pen or pencil till the end of his life. I have many samples of his handwriting still on hand; and their comparison with mine is enough to convince me that the art of penmanship must have become extinct with his generation.

Memories of life with grandfather from the time when I was three or four years old are already numerous in my mind; and some of these should perhaps be saved from oblivion. Grandfather was still in active service in those days; but his educational duties were far from over when he left his school: for there was still myself to be educated. He did not like to sit while teaching, and was (it seemed to me) always on the move. And how well do I remember all I learned when he took me for a walk! Not that it was easy to walk with grandfather! He was a tall man (though, in later years, he did not hold himself any too erect); and when he got going, he marched like he had done to the Holy Land in the days of his youth, oblivious of whether his little grandson could keep pace. Often he did not, but could catch up when grandfather stopped to point out some unusual plant by the road-side, a butterfly fluttering in the air, or a beetle crawling in the grass (which, to make sure that I would remember, he identified also with their Latin names); or again pointed out on the horizon distant hills or mountains visible on that day and identified them with their names. Those were the days when wildlife still abounded in the vicinity of human habitations and smog was still a curse of the future.

It was at that time that grandfather taught me also to read and write. It must have come very naturally to me because I cannot remember any problems†; though (I was told later) grandfather used a 'global' method for this purpose—he taught me to understand words as a whole, rather than individual letters, a method then rarely used, but later to be adopted in many schools. Before I was four, grandfather also taught me to understand the phases of the Moon; and I still remember my intense disappointment when told that the stars in reality are round, and not horny as on the Christmas tree.

On the whole, the attention I received from my grandfather no doubt accelerated my mental awakening; and our walks were certainly instructive, but sometimes physically tiring as well; and even grandfather noticed this on occasion. But what did he do about it at least

---

† Except that, when starting to write, I did so consistently with the left hand (and did not learn the use of the right one until I began to attend regular school). Although, for the most part of my life, I have been effectively ambidextrous, in the game of tennis I never learned to use the 'backhand stroke', but play a forehand with the left—an inclination which apparently goes through the family, since it was inherited by at least one of my children as well.

once? Just as I was unable to walk much further (and we were quite far from home), a train was coming from a distance; and its tracks were very close to our road. When the train came into view, grandfather waved to its engineer—and, behold, the train slowed down (they never went too fast in those days) and stopped for us to get on to it. This, to a little boy, looked like a miracle, and grandfather's hand like a magic wand stopping the mighty engine, heavily puffing. I wanted to learn to do the same, and asked grandfather how he had done it. 'Oh', replied he, 'its engineer was once my pupil.' This did not teach me quite what I wanted; but I learned from the incident something even more valuable: namely, the respect which teachers enjoyed in their communities in those days (now largely gone) and the teacher–pupil relationship which can last a lifetime. I never managed to stop a train myself by such means in any part of the world that I had a chance to visit, but I saw it performed at least once more by someone else: namely, by the late Professor Henry Norris Russell, who stopped a local train from Princeton to Princeton Junction (on the main New York–Philadelphia line, in New Jersey), by a wave, not of his hand, but of his umbrella!

In later years, as I grew into maturity, and after grandfather had gone to his reward, I often thought of his educational methods and used to compare them with those usually followed by my father. Thus whenever I asked grandfather a question, he always seemed to know the answer—at least he would never say he did not know. Father was much more open-minded in this respect, though not above poking gentle fun on the unsuspecting child. Thus when I asked him for the name of a bird (which may have been a sparrow, and which grandfather would have recognised at a glance), father's response was that he could not recall its Czech name; but in Latin it was called, he said, 'avis admirabilis'; and not yet being knowledgeable in Latin, I took it for gospel truth.

Also, as far as I remember, grandfather possessed but little sense of humour, which father possessed in a measure that endeared him to his pupils and colleagues alike. Grandfather would also never praise anyone's accomplishments, in the classroom or at home; his classroom must, I suppose, have been like a ship on which 'excellent performance was standard'. My father, on the other hand, was always generous with praise when it was due (albeit modest about his own accomplishments), and always ready to help when help was deserved (often at considerable expense of his own time or money).

During the high noon of his busy life (partly described already in the preceding chapter), father's preoccupations with the affairs of others left him but little time for his own family; and his principal

*Early years in Jičín*  41

(delayed-action) contribution to the education of his sons was his personal example which we instinctively learned to imitate. Admonishments were seldom heard from his lips (at home, these were mother's department). Perhaps he felt (rightly) that they would not do much good; and he left his sons to follow largely their own way.

How did such an educational system work? My older brother Miloš (born in 1910) eventually went up to the University to study law—a safe though rather unexciting career—and graduated in 1934. Since through all his school-years Miloš was a model student, his future never gave his parents (or teachers) any reason for concern. However (as I shall have to confess later in this chapter), with me it turned out to be a somewhat different story; and it was said (playfully by some, but by others with more concern) that in our family's nest I must have been born of a cuckoo's egg! My desire, in later years, to embark on a course of science at university level brought an anguished cry from father's heart: 'But what will you do with a science degree, my boy?'

Fortunately, he lived long enough to see! Natural science, to tell the truth, was not my father's *forte* (grandfather Lelek certainly knew more); but yet father too left me a priceless heritage which I never forgot: an example of what the life of a scholar (of any field) should be. Like he, I too have never ceased to be a student of my subject; and I hope to remain one till the end of my days.

But, to return to the war and the years of my early childhood, my recollections of its last years already become numerous; and not all of them are pleasant in retrospect. As the Great War was slowly approaching its end, the continental blockade was seriously undermining supplies of food; and, what was worse, the (rightly) anticipated bankruptcy of Austro-Hungary made farmers very reluctant to sell any food for money. This situation was worst in large cities and relatively not so bad in country towns like Jičín surrounded by fertile lands; but even so it became increasingly difficult to make farmers part with their food for paper money.

The consequences of food shortage were detrimental mainly for children of tender age; those who wielded power had enough for themselves till the end. However, the numbers of those who did not were steadily increasing; and, small as I was at that time, I took pride in the part which I was able to take in the family food-gathering efforts—by singing! When mother or our nanny set out on their perpetual journeys to the countryside to obtain food from the farmers, and when their efforts to bargain were not getting anywhere, I was usually called upon to sing ... the *Marseillaise*! I never had to be asked twice, and once I started, the audience gathered rapidly from all corners (sometimes I was even called upon to give an encore). When I finished, the desired food rapidly materialised; and we brought it home in triumph to grandmother Marie who was thus spared worries about what to feed her family with, at least for a while!

With a child's naivety I attributed my success to the strength of my voice (of which I was then very proud); but I learned better later. What I was called upon to perform, to sing the French national anthem in the land of 'Central Powers' which were at war with France, amounted, strictly speaking, to arch-treason; especially when sung in the enemy language! My father, by all his life's work a convinced francophile (a faith which was shattered, but not completely destroyed, by the Munich betrayal of 1938), taught me to sing the *Marseillaise* in French during his infrequent leaves at home—a song which no adult could risk singing in public without serious consequences. However, a child could; and the rural audiences (exclusively Czech around Jičín) traditionally sympathised with France. The acclaim (and reward) which the little singer received for his performance was, therefore, quite disproportionate to his art—as he learned only later and somewhat to his disappointment. At that time, I felt like a little urchin on the barricade of the famous canvas of Delacroix, following the flag of Marianne; and often kept singing the *Marseillaise* all the way home, somewhat to the apprehension of my female escort, who were continually on the lookout for gendarmes who would put a stop to it. But I, the only male (albeit little) of the group, felt full of courage, especially when thinking of the food which would await me on the return from a successful expedition. I may add that, during World War II twenty years later, such escapades could have had a swift and tragic end; but in the course of World War I, combatants on the domestic front were still a relatively civilised lot.

I do not know how often I took part in such foraging expeditions; but the end of the war came for the Czechs on 28 October 1918, when they at last shook off the yoke imposed upon them by the Habsburgs in the seventeenth century as a result of the Thirty Years' War, and once more regained (after almost exactly 300 years) their national independence—alas, not for too long! But I do remember very well the first Independence Day in 1918 and the delirious celebrations in the Jičín town square, all the windows full of (sometimes hastily improvised) flags of the new nation, and music—music everywhere! It was a sunny day in Jičín; and sunshine was in all the people's hearts; only father was not with us to share it on that day, as duty called him to Prague as an officer of the newly-forming Czechoslovak army.

He did return home later that year, none the worse for wear, and was demobilised by the end of the year. The Christmas of 1918 we still celebrated all in Jičín, the last Christmas in our grandparents' home;

*Primary school years*

for grandmother Marie (already then afflicted by cancer which she tried to conceal) had only little more than one year to live (she passed away in March 1920); and after her death only infrequent visits brought me back under the ancestral roof again.

In January 1919 our family returned from Jičín to Litomyšl (which we had left before the commencement of the World War), where father resumed teaching at the *Gymnasium*. That summer we visited both grandparents, in Jičín, as well as Hřmenín (where grandfather Kopal lived for four more years; though grandmother had already died in 1909).

By the end of the summer in 1920 (from 1 September of that year, to be precise) I began to attend primary school in Litomyšl, and thus embarked on the career of pupil and student, which I have made my own ever since. To begin with, I did not have to learn very much; for thanks to the previous efforts of grandfather Lelek, I could already read and write without much trouble; and spent most of my time (I was told) day-dreaming, or looking out of the window where more interesting things were going on all the time. But we had a kind and understanding teacher, Alois Vaníček by name, who made us all like school as a place of work and play, so much that we did not want to go home afterwards (in the first grade, instruction lasted only two hours a day). My playmate sharing the same bench, Mirek Šťastný, was nobody else than the future Professor of Civil Engineering at the Technical University of Brno, who was to become nationally known as an expert on large-scale ferro-concrete structures (and is now also emeritus); we have remained in touch with each other ever since.

By the time I spent my early years there, Litomyšl was a town of some 8000 inhabitants, barely larger than a century before when Smetana was born there, and although it lived largely on its national renown in the past, it was also full of life, intimately connected with its earlier humanistic traditions. Since the town's administration had long ago declined to have anything to do with railways, Litomyšl was largely by-passed by the 'industrial revolution' in the nineteenth century. It remained a town of schools, libraries, printers and booksellers, and of artistic life in general—facts reflected also in the daily life of its citizens. Professional sport was unheard of within its confines (to this day, Litomyšl does not have a football club worthy of the name), but amateur gymnastics flourished among citizens of all ages, from six to sixty, in the branches of 'Sokol' and other national organisations (of which I too was a member among the juniors; and performed with them on a few public exhibitions). Radio—let alone television—was yet to be invented; but local cultural activities—musical, literary, theatrical—flourished; with everybody taking part as well as they could. I myself was directed twice to try my luck on the stage (in roles appropriate for my age); but it soon transpired that I inherited none of the theatrical talents of my grandmother Marie; and the experiment was not repeated; nor did I show any sign of being a musical prodigy; so that my place in all such productions was henceforward relegated to the audience.

But how much there was to watch from this vantage point! I still recall today the *Maiales* (a traditional festivity which was a privilege of the 'sixth-formers' of the *Gymnasium*), performed on 1 May 1920 in nineteenth century costumes in the spirit of that delightful novel by Alois Jirásek; and remember vividly the day in 1921 when that great master of Czech historical prose (who between 1874 and 1888 taught at the local *Gymnasium* himself) re-visited the town on his seventieth birthday, to witness the unveiling of a memorial plaque in his honour (at 'Buřvalka', where he lived as a young man). When his attention was called to the fact that a seven-year old son of one of the present professors of the old *Gymnasium* had already read many of his books, the old gentleman was very kind to me for a few moments; and autographed for me a photograph which is still in my home in England.

Perhaps the greatest fête which I remember from my years in Litomyšl was, however, the unveiling of a monument to Jan Amos Komenský (Comenius), the last bishop of the old Unity of Bohemian Brethren, and (his chief claim to fame) one of the founders of modern pedagogy. On 19 June 1921, two monuments (erected by public subscription) to Comenius were unveiled in Litomyšl—one in the town, and the other some distance away, at a place called 'Růžový palouček' ('Rose Field') where (according to tradition) the local Brethren, on their expulsion from the region (by orders of the new Catholic rulers) buried their chalice. Whether or not Comenius was there in person is uncertain (though he may have been; for he is known to have lived at that time in the neighbouring town of Brandýs nad Orlicí, under the temporary protection of the Lords of Žerotín); and nobody has searched for the legendary buried chalice in modern times. However, the field of wild roses (which, according to the legend, grew up from the tears of the Brethren leaving their homes) were still there in full bloom (as they no doubt are up to the present time) and so is the memorial unveiled on the same day (though the one erected on Smetana Square in Litomyšl was destroyed by the Nazis during the Second World War).

I had a chance to attend (with my parents) both these occasions; and there saw Alois Jirásek (who delivered the valedictory address at the Rose Field) once more—and for the last time. He still had almost nine years to live (he died on 12 March 1930); but after June 1921 he never visited Litomyšl again. Only once more did I come close to him, in March 1930, when his coffin was driven from the Pantheon of the

National Museum through the streets of Prague, lined with schoolchildren of all schools of that city, to pay a last farewell to the old bard who was—and still is—held in the affectionate respect of the Czech nation, as Walter Scott is by the Scots, or Henryk Sienkiewicz by the Poles.

But it was not only Jirásek whom I had a last opportunity to see at the Rose Field on 11 June 1921; but also the last male descendant of the Comenius family whom the town-fathers of Litomyšl invited (and brought over) for this occasion—all the way from South Africa! Political upheavals of his times forced Komenský (1592–1670) to spend most of his life abroad. He also visited Britain more than once: in 1641, by invitation of Parliament for consultations (whose delayed action was the foundation of the Royal Society of London); and earlier (in 1637) he was offered also the presidency of the newly-founded Harvard College in Cambridge, Massachusetts†. The last years of his life were spent in Holland, and he was buried in Naarden. He left no sons; but some of his daughters married Dutchmen who later emigrated to South Africa; and their offspring continue to live there up to the present time.

The name of the gentleman who came to Litomyšl from the south to visit the land of his forefathers was Captain G Victor Figulus OBE of Johannesburg. He knew, of course, no Czech (nor German); and as no one in Litomyšl could speak Afrikaans or English with sufficient fluency, conversations with Captain Figulus had to be conducted in French (which the good captain probably learned while serving with the British Expeditionary Force in France during the Great War), with my father acting as interpreter. The captain also visited our home (I remember him to be a very tall man); who as a souvenir (still in my hands) left behind a visiting card with the Czech inscription *Nazdar* ('cheers'). The last national occasion in Litomyšl at that time, celebrating the hundredth anniversary of the birth there of Bedřich Smetana on 2 March 1924, I could no longer attend; for by that time we were already in Prague.

Not all the time in Litomyšl was, of course, equally festive; for the years immediately following World War I in Czechoslovakia were economically no more prosperous than (say) Great Britain was for several years after 1945. But in the time which followed World War I we did not notice this much, and minded still less; for the recovery of political independence in 1918 more than compensated for such

---

† Comenius did not accept this invitation (tendered to him at that time by Governor John Winthrop of Massachusetts), because he was afraid to cross the ocean. However, as a token of his appreciation, he prepared for Harvard a tract on how to teach Latin to the American Indians! A copy of this tract is still extant in the Widener Library.

vicissitudes which came in its wake; and the uphill struggle of the new republic was considered an inevitable fact of life.

Essential food was not rationed; and soon everyone had enough to eat. Luxuries were scarce; but were not missed by those who, like most children of my age, were not brought up with them. Thus for the better part of the first decade of our lives, bananas or oranges (not to speak of pineapples!) remained for most of us only pictures in books, and we had no idea of their taste; only later did we learn that such exotic fruits had to be paid for in foreign currencies, for which the young republic had other (and more pressing) needs.

The reader may also be amused to hear that another commodity which schoolboys regarded as valuable was clean sheets of paper, especially those suitable for painting; and to receive some (from grandfather, for instance) was a real gift! It is in those years that I learned to respect properly paper as a vehicle of self-expression; and even today—to the amusement (and, sometimes, exasperation) of my secretaries—I do not throw lightly into the waste-basket as scrap those sheets of paper which have not been utilised on both sides. It is also no joke, but the truth, that several of my books of more recent vintage were written on the reverse sides of IBM computer printout records, which my students waste in ever-increasing tonnage (sometimes I think that if the numbers printed on one side are all nonsense, one could at least do something more useful on the other).

Moreover—unbelievable as this may sound to our juniors today—the years which followed World War I were the last time when one had some privacy at home, before this was invaded by the telephone, radio and (since World War II) television. Cars (or, as we then called them, 'automobiles') still made but rare appearances in the streets of Litomyšl, and were viewed as a rarity of little significance for everyday life. I had my first ride in one (belonging to a manufacturer in Pardubice who came to visit us, and whose factory was commanded by my father during the war), little knowing that in the next sixty years I should travel behind the wheel of this 'horseless carriage' a distance exceeding a round-trip to the Moon!

But the lure of distance already began to beckon me at that age, not via the 'horeseless carriage', but via the 'iron horse' which could run much faster on the rails than any car could do on the road at that time. Litomyšl itself was not on the main line of rail traffic; and was connected with it only by a local doodlebug which grandfather Lelek could stop with one finger of his hand. However, a part of our summer vacations in 1920 was spent in the town of Ústí nad Orlicí (in a railway hotel which belonged to another one of father's war-time fellow-combatants); and, although not all trains passing through stopped there, an international Paris–Warsaw express thundered through the station at high speed several times a week. This train held a particular fascination for a boy of my age. I wondered what it would feel like to be aboard, and what would I have to be to make it—an engineer, steering the iron horse, was too much to hope for; but perhaps I could go aboard as a train guard. Little did I know that, only sixteen years later, I should be a passenger on the same train, going not only to Warsaw, but to Moscow and across Siberia to Japan, and returning via China—perhaps the longest train ride one could take on the surface of our planet. Moreover, to go then I did not have to travel in the guard's uniform (although these were very impressive in those days) but as an astronomer, *en route* to observe the total eclipse of the Sun on 19 June 1936 at Nakatombetsu, not far from the shores of the Okhotsk Sea.

More will be said on this long journey in the next chapter. In the meantime, we are still in the primary school at Litomyšl and between events which one cannot forget, everyday life went on as usual, at school as well as outside. Some of the influence of grandfather Lelek still persisted (my interest in collecting beetles commenced in Litomyšl); but more often I went to the town library (then housed in the Old Town Hall going back to the early years of the fifteenth century)—one of the largest in the country—to borrow from the kind old librarian Antonín Brachtl the books which interested me at that time. 'Cowboys and Indians' stories had not yet been invented; but the novels of Fenimore Cooper or Karl May took me to Canada or the Far West of the United States, to roam with Old Shatterhand and his Indian friend Winnetou through the endless plains of Arizona or Nevada. It was safe reading, for one was sure that the good cause would always triumph in the end, and the bad guys would be suitably punished—agreeable fiction, rather different from the reality with which I became acquainted at closer range thirty years later.

But the novelist who enthralled me most in those days was Jules Verne. How many hours did I spend then on the *Mysterious Island*, in the company of engineer Cyrus Smith, Gideon Spilett of the *New York Herald*, and their companions who escaped by balloon from Richmond during the American Civil War to get stranded in the South Pacific, and who lived there till their dramatic rescue years later—French literature (in Czech translation, of course) which I was encouraged to read from childhood by my father. It was from Verne's *Mysterious Island* that I learned to respect President Lincoln (and was so sorry that he lost 'his' island as a result of a volcanic explosion).

But even this was nothing in comparison with the impression left in my mind by another Verne novel, *From the Earth to the Moon*, which I must have read endlessly at about the same time! How I loved Michel Ardan (more, I confess, than that typical Yankee President Barbican or Captain Nicholls) and enjoyed his stunts in the weightless state; or the delightful zeal and belligerence of unemployed cannoneers like the worthy T J J Maston! Although I could then have no inkling of the extent of serendipity with which Verne sometimes anticipated the shape of things to come (or again went hopelessly wrong with others), the book has remained very dear to me since that time.

When, therefore, much of what Verne foresaw a hundred years ago came to pass within our lifetime, I applauded our Russian colleagues for bestowing Verne's name on one of the large craters discovered by them in 1959 on the Moon's far side; but thought that I also had a personal debt to settle with the great novelist for the inspiration received from him during my childhood. As is well known, the Apollo missions of 1969–1972 to the Moon returned with a cargo of almost 400 kg of lunar rocks, the bulk of which is still kept at NASA's Johnson Space Flight Center in Houston, Texas. And once, when I received a tiny amount of this precious soil (already used up for scientific purposes), I brought it to France and (during one of our summer trips to Switzerland) mixed it with the soil of Verne's tomb at Amiens. It may not be the only lunar soil which found its way to France; but it was the first to belong to Jules Verne himself.

With such thoughts wandering through my young mind, the happy years in Litomyšl were, however, rapidly drawing to a close. As was mentioned already in Chapter 1, some time in 1920, when I began to attend primary school, my father decided to return to his academic career interrupted by World War I; and the first necessary step was to return to Prague. From the commencement of the academic year 1921/22 he was indeed transferred from Litomyšl to a secondary school in Prague, and he moved there alone at first, as the family could not follow him at once because of the acute housing shortage in the capital. It was, indeed, not till two years later that the building of our family home was completed—in a suburb on the outskirts of the city (it still stands there, although engulfed now by the growing metropolis). It was to become my home for the next fifteen years, where I grew up into manhood; and still remains one during my infrequent visits to my native country. In Litomyšl (my birthplace) I lived barely four and a half years between the ages of four and nine—a much shorter time than I was to spend in Prague (and only marginally more than I spent in Jičín). In Jičín, however, I was still too young to remember much beyond what I have recounted in these recollections; while in Litomyšl I was already influenced very much more by the life around us, though it was not till in Prague during my teens that I had the first rendezvous with what proved to be the fate of my life.

But this was still some time in the future when we left Litomyšl on 5 May 1923, in the midst of a glorious spring morning; and an express train (what a thrill it was to mount one for the first time!) transported us within two hours from Choceň (the nearest stop on the main line) to the capital. It was a beautiful day as we approached Prague around noontime, when I beheld for the first time that architectural jewel of the land on the Vltava river with its eleven centuries of history, its medieval city and royal castle on the hill with the magnificent Gothic cathedral in the background (little knowing that fifteen years later I was to get married there; as 44 years later did also our daughter Zdenka). The beauty of this ancient city, which had survived so many wars in the past millennium, has been attested by more competent observers than I was myself at the age of nine; and nobody less than Alexander von Humboldt, a great traveller of the first half of the nineteenth century, bracketed Prague with Istanbul and Lisbon as the most beautiful cities of Europe. 'Golden Prague' it is still called today, or the 'city of a hundred spires' (an obvious understatement, as there are actually more than 340 of them).

This all was not yet known to a young immigrant of nine years, who could perhaps be excused for not having counted the 340 spires correctly to begin with. It took me several years, coming as I did from a provincial town (albeit of distinguished history and almost as old as Prague) of 8000 occupants, to get acclimatised to life in a metropolis with a hundred times as many inhabitants. This was perhaps the greatest such discontinuity I experienced in the course of my life, and several years elapsed before I came to terms with my new surroundings and the wider world.

Incidentally, our removal to Prague in the spring of 1923 meant also the parting of the ways with our family's older generation. Both grandmothers were already dead for several years; and although both grandfathers came to see us safely installed in our new home, their first visits also proved to be the last. As was mentioned already in Chapter 1, grandfather Kopal (who visited us in June) passed away in September of the same year, while grandfather Lelek (who spent the summer with us) suffered soon after another stroke which prevented him from any further travel. Although this (second) stroke did not affect his mental faculties to any noticeable extent (and he remained as avid a reader as before), he was thenceforward confined to his flat in Jičín, and cared for by local relatives, until his passing in April 1930 as the last one of the old generation. By that time I was, however, already sixteen years old and well launched on pursuits which eventually became the aim of my life. Grandfather Lelek already knew about it, and approved: his last gesture of encouragement was a small present for which I bought myself a copy of Schüller-Novák's *Atlas of the Heavens* (published in Prague around that time and still with me in our house in Wilmslow where I am writing these recollections).

But this is again getting far ahead of our story. Yet turning back to our first years in Prague, two distinct memories remained engraved in my mind since that time. In the summer of 1923 (and again a year later) I was sent, together with a group of other children in need of physical build-up, to Yugoslavia for one month of vacations; and thus beheld for the first time the Adriatic Sea. And, secondly, on the way there (by train) I got the first glimpse of high mountains (the Austrian Alps), the view of which so impressed a child born in flatlands as not to fade from my mind ever since.

Next, having completed in June 1925 the five-year primary school in Prague, I passed the entrance examination for the *Realgymnasium* at Smíchov (the part of Prague where we lived), and entered that school for an eight-year course; an *Abitur* from which would eventually qualify me for university entrance. This proved to be an important event of my life; and in putting me up for this school my father made a good choice. In the *Realgymnasium* which I entered the teaching of classical languages was de-emphasised in favour of mathematics and natural sciences. In the 'classical' *Gymnasium*, like the one which father attended in Jičín, or which my brother entered in Litomyšl, both Latin and (ancient) Greek were taught as the main subjects. At *Realgymnasien*, Greek was dropped altogether; and although Latin lessons continued to be taught six times a week (including Saturdays!) for eight years (so that, in upper sixth, we could tolerably translate from sight both Tacitus and Horace), two modern languages were added: namely, German (for eight years) and French (six years), in addition to the Czech language, literature and history; and, of course, science (mathematics, physics, biology) which occupied about one-third of the entire curriculum.

The English reader may perhaps be surprised to learn that sports or games were no part of it—almost none at all (save simple gymnastics for one hour per week). The days were, to be sure, long gone by when merely to be seen in public in the Vltava river swimming with nothing on but bathing trunks could provide the reason for expulsion from the school (as had almost happened to my father thirty years before); but athletics was regarded still in my time as a strictly extracurricular activity; and secondary schools maintained no organised (let alone supported) teams. My favourite sport since the time of my youth has been mountaineering; and still today, at well past 70 years of age, I can stay on my legs (though no longer climb) in the Alps for twelve hours a day or more; but when I began to play tennis in my teens with some regularity, I had to book time on courts maintained for university students.

The classroom instruction itself in our *Gymnasium* occupied thirty hours per week—five hours each day (including Saturdays); and we were fortunate to have come under the influence of several outstanding teachers. Four in particular stand out vividly in my mind—all gone now to their reward, but living in the grateful memories of their pupils. Dr Josef Klik, who taught us history for seven successive years, was at the same time assistant and right-hand man to Professor Josef Pekař of Charles University, the greatest historian of the Czech nation after Palacký, and as such opened to us horizons way outside ordinary textbooks. Professor Vojtěch Prokeš, brother of one of the Czech chess champions of that time, opened for us a knowledge, not only of modern languages, but also of their literatures; and he did so in a way which often kept us spellbound for hours at a time. And what Prokeš did for us in modern literature, Professor Augustin Wolf did for the classical languages and literature. In his university years, Wolf was a pupil of Thomas Masaryk; and his main contribution to our education was the fact that, under his guidance, we became acquainted with the history and culture of the ancient world, interests which some of us never lost in the course of our lives.

To turn from arts to sciences, Professor Václav Breindl, already then a lecturer at Charles University, and destined to become full Professor of Invertebrate Zoology at the University's Faculty of Sciences not too many years later, was likewise an excellent teacher, who paid little attention to prescribed curricula and, in more senior forms, often mistook us for his university students. The only disadvantage of such an advanced mode of education was the fact that, whereas the general plan of instruction called for a systematic coverage of the animal kingdom, from protozoa to *Homo sapiens*, we did start from the former (which were Breindl's professional speciality); but in our subsequent climbing of the animal tree we did not progress—if I remember rightly—much beyond the spiders! Whether this was the right way of teaching I do not know (no headmaster or school inspector would have dared to tamper with Breindl's curricula!); but it was under Breindl's microscope that I saw for the first time the malaria-causing parasites which were to give me some trouble during my first trip to the Far East in 1936; when it came to apes or elephants, one just had to go and make their acquaintance in the zoo!

However, the teacher dearest to my heart during the upper forms of our *Gymnasium* was Dr Ladislav Klír, who educated us in mathematics for four years, and who did so in a way I never saw surpassed in my life. Ever cheerful and friendly—and equally so with his pupils as well as with our head (he was assistant headmaster himself in those years)—Klír made mathematics effortlessly intelligible to all. One could never forget the way in which he used to encourage those who were not sufficiently quick on the uptake with the arcana of his science. When anyone (in distress) asked how he should deal with a particular problem, Klír's usual answer was an unruffled: 'But cleverly, my boy; cleverly!'; and with but a few words he usually made the subject obvious to us all. His constant message that mathematics was 'truth and beauty' I have always tried to imitate and pass on to others in the course of my own teaching career in different parts of the world. To be inspired and guided by such a teacher in the formative years of one's life was a gift which can never be repaid in full—the only way to do so by those of us who embraced teaching careers ourselves is to return to the next generation at least a part of what we received, at the time of need, from our own masters.

To enter the *Gymnasium* in my time for anyone of the requisite age who passed (not too difficult) an entrance examination was relatively straightforward, with none of the social barriers in the way that still survive in some cases in England; but to remain there required greater effort. The lower third form (*prima*) which I joined in September 1925 had 64 pupils enrolled in it; while in the senior sixth (*octava*) our number was reduced by exactly one-half; the rest had fallen by the wayside in between for not being able to keep up with the requisite scholastic standards. I never found this task difficult, and did, in fact, little school-work at home (at first somewhat to the concern of my parents; but gradually they got used to it).

Although my report cards (still in my hands) show that I maintained consistent progress 'with distinction' (a classification requiring top marks in the majority of subjects, and not more than four of the next best grades), I was seldom—if ever—the best pupil of the form, partly because of habitual unwillingness to over-exert myself in any subject of no particular interest to me at that time; but mainly because in subjects which were of interest, I soon developed a habit of asking embarrassing questions. Not all teachers take kindly to teasers of this kind, especially from those still wearing short trousers, and some of ours (not mentioned by name in this chapter) were no exceptions. For my part, the habit of raising embarrassing questions, evident already in my early teens, has only grown with age, and has remained with me all my life.

By all instincts of my personality I have been an 'inner-directed' rather than 'other-directed' creature—a 'lone wolf' inclined to bark whenever the spirit moves him to do so, rather than an 'organisation man' who should know better when to keep his mouth shut. The rest of my life's story which will unfold in subsequent chapters will support these confessions with more concrete facts. In fact, when I count the blessings of my life, not the least of them is the fact that fortunate circumstances permitted me to lead such a life as I enjoyed, and get away with it!

But there was still a long way to go before reaching that stage; and on the preceding pages we left our young subject in the lower third

*Extracurricular interests* 53

form of his new school, with eyes gradually opening to new adventures. The school and the subjects taught there soon ceased to claim all my attention; and the seeds planted in my mind by grandfather Lelek in earlier years began to assert themselves. For some years, I recall, I was a keen collector of beetles and butterflies—especially of the former (and grandfather, becoming physically incapacitated, gave me his own collection of these around that time, containing many species which, I am afraid, may now be extinct). The transfer from Litomyšl to a city of the size of Prague limited these collecting periods to summer vacations (which, in those years, we used to spend in Harrachov in the border mountains of Krkonoše).

Prague is, however, the centre of a geologically very interesting region, where shallow seas abounded in early Paleozoic times (long before William Shakespeare, in his *Winter's Tale*, made the King of Bohemia a sea-coast ruler); and localities exist within almost walking distance from the termini of some tram lines where rocks replete with fossils of the Silurian and Devonian periods could be found. Soon I became an enthusiastic collector of such fossils; and several times each year I took my precious collection of trilobites and ammonites to the National Museum at St Wenceslas Square in Prague to identify them by comparison with specimens there on exhibit from the world-famous collection of Joachim Barrande (1799–1883), a French paleontologist who dedicated his lifetime to the exploration of the Silurian system of Central Bohemia. (My collection of fossils, which eventually numbered several hundred species, was later accepted by Professor Breindl for the natural-history cabinet of our *Gymnasium* at Smíchov.)

Extracurricular pursuits of this kind claimed most of my free time during the first two (and partly the third) years at the *Gymnasium*. But, in the lower fourth form (*tertia*) experimental physics began to claim its place in the interests of young scholars of my age. I vividly recall the excitement created in our form by laboratory experiments with electrical discharges in vacuum, or with cathode rays and x-rays, which then formed a part of our educational curriculum. In order to duplicate these at home, I remember I set out to build for myself an induction coil capable of generating a sufficient voltage; but had to abandon the effort when the pocket-money available for the purpose proved sadly insufficient for the requisite amount of copper wires. Or, that spring, I switched my interests from dead fossils to living creatures, and set up an aquarium at home with tropical fish, a project which flourished for at least a year, until I had to dispose of it by selling it to a dealer (we shall soon see why).

But it was not only natural sciences which were then attracting my mind with a promise for the future. For even before entering the *Gymnasium* in 1925 I had started to write—possibly a budding inclination inherited from my father—with results which earned greater approval from the family than my collecting activities. The first such effort I recall described my itinerary to the Adriatic Sea in the summer of 1923 for my grandfather; and although its original has long been lost, I can still recall several sentences from it—possibly because these were read to some family members in my presence (a fact which then filled me with great pride).

Some years later, when I entered the *Gymnasium*, indications that I could perhaps write better than most of my contemporaries took a more definite form, when in stylistic exercises at school I was more than once summoned to the rostrum to read my exercise to the rest of the class—an exhibition which the audience may not have enjoyed, and neither did I; for although the text may have passed muster, my ability to read it aloud did not; and it took several more years before I could express myself orally as well as in writing in a passable manner. That the ability to write comes first is only natural, for one learns writing essentially by imitation; and (thanks to grandfather Lelek) I had learned to read before attaining school age.

One more nascent literary effort may perhaps be mentioned in this place, because of its connection with what was to come later in my life. In 1927, my classmate Zdeněk Němec (a very talented musician) and I decided to start 'publishing' a class newspaper. The idea was neither new, nor original; for several forms in several schools were doing the same at that time; so we simply decided to 'step in the queue' and attempted to do likewise. It went without saying that, save for (doubtful) sales of the finished product, our undertaking could not rely on any regular source of income; and all typing as well as duplicating would have to be done by volunteers (or, in practice, the 'editors').

Mindful of the tribulations my father had with a similar flyer at Jičín thirty years before, we first sought to obtain a blessing for our undertaking from our headmaster. He (a historian by profession) granted such a permit very willingly; and, in fact, volunteered to become the first subscriber to the broadsheet—on condition that all contributions appearing in it would be written by the pupils. Who else would have written anything for us? And so the new 'journal' saw the light of day—the first periodical and publishing venture with which I became associated—and kept appearing (at irregular intervals) until drifting interests of both 'editors' made it expire by natural death. I no longer have any of its issues on hand to refresh my memory on its contents; nor do I remember any more its name (I believe it was *Omnia*); but at least we did not end up with any deficit.

The reader who has followed us so far in this chapter of our recollections may stop at this stage to wonder: is the world described here only sixty years behind our times? The years which elapsed since then have indeed changed life almost beyond recognition. But to appreciate the reasons, let us recall that, in those times, radio was only a recent newcomer (the only receiver we had at home at the time of my youth was a small set with crystal rectifier), that television was not yet invented, and aeroplanes were for all practical purposes non-existent. Horse-driven city transport disappeared only twenty years before to make room for electrical tram-cars; but long-distance transport was still dominated by the 'iron horse' on steel rails moving only at an average sedate speed of 40–50 miles per hour for express trains, and half that speed for locals. Telephones were still almost exclusively reserved for business use; and private automobiles were manifestations of luxury.

On the credit side, communications by surface mail were considerably faster than they are today; and films were (occasionally) still works of art. Legitimate theatre lorded over the world of entertainment; and great actors (of stage or silver screen) were enjoying prestige, or even adulation, exceeding that of any pop-singer of more recent times. The politicians who still, on occasion, spoke what was in their minds were not yet completely extinct. Professional sport had not yet taken over its principal (and lucrative) role of dispelling the boredom of stay-at-home armchair 'supporters' (television had to come first to make this possible). Football fans were still regarded by most of the uninvolved population as a particular species of naive simpletons; and outbursts of violence on such occasions, as the social phenomenon they were later to become, particularly in England, were at least forty years in the future.

In the days of my youth, the schools of the type I used to attend were still the places to which one went with the hope of learning something of value for one's own future; and teachers, as fonts of wisdom, still enjoyed social prestige in the community not measured by their income alone. Indeed, few people doubted then that this world of ours had a future, and one had to prepare to play one's part in it. Books or journals were relatively cheap; and the press (in any form) still served as the principal vehicle for the sharing and dissemination of information. All this may seem almost unbelievable to our grandchildren today; and the main aim of our recollections is to bear testimony that our story is not fiction—just recollections of the times which (like so much else) have since gone with the wind.

## Rendezvous with Destiny

But—after these encouraging remarks—let us return to the main theme of our narrative, and ask ourselves where and when we met our destiny which launched us on our life's mission. Sometimes, after a life spent in pursuit of a particular line of endeavour, it may be difficult to recapture 'how it all started'. This is, however, certainly not the case with me; for to this date I can specify the day and (almost) the hour when I received the call to become an astronomer, although I cannot recall any particular interest in the subject prior to that fateful 'moment of truth' which decided all my subsequent life.

It was on the last day of July 1928, soon after I had returned to Prague in the middle of the long summer vacations with a harvest of newly-collected beetles; and as was my custom in those years, I took the new acquisitions to the National Museum on St Wenceslas Square to compare them with the specimens on exhibit there for identification. The Museum was about three miles from our home; and as it was a warm sunny day, I decided to cover this distance on foot.

The return journey called for the crossing of the Vltava river by the bridge leading to the National Theatre, and on that corner (from which there is one of the most beautiful views of the city; see figure 2.3) I saw a portable telescope placed there, attended by a commercial optician who was offering to show sunspots to any passerby for one crown (about the price of the tram-car ticket from the Museum home). This arrested my attention; and coming closer I saw a classmate of mine, Václav (Venda) Izera, together with a fellow-student from another school in the same district (Miroslav Stelčovský by name) who had just had a look through the telescope themselves. I followed their example; and marvelled—I do not know for how long (there was no queue of Prague citizens behind me lining up to see sunspots); but my decision to forget all about beetles or fossils and to become a student of the heavens was probably formed during those fateful minutes, and was reinforced by the decision of my friends standing there with me to build a telescope to see the celestial phenomena on their own. How could I stay behind? I decided, on the spur of the moment, to do likewise; and this was, in effect, the commencement of my astronomical career.

That it was the first flicker of interest in astronomy in our family I would not dare to say. My father, otherwise a dyed-in-the wool arts man, learned to know at least the names of the brightest stars in the sky or of their constellations in his youth, and remembered them well till he was past eighty (as I had an occasion to verify); while my elder brother Miloš (1910–1948) exhibited a keen interest in astronomy in his mid-teens, going so far as to spend most of his pocket money on the assembly of a small library of books on the subject (of which I was to become the principal beneficiary), but his interest in the stars waned in his late teens about as rapidly as mine did in beetles or fossils. And, interestingly enough, his only (posthumous) son, Miloš junior (born in 1948), who later became an accomplished linguist, went through the same period of intense interest in the sky in his early teens, which was

## Rendezvous with destiny

also deflected to other channels (but not extinguished, as is attested by his translations of some of my astronomical books from English into our native vernacular). In brief, there is some evidence that interest in astronomy was being transmitted in the family tree in the form of a hereditary recessive characteristic, only occasionally (or fleetingly) coming to the fore, and it was at first thought by my parents that its emergence in their second son would also prove temporary. But, for better or worse, this was not to be the case; my interest in the stars was to remain dominant—and more: for it provided sufficient incentive for me to remain a student of astronomy for the rest of my life.

**Figure 2.3** The place in Prague, by the Vltava river in front of the National Theatre (to the right, off the picture) where I first saw sunspots through a telescope and decided to become an astronomer.

This could not, of course, have been foreseen yet by a young boy of fourteen, walking home from his rendezvous with destiny in front of the National Theatre of Prague on 31 July 1928. When I reached home, I quickly looked up, and ran through, all the astronomical books accumulated by my brother some years before. Among them the Czech translation of a book *Astronomy for Everybody* by Simon Newcomb (1835–1909) particularly attracted my attention. This book, written by a leading American astronomer of his time, appeared in New York in 1902, and was translated into Czech in 1909. By the time its message reached me, its contents were already dated (concerned as it was mainly with solar-system topics, it had very little to say on the stars). There were, however, other books around of more recent vintage whose message I proceeded to absorb. I was alone at home that enchanted day; with nothing to disturb the lure of the heavenly kingdom opening up for my eager mind by the perusal of their contents; and before the Sun set that evening I was completely under their spell.

The month of August still took me away from Prague (to a summer camp in the Low Tatra mountains, I believe); but all previous pursuits had been summarily forgotten, and in my mind I dwelt only among the stars. Even before my return for the commencement of the new school year, I started correspondence with Venda Izera (another participant of that brief conjunction before the National Theatre on 31 July) as to the best way of acquiring a telescope of our own, with which to behold the other celestial wonders that we read about in books. To acquire one by purchase was wholly beyond our means; therefore, we had to resort to self-help. It did not take us more than a couple of weeks to construct our first telescope. One could only with some blushing call it a 'refractor', but it had a lens: a (chromatic) objective of 40 mm diameter and 1 metre focal length (a 1-dioptre spectacle lens, which we bought from the optician); while the eyepiece was a magnifying lens of about 2 cm focus. The lenses were fixed at opposite ends of a 1 metre long paper tube, and mounted (azimuthally) on a portable tripod with the assistance of a local tinker!

Primitive as was this instrument, it provided us with an unending source of joy. Turning it on the Sun in daytime, we verified that the spots we saw on its disc were no mirage (the solar activity was then close to its maximum); while as the sky grew dark after sunset, our telescope disclosed to view many craters on the Moon, or the satellites of Jupiter dancing around their central planet; and (somewhat later at night) also the disc of Mars. Venus was not visible that autumn in the evening, to show as a crescent in solar illumination (and next spring we were already able to observe its phases in much superior circumstances).

Since that time, I have been privileged to observe occasionally the same targets with telescopes of vastly greater optical power, at Mt Wilson in California or Pic-du-Midi in southern France; but at no later time do I recall having been more exhilarated at what I saw than the excitement provided by our home-made telescope in the autumn and winter of 1928–1929 in the days of our youth. All the adventures of Julius Caesar and of his legions in Gaul (as described in his *Commentaries*, which we had at that time to read at school in the original Latin) were nothing for us in comparison with the Odyssey we could wage through space with our home-made optical contraption. Venda Izera—my bosom friend and faithful companion on such trips—no doubt agreed even more; for Latin was not his *forte* (and caused him, in fact, to leave our *Gymnasium* after the first four years for more practical pursuits).

As the years went by, Venda and I had less and less chance to see each other; for the Atlantic Ocean separated us for many years. Now, much to my regret, he will be unable to read these reminiscences; for he passed away in Prague on 26 August 1983, aged 69 years. At least his younger brother George and sister Libuše were destined to find their way to England, and become British by naturalisation as well as by marriage. But, I am sorry to say, neither shared with me my celestial preoccupations to the same extent as their elder brother had done.

In the autumn of that year an event took place which was of fundamental importance for my whole future: I discovered in Prague the existence of an Astronomical Society, founded in 1917, which provided a national centre for amateur astronomers of the country. By 1928 it already had over 800 members; and its Chairman since 1926, Dr František Nušl (1867–1951) (see figure 3.1), was Professor of Astronomy at Charles University and Director of the National Observatory of Czechoslovakia and between 1929 and 1935 served also as one of the Vice-Presidents of the International Astronomical Union. If only my brother Miloš had become aware of the existence of this society four years before, two Kopals, rather than only one, could have become students of this subject in their turn. This society has since 1920 published a monthly journal called Říše Hvězd (*The Realm of the Stars*), which by 1928 was in its ninth volume (and to which I was soon to become, not only a subscriber, but a regular contributor). My father presented to me for Christmas that year some of its past volumes, from which it was easy to learn about the Society's activities. My friend Izera and I applied for membership, and were accepted as from January 1929. This marks the beginning of my association with this Society which has lasted up to the present time (since 1967, from the Society's fiftieth anniversary, as an Honorary Member).

The year of 1929 was a particularly auspicious one for a young tiro to join the Society; for after several years of effort the Council managed to raise sufficient funds to open up on the Petřín hill in Prague an observatory for its members as well as for the general public, equipped not only with instruments vastly superior to what a young amateur was likely to own, but also with an astronomical library of several hundred volumes which contained not only popular books, but also professional literature purchased from the estate of Dr Ladislav Pračka (1877–1922), a prematurely-departed keen student of variable stars, trained at Bamberg and Potsdam. It is in this library that I made my first acquaintance with BD or AG catalogues, or beheld the imposing serials of *Annals of the Harvard College Observatory*, containing the work of astronomers who were to become so well known to me only ten years later. I may add that, of the Society's instruments, an 8 inch Zeiss comet seeker was already installed in the observatory's eastern

**Figure 2.4** (above) The Štefánik Observatory on Petřín hill (327 m above sea-level) overlooking Prague (and administered by the Czech Astronomical Society) which was my astronomical kindergarten between 1929 and 1933. (below) The central dome of the Štefánik Observatory (housing the 8 inch double refractor seen in figure 2.6), with damage suffered from artillery fire during the last days of World War II in May 1945.

dome (figure 2.5); while a twin photographic Zeiss refractor with two 8 inch lenses (one visual, the other photographic) of focal length 343 cm (equivalent to the standard Carte-du-Ciel refractor), and formerly in possession of the Austrian selenographer R König, was installed a year later (see figure 2.6). After a short period of initial training I was allowed free access to both these instruments (and, in fact, the photographic objective of the twin refractor was to accompany me to Japan in 1936 to observe the total eclipse of the Sun on 19 June of that year).

**Figure 2.5** The Zeiss 8 inch comet seeker in the eastern dome of the Štefánik Observatory (to the right on the top photograph of figure 2.4). This was the first larger instrument I was able to use for astronomical observations.

But this is getting ahead of our story. The observatory just described was formally opened to the public on 4 May 1929; and given the name Štefánik Observatory in honour of General Milan Rastislav Štefánik (1880–1919), the first Czechoslovak Minister of Defence, who started his professional career as an astronomer, and who had perished in a plane crash exactly ten years before. This observatory proved to be my astronomical kindergarten before I joined the University, and had it not existed at the right time, my own whole subsequent career might have been very different.

These newly-opening facilities for observational work revived, and greatly encouraged, amateur work in three distinct fields of amateur endeavour—solar activity, meteors, and variable stars. A continuous tracking of solar activity (then near one of its maxima) never held out

**Figure 2.6** The double astrograph housed in the central dome of the Štefánik Observatory (once the property of the well-known Viennese selenographer R König (1865–1927)) with its occasional youthful observer (inset). The visual objective of 18 cm aperture and the photographic objective of 21 cm aperture both have a focal length of 343 cm. The third telescope is a coronograph (added after the war) of 16 cm aperture and 296 cm focal length, used in conjunction with a polarising filter of 5 Å pass band at the wavelength of Hα for observations of solar prominences. The 21 cm photographic objective of this instrument travelled with me to Japan in 1936 to observe the total eclipse of the Sun.

much attraction for me (after that traumatic experience in July 1928 in front of the National Theatre); and visual observations of meteor showers I also never much indulged in; but variable stars attracted me greatly from the beginning. Their systematic observations (by Argelander's method) in Czechoslovakia had a tradition going back to the nineteenth century, when Professor Vojtěch Šafařík (1829–1902), son of the famous Slavist Pavel Josef Šafařík, commenced systematic observations of long-period variables, with an 8 inch refractor whose objective was one of the first to come out of the hands of Alvan Clark†. He made about 20 000 observations, at a time when very few astronomers paid any attention to them, which were subsequently prepared for (posthumous) publication by Pračka.

Some years after the foundation of the Czech Astronomical Society, variable star observations were resumed by its members under the leadership of Dr Bohumil Hacar (1886–1974) whom only a shortage of suitable posts prevented from following a professional career; and who contributed a number of inspiring papers on variable stars to the Society's monthly journal *Říše Hvězd*. Distance from Prague prevented him, however, from discharging the tasks of a group leader in this field effectively; and on his resignation he was replaced by F Schüller who, shortly before, had acquired local renown with the publication of a truly excellent *Atlas of the Northern Hemisphere*—one of the best works of its kind then available, which the French Astronomical Society honoured with a Jean Rey Prize.

Among the members of the revived group the most active one proved to be Rostislav Rajchl, a university undergraduate four years my senior; and when Schüller was soon obliged (for health reasons) to resign the leadership of the group, it was Rajchl who succeeded him in this function—but again only temporarily; for when increasing pressure of academic work prevented him from discharging his duties as he would have liked, the chair of the section devolved on—myself! I was then barely more than sixteen years old, and very probably its youngest member, when I took the chair on 6 September 1930; and I guided the group for some years.

How did it come about for someone so young to be called upon to fill such a function? It is true that my name had begun increasingly to appear in the Society's journal; and my first articles began to appear there (in Czech, of course) around that time (the first one, on $\rho$ Persei, in 1930). But the main reason may have been the fact that I had

---

† This historic lens was originally made for the British double-star astronomer W R Dawes (1799–1868), and acquired from his estate by Šafařík; on his death in 1902, Šafařík's widow donated it to the Ondřejov Observatory. In my student days it was still in use there, but has since given way to larger instruments.

enough time to prepare star charts of the surroundings of the respective variable, and to look up the brightness of the comparison stars in the *Harvard Photometry* catalogues and to distribute them to members. This became almost solely my responsibility, and with it (largely) the choice of stars selected for observation. Another function which went with the job was the duty of publishing a (duplicated) *Circular* to serve as a vehicle of information between members on what was going on in the field of our activities at home (and sometimes abroad). With my previous experience with our school journal I was fairly well qualified to do so, and eventually produced three volumes of such circulars between 1931 and 1933, as a foretaste of the more serious editorial work awaiting in the future (see Chapter 5).

Some time that winter I made, moreover, my debut at the rostrum of one of the monthly meetings of the Czech Astronomical Society in Prague. My own memories of that occasion are blurred with the years; but were recently refreshed by an article which appeared on 27 June 1982 in the Chicago *Hlasatel (Announcer)*—the largest daily newspaper in the Czech language published in the United States—by someone who attended that lecture and recorded his reminiscences for the press. The author was Ing Ladislav Matoušek, my senior by some years, whom I knew well from the Štefánik Observatory in Prague in pre-war years; and who after the war emigrated to the United States.

> In the early 1930s [to quote from Matoušek's article] I once went to attend a meeting of the Czech Astronomical Society, to hear a lecture by its Chairman, Professor František Nušl. Nušl's name was an attraction; and the lecture hall at the Technical University was full to overflowing. However, the announced speaker did not show up. The Vice-Chairman announced that Professor Nušl was laid horizontal with a flu, and would be replaced by a substitute.
>
> To the great surprise of us all, this substitute turned out to be a slim young boy, about sixteen years old. I leaned to my neighbour in the audience (it was the Society's Secretary, Mr. Kadavý) and asked in a whisper why they did not procure a more significant substitute for Professor Nušl. Mr. Kadavý merely smiled and whispered back: 'just wait and see!' And, indeed, the young boy's performance at the rostrum just kept us spellbound. For about an hour or so we heard a wonderful lecture on binary stars, which under certain circumstances can become variables; and on what we can deduce from their observed changes of light.

This report brought back to mind ancient memories which have long faded out of my mind; while I remembered the occasion, I could no longer recall the topic. I was, however, gratified to learn that already it dealt with eclipsing variables, which (as we shall see in subsequent chapters) were to become the fate of my life. I doubt, however, if my performance was as good as described in Matoušek's report quoted above. For I, too, had come to the meeting to listen to Nušl's lecture, and learned about my role as his proposed substitute only about ten minutes before the commencement of the actual performance, without the benefit of any syllabus or notes—wasn't this bound to make me nervous? Moreover, in order to justify my place at the rostrum I probably tried to say too much in too short a time—in other words, spoke too fast. This has, indeed, been my initial failing; and it took me many years to get over it.

Another occasion which I fondly recall from those days was the visit to Prague of Professor Giorgio Abetti (1882–1982), then director of the Osservatorio Astrofisico di Arcetri near Florence, who was invited to deliver guest lectures at the Charles University in March 1931. The subjects of these dealt mainly with the Sun, but also with stellar parallaxes, and the history of astronomy at the time of Galileo Galilei. These lectures (in French) were open to the public; and as they were scheduled in the afternoons, I could (and did) attend them all.

A very pleasing feature of Abetti's lectures was the fact that, after each lecture, the speaker was willing to answer questions from the floor; and as they were many, the questioners had to line up to ask questions privately one by one. Encouraged by Abetti's kindness, I too ventured to step in this queue one day to raise some questions concerning statistical parallaxes. Abetti listened with patience to my school-book French; and thought that these were indeed interesting problems; but to discuss them at that time would have taken too long; could I come to see him next morning to continue our discussion? I said I could not, because I had school. 'Where do you teach?', asked Abetti. I remembered this; and many years later (in September 1952, at the meeting of the International Astronomical Union in Rome) I told Abetti that (being already Professor of Astronomy at the University of Manchester and President of Commission 42 of the IAU) I could then answer his question to his satisfaction, which I was unable to do in March 1931; but added that his hopeful anticipation of my future was a source of great encouragement to a young budding astronomer in his teens. Abetti laughed; and insisted that we celebrate our reunion with a drink!

But to return to the variable stars and their observers in Czechoslovakia in 1931: which stars principally attracted our attention? Eclipsing variables were not yet among them (as they became later). Instead, we were mesmerised by those which, in the variable star catalogues published yearly by Professor Richard Prager from Berlin-Babelsberg, had their type described as 'unknown'. I vividly recall the thrill we all shared in reading Dr Cuno Hoffmeister's stirring address delivered before the Verein der Sternfreunde in Stuttgart, 1928 (and reprinted in *Die Sterne*): 'A very rewarding field is opening up before us; and with zeal and perseverance we must reach our goal. Whose heart would not beat faster at the thought that he may discover great things? Volunteers, forward!'

We felt indeed like volunteers answering the call; and some of the stars we selected then for observation did make history later; but unfortunately, techniques other than visual observations were necessary to impel such stars to share their secrets more fully with us; and these were still some time in the future.

At about the same time (or shortly thereafter) I joined the American Association of Variable Star Observers (with headquarters at Harvard Observatory and Mr Leon Campbell as its Recording Secretary), as well as the Association Française d'Observateurs d'Etoiles Variables (with headquarters at the Observatoire de Lyon). My connections with the AAVSO were, however, difficult to maintain effectively because of my ignorance of the English language at that time. It was only since 1938 that I learned to know Mr Campbell at close range; and it was he who, years later, served as one of my sponsors when the time came for me to become an American citizen. My contacts with the French Association, and with its genial Secretary M Henri Grouiller, were much easier, not only because of shorter distance, but also because my knowledge of written French was by then sufficient for professional correspondence with the headquarters. M Grouiller was a true friend of all students of variable stars; and the French Association represented a forum for international cooperation of like-minded astronomers (mostly young) in the European area of the globe. And more: publications of the Lyon Observatory served as an instrument where the observations of the members of AFOEV could be published; and a subsequent serial, the *Bulletins* of AFOEV, offered a medium in which longer papers by its members could appear in print in French. It was in this periodical that my first longer study on the variability of AF Cygni and RR Coronae Borealis saw the light of day in the autumn of 1932 (neither of these stars was, incidentally, an eclipsing variable).

These were indeed good times, stimulating for young astronomers of that age, but were, alas, not destined to last. For the respite between the two World Wars was, unfortunately, inexorably running out; and the storm which was gathering on the horizon cast its shadow also on the AFOEV before it broke out in 1939 in all its fury. Today, 50 years later, some of its former members are still around; but others are not. Thus Eppe Loreta, a keen Italian observer of variable stars in the 1930s, got mixed up too much for his own good with the Fascist regime of Benito Mussolini during World War II, and perished towards its end as a consequence; while N Florja, a Russian, lost his life on the eastern front in the ranks of the Red Army in defence of his country. B V Kukarkin and P P Parenago, both as brave in their officers' uniforms at the front as they were diligent at the telescope, survived the war, but they have left us since at different times. The only survivors of our once young band are Professor Vorontsov-Velyaminov in Moscow, Dr Luigi G Jacchia (then of Bologna, and now at Harvard), and myself—septuagenarians all—still awaiting our call to join the Grand Army.

But to return back in time to our (then) relatively safe homes, the year 1931 saw also the publication of my first book on variable stars (the first ever to appear in Czech), written jointly with František Kadavý, in which I outlined their physical properties; and he provided an introduction to the methods of their observation (Kopal and Kadavý 1931). Two years later, I expanded my part of it into another, larger, fascicle published likewise by the Czech Astronomical Society when I was in the upper sixth form (Kopal 1933). This was a book which appeared in print solely under my name, as a forerunner of others which have been written since that time.

Both these booklets, being written in Czech, were of interest only to local amateur astronomers. This was, however, no longer the case with a third (larger) work going back to these days—an *Atlas d'Etoiles Variables*, containing 28 charts for 28 variables (mostly new), together with sequences of comparison stars whose brightness was established photometrically by means of the 8 inch refractor of Štefánik Observatory; which we published in 1933 jointly with Vladimír Vand, a member of our group, who succeeded me in the chair of it when I joined the University in 1933. The appearance of this work was already noted on the international forum, and was commented on favourably in several reviews (cf, e.g., Nielsen 1934, Plassmann 1933, or Prager 1935). Unfortunately, it had no sequel; for soon thereafter both of its authors turned their interests to other problems.

Vladimír Vand—another one of the dear friends of my young days—deserves more than a passing mention in these reminiscences. Born in 1911 in Sumy, Russia (where his father was then employed as a chemical engineer in the sugar industry), he returned with his family to their ancestral country after the Communist revolution; and was educated in Prague, where he received his doctorate in physics from Charles University in 1934. Engaged as he was during a subsequent term of national service in military research, he had to leave Czechoslovakia (via France) for Britain which he reached in the summer of 1940; and for the next fifteen years he was engaged in postdoctoral research (in solid-state physics) at the University of Glasgow. When I came back to England myself in the 1950s, steps were initiated to bring Vand from Glasgow to Manchester and attach him to the astronomy department. However, before these steps could be consummated, Vand was offered (and accepted) the Professorship of Physics at the Pennsylvania State College (later to become the Pennsylvania State University), with which he remained associated until his premature death in 1968 caused (I believe) by rapid cancer. Although, since he left Czechoslovakia in 1936, Vand could only occasionally return professionally to astronomical subjects, his premature passing was a source of sincere regret to his many friends on both sides of the Atlantic as well as in his home.

But—to return again to the old times—the organisational and service work connected with my position as Chairman of the variable-star group of the Czech Astronomical Society between 1931 and 1934 was not the principal part of my preoccupations; and I was increasingly driven to try to make original contributions to the subject. Articles of my own on different aspects of variable stars began to appear with increasing frequency in the *Říše Hvězd* journal from 1930; and since 1931 I ventured to send the results of my observations abroad—to *Astronomische Nachrichten* in Kiel, in German; and to the *Bulletin de l'Observatoire de Lyon* (later renamed *Bulletin de l'Association Française d'Observateurs d'Etoiles Variables*) in French. Professor Hermann Kobold (1858–1942) very kindly accepted for the former journal such contributions as I was then able to submit from 1931; and in looking over their subjects, I note that two of the papers submitted in 1932 were already concerned with eclipsing variables (cf Kopal 1932 a,b). They reported no new observations, but rather solutions for photometric elements of several eclipsing systems observed by others—mainly the late Professor Nijland from Utrecht, a keen observer of variable stars, who preferred, however, to publish his observations as they stood, and abstain from any analysis. I applied to his observations the Russell–Shapley methods (whose rudimentary knowledge I acquired privately around that time; cf Chapter 6); and although my results could not claim any particular significance, they were perhaps the earliest pointers of the direction in which my interests would develop in the future.

But I recall, in this connection, one amusing incident which it may be of interest to invoke from the distant past, mainly as a reminiscence of the gentlemanly ways then observed between editors and contributors to astronomical literature. Professor Kobold, the editor of the *Astronomische Nachrichten*, was already past eighty years of age; and, in his correspondence, he used to address the recipients of his missives as 'Sehr geehrter Herr Kollege'—a term almost untranslatable into English in its full politeness. I was similarly favoured by him, a fact not altogether pleasing to some more senior Czech astronomers, who already possessed doctorates. Once, when Professor Kobold acknowledged one of my contributions in his usual style, he added the following note: 'Incidentally, I recently received a letter from one of your colleagues in Prague, which I enclose for your information', signed as it was by one recently graduated astronomer whom I knew very well indeed. What the letter did—no less—was to call Kobold's attention to the fact that he was accepting, and printing in his journal, the contributions of a grammar-school boy! Naturally I was in no position to deny the charge; but was deeply gratified that the old professor, then in full knowledge of the facts, continued to address me as 'Sehr geehrter Herr Kollege'; and did so for the rest of his life! He passed away in 1942 in the midst of World War II, aged 84 years. I never met Kobold in person; but understand that (true to his family name) he was indeed of diminutive size.

And in the midst of such preoccupations, both serious and on the lighter side, the time approached when I completed the last form (*octava*) of our *Gymnasium* and was to take my *Abitur* examination (or *matura*, as it was then called). Such an examination was (at least at that time) more difficult than British GCE at A-level, if only because of the fact that specialisation to science or arts subjects, permissible on the British system at a fairly early stage of one's scholastic career, was not permitted on the Continent. In order to pass the *matura*, one had to satisfy the examiners, orally as well as in writing, of the requisite knowledge in at least four subjects which had to include science as well as arts; and only after passing the oral (which lasted a whole day, and was presided over by the headmaster of another school) was one given a certificate which entitled one to admission to the university or other institution of tertiary rank.

Truth be told, I did not give the forthcoming *matura* examinations much thought; for their outcome appeared to be a foregone conclusion. And so it came to pass; on 9 June 1933 I received my *Abitur* 'with distinction' (for what I used to look like at that time, see figure 2.7). However, no sooner done, then my mind was already on my university course due to begin in October. This will, to be sure, be the beginning of another story on which we shall embark in the next chapter. In concluding the present one, it is time to say farewell not only to one's youth, but also to the companions who shared those years with me, as well as to all our teachers whose efforts laid the foundations for our future. Of the latter, none are alive any more as I write these lines; and of the classmates who took their *Abitur* in June 1933 more than half are likewise dead now. Moreover, some departed from this world under tragic circumstances which deserve to be remembered by the survivors.

Perhaps the most tragic was the fate of Zdeněk Němec (1914–1945),

70  Awakening

an outstanding musician and literary critic, with whom we set out to publish our first school journal in 1927 or 1928. In subsequent years, our interests and inclinations diverged: while mine took a scientific path, Němec perfected his musical talents by studies at the Czech Academy of Musical Arts (once directed by Antonín Dvořák) as a violinist—an art in which he became a veritable virtuoso—and also at the Faculty of Arts of Charles University where he attained a doctor's degree in the history of music. In 1941, he published a valuable monograph on the Czech musician V J Tomášek (1774–1850) of the time of Mozart. Moreover, as a performing musician, he soon reached the top of his art to be appointed leader (i.e. first violin) of the world-famous Czech Philharmonic Orchestra. In addition, his literary talents predestined him to the role of one of the best music critics of his generation; but this combination of gifts proved, unfortunately, fatal.

**Figure 2.7**  A graduation photograph of the author, taken on 18 May 1933 at the time of his *Abitur* from the *Gymnasium*.

In February 1945, when the bloody World War II in Europe was approaching its end, Němec wrote a review of a performance then given by the Philharmonic Orchestra of Smetana's symphonic poem *Má Vlast*, whose last two parts glorify the Hussite warriors of the fifteenth century (in *Tábor*) and the mythical knights of the Middle Ages (in *Blaník*). Němec's review of this music and its performance, which appeared in the press, did not try too hard to conceal an analogy between the Czech medieval warriors of old and the Soviet armies which were then rapidly approaching Czechoslovakia from the east. In

*Outstanding schoolfellows*

intimating this Němec spoke no doubt for many, but the Nazis were no fools: the day the review appeared Němec was arrested and dragged from the concert stage (still in his evening dress) to the dungeons of the Gestapo, where he was accused of 'spreading propaganda inimical to the Third Reich' and murdered without trial before daybreak on 8 February 1945.

**Figure 2.8** Dr Zdeněk Němec (1914–1945), leader (first violin) of the Czech Philharmonic Orchestra in Prague. He was murdered by the Nazis on 8 February 1945.

Such was the untimely end of one of the most talented Czech musicians of his generation, with whom we were once closely connected by our literary inclinations. I recall that it was Němec who urged me in the lower fourth form (*tertia*) of the *Gymnasium* to read Nietzsche's *Also Sprach Zarathustra* (which he brought me in Czech translation)—a fact which did not save his life from the Gestapo. Whether or not Němec, had he lived, would have welcomed the Soviet armies on Czech soil in the summer of 1968 with the same expectations as he may have done in the spring of 1945 is another question which no-one can answer today. In mourning his loss, I wish by these recollections to place a few flowers on his (unknown) tomb.

The second outstanding musician who studied in the same form with us at our *Gymnasium*, Josef Páleníček, was luckier in life. An outstanding pianist almost from the cradle, he studied at the *Gymnasium* while receiving private instruction from professors of the Conservatory of Music at the same time. In the third form (*secunda*) we sat on the same bench; and while I helped him occasionally with mathematics, Jožka came sometimes to our home to play for us on the piano—he was already wonderful then at the keyboard! We also went out together to collect fossils; and (in later years) Jožka's younger brother became a member of the Astronomical Society as well.

In the fullness of time, Josef Páleníček (who later studied with E Cortot in Paris) became professor at the Prague Academy of Musical Sciences and an outstanding pianist, especially as an interpreter of Leoš Janáček (Janáček and Páleníček were fellow-Moravians). On the European scale, Páleníček is at present probably a pianist of the same class as Rudolf Firkušný (another Moravian and pupil of Leoš Janáček); the main difference between them goes back to the fact that while Páleníček has spent his lifetime largely at home, Firkušný has for many years been performing on the stage of the world and, as such, is better known to the general public.

Of other classmates who carried their talents (and bodies) abroad as a consequence of World War II, I should name Karel Brušák, perhaps my closest friend in the upper forms of our *Gymnasium*, who did not, however, manage to complete his university course at the Faculty of Arts before he was obliged to flee from his native country in the spring of 1939 to France, an Odyssey which took him after the French defeat in 1940 to Britain, which has become his home ever since. For most of the past years he has served Her Majesty as Chief of the BBC shortwave broadcasts in Czech (I sometimes took part in these myself); and, although now retired, he still continues to teach Czech language and literature at Cambridge.

A second close friend, Jaroslav Kozák, had very different fortunes. He survived the German occupation in Prague; and due to his 'correct' orientation in post-war Czechoslovakia he rapidly made a political career, first in the Communist Party organisation, and later in the Czechoslovak Foreign Office. In this latter capacity he was assigned senior diplomatic missions to several African countries; and was reported shot down with his plane over Angola during their civil war in the 1970s—the second one among our classmates whose final resting place will probably never be known.

The third 'musketeer' from our class who saw military service abroad during World War II is Gustav Jelínek, who almost completed his medical qualification at Charles University before the Germans moved into Prague in the spring of 1939. Obliged to flee from the Gestapo, he escaped from the country in time, first to France where he joined the re-forming Czech Army in exile; and with the remnants of this army he moved to England in the summer of 1940. He completed his medical qualifications at Oxford, and promptly joined the British Army Medical Corps for field duty; but being always of an adventurous bent of mind, he volunteered for overseas service in the China–Burma–India theatre in the Far East. On his return in 1945 (with the rank of Major, RAMC), he settled in London and became a well-known consultant at St Bartholomew's Hospital in London. However, far-away lands never ceased to excert their lure on his mind; and in spite of his years he has remained an inveterate traveller. Now retired, he can indulge in this hobby to his heart's delight; and, like most old soldiers, is always glad to recount his adventures (in fighting not only the Japanese army, but also monkeys in the jungle) if you give him a chance.

None of us knew, of course, what the future held in store as we parted from each other after the *matura* examinations in June 1933, to follow our own stars. In looking back at those years more than half a century later, I can see only too clearly that while we were all good friends, my relations with most of my fellow-students were on an intellectual rather than emotional basis: I knew, on the whole, more than any one else in the form—a fact accepted by our teachers and fellow-pupils alike—probably because my mind was centred on things rather than people (and this at times did not earn me any particular popularity). Moreover, a certain degree of precocity in one sense probably implied an immaturity in other directions; for while I may have known more facts, I no doubt knew less of life than many of my contemporaries at that time; and it took the experience of many years to restore the balance.

In conclusion, one more farewell remains to be bid before this chapter can really come to its end: and that is to the Astronomical Observatory in Prague, and its Štefánik Observatory at Petřín, which were my astronomical kindergartens in my teens, and nurtured my interests in astronomy until the time came to join the University, and later to take off for faraway lands. Although it soon transpired in the course of this metamorphosis that my principal contributions to astronomy would belong to the theoretical, rather than practical, aspects of our subject, after the tussle of daily life I often return in my thoughts to the observatory at Petřín hill, among the amateur friends of my youth (now also largely gone by), of whom Mr Josef Klepešta (1895–1976), for many years the Society's indefatigable Secretary (figure 2.9); and Mr František Kadavý (1896–1972), the observatory's Superintendent (figure 5.37), have an especially warm spot in my heart. In due course, both of them co-authored some of my books; and we watched together the stars go by when the century was young; for us the passing years detracted nothing from their lure. Although it has been my good fortune to be able to remain Urania's full-time servant for a major part of my life, astronomy has never been a mere profession with me to be pursued during office hours, but (like for them) a life-time liaison which can be dissolved only by death.

And the fellowship bond of like-minded people, wherever they live and whichever language they speak, should also remain equally undissolvable, whether astronomy is their vocation or avocation. My continuing association with the Czechoslovak Astronomical Society has already been mentioned. Let us add that, since 1952, I have been an honorary member of the Astronomical Society of Manchester; and since 1967, at Liverpool; while, in 1973, the Salford Astronomical Society elected me their Patron. For many years now, no season has gone by without my giving these societies at least one lecture each session, to bring their members up to date with the forward march of science; and I only hope to be able to do so still for at least some years to come.

**Figure 2.9** Mr Josef Klepešta (1895–1976), one of the founders, and for seventeen years (1926–1943) Secretary of the Czech Astronomical Society in Prague. A well-known publisher of astronomical literature, during his lifetime he was one of the best-known amateur astronomers in central Europe.

## References

Kopal Z 1932a *Astron. Nachr.* **245** 335–8
—— 1932b *Astron. Nachr.* **247** 117–120
—— 1933 *Physics of Variable Stars* (in Czech) (Prague: Czech. Astron. Soc.)
Kopal Z and Kadavý F 1931 *Variable Stars and how to Observe them* (in Czech) (Prague: Czech. Astron. Soc.)
Kopal Z and Vand V 1933 *Atlas d'Etoiles Variables* (Prague: Czech. Astron. Soc.)
Nielsen A V 1934 *Nordisk Astron. Tidskrift* **14** 77
Plassmann J 1933 *Die Himmelswelt* **43** 113
Prager R 1935 *Vierteljahrsschr. Astron. Ges.* **70** 11

# Chapter 3

## Lehr- und Wanderjahre

*Per Ardua ad Astra*

If I have chosen to commence this chapter with my *Abitur* in June 1933, and not with my university entrance in September of that year, there is a good reason for this (without which, in fact, this book could never have been written): namely, the fact that it was then that I learned to read (if not yet speak) the English language!

Up to the nineteenth year of my life the English language was for me a treasure which was very largely closed. German (some) I knew almost before I went to primary school; and French we were systematically taught at the *Gymnasium* from the lower fourth form (although never enough for my father's liking!). It was, to be sure, largely bookish French which was taught to us, by teachers who learned its academic (rather than practical) aspects as a part of their own university course; but it was sufficient to acquire a reading knowledge good enough to enable me to correspond with my colleagues in AFOEV in that beautiful language, as I did with Professor Kobold in German; and when I visited France for the first time in 1934, French heard in Paris did not exactly sound like a foreign language.

But English was not taught at secondary schools in Czechoslovakia in the days of my youth, a fact which I found increasingly exasperating, as a large part of the original astronomical literature which I wanted to study was written in that language. With some previous knowledge of French and German (as well as Latin) it was often possible for me to gather at least the gist of the meaning of English texts; and I was told that the grammar of the English language is relatively simple. However, the pronunciation was certainly not; and I realised by that time that a good knowledge of English was essential for every student of astronomy.

This was the intention; but how to implement it? In order to embark on this task, I chose the way which philologists would certainly have condemned as unorthodox: namely, to begin by translating a book from English into Czech! The chosen victim of my efforts was to be *The Mysterious Universe* by Sir James Jeans, the first English edition of which appeared in 1929 and immediately became a great success, as perhaps the most popular book on astronomical topics since the days of Camille Flammarion half a century earlier. The Czech translation of Jeans's preceding book, *The Universe Around Us*, appeared in Prague shortly before (in 1931); and the enthusiasm with which it was received by Czech readers augured well for his next book.

And as we shall see later in this chapter, my hope was indeed not disappointed. My translation of *The Mysterious Universe* appeared in print, to be sure, only three years later (1936); but although somewhat delayed, it came out in time for the proceeds from it to enable me to travel to Japan in 1936 to observe there the total eclipse of the Sun on 19 June of that year.

But this is getting ahead of our story; for before the book could be published, it had to be translated; and this I accomplished in the following manner. The library of the Czech Astronomical Society possessed already at that time the French translation of Jeans's book, as well as its English original. I borrowed both, and compared their texts sentence by sentence. With an English–Czech dictionary constantly at my elbow, I managed to complete the translation (which by the time it went to the press, three years later, needed very little change) in less than two months of my summer vacations, spent that year in a village called Raspenava in the border mountains some hundred miles north of Prague. The avowed aim of that stay was to improve my fluency in German (Raspenava was in the border district inhabited mostly by the Germans), which I suppose I did (I remember very fondly the family I stayed with at that time). Little did I know that these were to be my last vacations at home, which a few years later I was to leave for good.

It was an enchanted time, translating Jeans in the daytime, and observing variable stars at night, and not alone, but in company of my friend Vladimír Vand (mentioned already in the preceding chapter), who came to Raspenava at the same time and for the same purpose. The telescope with which we observed the variable stars at night (cf Kopal and Vand 1933) was his; and he also built the photometer attached to it. During the daytime, Vláďa was watching over my shoulder the gradual growth of the translation of Jeans's book. I should add that, at that time, Vláďa knew no more English than I did; but

*Preparations for university* 77

before too long, English was to become our second mother-tongue (especially for Vand, who during his long years at Glasgow married also a nice Scottish lass from Milngavie before decamping for the United States).

As for myself, I learned to speak English too late in life (after twenty) for anyone not to notice that I had not acquired its knowledge at my mother's knee. However, in the course of time I adopted it as my literary language, in which I have since written over 350 individual papers and 40 books; and, especially in the latter, to use the kind words of some of their reviewers (cf, e.g., Fletcher 1957), I have 'enriched the English language with several new words which are not standard on either side of the Atlantic'.

But such feats were still a long time in the future. Since the commencement of the 1933/34 academic session, I pressed hard with further studies of English, and enrolled in evening courses (between 7 and 9 PM) given twice a week, by native Englishmen (the headmaster was a Cambridge BA) resident in Prague. I enjoyed these greatly, and attended for several years: the classes were small (six to twelve pupils, mostly my seniors); and in them I learned to know not only the English language, but also the English people—their little idiosyncrasies as well as their sense of humour, or their social and even political views. I felt the atmosphere of these classes very congenial; and when I visited England for the first time in the summer of 1934, I was not too surprised to be addressed in London (occasionally) as 'guvn'r' any more than at being called 'my luv' in Manchester in later years.

Let us return to the summer of 1933 and my university entrance in September of that year; but first one brief recollection. Before that memorable step was taken, I was summoned to the presence of old Professor Skála (see p. 23) who played an important role in the early steps of my father's academic career; and who (now completely blind, and with less than one year to live) wanted now to meet the son also. The main aim of the interview was, of course, to see if I could be induced to follow in my father's footsteps to study philology; but Skála was too intelligent not to see that there was no hope of such a conversion; and accepting the situation in good grace he gave me much advice for the years ahead—words which may not have been fully appreciated at that time, but whose significance has grown with the years: 'Never waste your time', Skála urged me; 'for it is a limited commodity; and time lost can never be recovered... Do not fritter it away on irrelevant pursuits—such as the reading of second-rate literature, in science as well as for relaxation—or other pursuits in which you can be only at your second-best... If you do not heed this, the work left undone of what you once hoped to accomplish in your life would become a source of continuous frustration, for which there is no reprieve.'

These words have been in the back of my mind ever since; and their gravity has only grown with time. In this one interview Skála taught me more than many other professors did in a whole semester.

## Charles University

On 6 November 1933, with the tacit acquiescence of the family counsels, I formally matriculated at the Faculty of Science of Charles University. For those who passed the *Abitur* examinations from the *Gymnasium* with distinction, entrance to the University was free (and without any further examinations at that time). Every qualified entrant had his (or her) chance; but the process of attrition started with the first semester; and the number of those who (for one reason or another) fell by the wayside each year was certainly larger than at the preceding grammar-school level.

Before we follow my own footsteps on this pilgrimage for the next four years, let us describe first with a few words the institution whose academic citizen I became at that time. Charles University of Prague, my *alma mater*, founded in 1348 by the Emperor of that name, is the oldest university on the European continent north of the Alps. There were ups and downs in its fortunes (caused mainly by religious strife† in the past, for which Master Jan Hus, its rector in 1409, and M Jeronym of Prague, fifteenth-century forerunners of Church reform, paid with their lives in 1415 at the Council of Constance) in the course of its history, which now extends over more than 600 years. However, over all these years the University has never closed the doors to its alumni (except during World War II, when it was closed by the Nazis); and it retained some of its medieval forms of academic life almost to my time.

Until 1918, the Chancellor of the University was *ex officio* the Arch-

† The first such strife was sparked off by the teachings of John Wycliffe (1320–1384) of Oxford, whose books were brought over to Prague from Oxford by Hieronymus (Jeronym) of Prague. In the last decade of the fourteenth century academic contacts between Prague and Oxford were especially lively (endowed fellowships in each institution for scholars of the other were maintained for some time); and one of the senior members of the Prague faculty of theology, M Peter Payne ('Master English' to the Czechs), was an academic refugee, obliged to flee from Oxford after suppression of the Lollards.

It may be added in this connection that Emperor Charles IV (the founder of the University in Prague) was father of the English Queen Anne of Bohemia (1366–1394; wife of Richard II), and it was her clerical retinue that brought over to England the Christmas carol *Good King Wenceslas*, well known up to the present. (King Wenceslas, the national saint of the Czechs, was a tenth-century ancestor of Charles IV and, therefore, of Queen Anne, on the distaff side.)

bishop of Prague; but *de facto* the University was headed by its rector, elected annually by the academic Senate among its professors largely by seniority (the same was also true of the deans of its five faculties: law, medicine, science, arts, and theology). Moreover, in accordance with canon law going back to the Middle Ages, all University grounds and buildings were (until 1939) solely under the jurisdiction of its academic authorities, and civil authorities were excluded from entering them. I recall from my student years that when police once inadvertently entered the academy the Police President of Prague had to present to the Rector in person his apologies for overstepping his authority. In those days, the University as a whole had approximately twenty thousand full-time students registered for a degree—i.e. it was approximately the size of large Mid-Western universities in the United States; and it has grown up since in almost the same proportion.

An annual change of elected academic officers would have made its financial administration effectively impossible within the University. This administration was firmly in the hands of the Ministry of Education, which (together with the Ministry of Finance) controlled the University's budget—as it did for all schools in the country. However, while the Government could in this way specify the number of posts available for each faculty, it could not appoint anyone without the recommendation of the Senate (at most, it could delay implementation of the faculty's candidate; but it could not substitute anyone else). As is true of most aspects of human affairs, in Czechoslovakia as well as anywhere else, this system had its good as well as bad aspects. It effectively prevented politicians from staffing the universities with their appointees (not that, to the best of my knowledge, they ever attempted to do so before 1948!) who would have been unsuitable on academic grounds; and this was on the credit side—aided perhaps further by the fact that, between 1918 and 1938, the majority of the Ministers of Education were university professors!

On the debit side, the 'academic freedom' insisted on by the University was more than once (within my memory) misused to protect incumbents whose appointments (made on the recommendation of the Senate) turned out to be mistakes, for the sake of 'union solidarity', and regardless of the interests of the subject or of its students. Inadequacies of this kind, our professors (or, at least, a majority of them) maintained, should (according to canon law promulgated in the *Statuta Arnesti* by the first Chancellor of the University in 1355) have come under the jurisdiction of the dean of the respective faculty. As, however, these changed from year to year, it was only too tempting to shelve unpleasant cases to be dealt with by his successor, a process which led to the academic tribunal being described as the strictest court of the world, one in which every case brought before it ended in death.

That is, it dragged on for so long that the delinquent could retire in peace and pass away in the fullness of years, twiddling his thumbs at the accusations!

I readily admit that (as far as I know) such cases were very rare; but, unfortunately, one of then concerned astronomy and the incumbent of the only chair of the subject at that time (Nušl's chair being an honorary one)! Half a century ago, this delinquent could easily have thwarted my astronomical career before it started; and it took the (inadvertent) intervention of no one less than Sir Arthur Eddington to prevent this from occurring!

But, by saying this, I do not wish to imply that the majority of the University's staff could be so described. The faculty of science at Charles was administratively separated from the faculty of arts (or, as it was then called, of philosophy) less than twenty years before my time; but already in the 1930s the science faculty managed to assemble a staff, some of whom would have been to the credit of any university in the world! This was certainly true of Professor Jaroslav Heyrovský in physical chemistry (1890–1967; Nobel laureate in 1959); Professor Bohumil Němec (1873–1966; renowned plant physiologist); Professor Eduard Čech (1893–1960; topologist) or Professor Václav Hlavatý 1894–1969; an expert in differential geometry). Čech did not, however, come to Prague till I was about to leave; while Hlavatý, a future collaborator of Albert Einstein, had to leave Prague for political reasons in 1948 to finish his life at the University of Indiana in the United States.

The *Index Lectionum* from my student years discloses that I took some courses with them all, but did not establish any closer personal contact with them or their subjects, being overawed at a distance. I learned much more from Professor František Záviška (1873–1945), a theoretical physicist whose research days were perhaps over by that time, but who was one of the best teachers I ever knew: a kind and understanding person and also a friend of Einstein from the latter's years in Prague. In the fullness of time, he was to be one of my examiners for the doctorate. But, unfortunately, my *viva* was also the last time I ever saw Professor Záviška; for soon thereafter I left for abroad, not to return to my native country for nineteen long years; and by that time Záviška was already dead. During the 1939–1945 war he was arrested by the Nazis as a hostage, and perished on the 'march of death' from the concentration camp at Osterode shortly before its liberation by the US Army in April 1945. Readers who happen to be philatelists may see Záviška's face on 40h stamps issued by the Czechoslovak Government in 1962.

The same was, moreover, true also of Professor Václav Dolejšek (1895–1944), the director of the University's Spectroscopic Institute, and an experimental physicist *par excellence* (a pupil of Manne Siegbahn, and discoverer of the N series of x-ray spectra). On the recommendation of Professor Breindl, Dolejšek very kindly allowed me to come and work at his institute in the afternoons (and over the weekends) while I was still a grammar-school boy—an act of faith in my good luck on both sides, for I was sometimes left alone for hours in a crammed laboratory abounding with high-voltage wires which often caused my hair to stand on end. But, fortunately, nothing untoward ever happened to me (or to the equipment entrusted to my care). In due course, Professor Dolejšek also became one of my examiners for the doctorate; but his end, only six years later, was tragic. During World War II Dolejšek (a keen reserve army officer) was entrusted to maintain radio links between occupied Czechoslovakia and the exiled government of President Beneš in London—a task which he carried out with bravery and skill until 1942; but the Gestapo finally closed in on him and his collaborators (many of whom were members of his university staff), and arrested them all. Dolejšek himself was taken to the Terezín concentration camp, and died there in unknown circumstances.

It is indeed sad to recall the personal losses suffered by Czechoslovak universities between 1939 and 1945 in active defiance of the Nazi occupation; particularly among the physicists, some of whom were killed with a sword in their hands, while others perished (like Záviška) as hostages in various concentration camps. My father was also on the list of the latter in 1942; and we have already mentioned how he managed to avoid that fate. On the whole, about one-sixth of the Czechoslovak university staff did not live long enough to see VE Day; and the losses among senior staff were disproportionately large.

But by the time I joined the University in 1933, this was all still in the (albeit not too distant) future, which very few foresaw at that time. Let me, therefore, end this section devoted to my *alma mater* in the last years of its academic autonomy with a few words on the life of its students. Unlike the situation I encountered later at most universities in England or the United States, the individual academic courses were not, in general, very closely supervised. A rigid *cursus studiorum*, where prescribed examinations had to be taken in the sequence specified by the regulations, applied only to professional courses (such as medicine or law, leading to degrees which recognised the recipient's right to practise publicly his profession); or to the candidates for 'state examinations', the passing of which qualified them to teach at state-supported secondary schools. But the time when the student presented himself for such examinations was largely left at the discretion of the candidate, another remnant of the academic ways of the Middle Ages, when many students came to the universities to have a good time in congenial company (a way of life encouraged by the faculties, for which 'eternal students' from the ranks of the gentry represented a steady source of income).

As to the University staff, the average teaching load amounted to 5–7 hours of lectures during term-time (which occupied only somewhat more than half of a calendar year). The size of the classes depended very much on the subject: several hundred may have been enrolled in courses compulsory for a professional degree such as law (a degree in law in those days offered the usual entrance gate to most branches of the civil service in Czechoslovakia, as the arts degree did in Britain). But only a fraction of students enrolled actually attended the lectures, and studied for examinations with the aid of professionally prepared class notes (called *scripta*) or published textbooks. In the majority of lecture courses in non-compulsory subjects, especially at the faculties of arts (called 'philosophy') and sciences, the audiences were usually much smaller, especially if the lecturer in question was not a good speaker.

According to the first statutes of the University (going back, I believe, to Archbishop Arnošt, the first Chancellor of the University, in 1355), *tres faciunt collegium*; or, in plain English, if at least two students come to hear the professor's lecture, the latter was obliged not to disappoint his audience; and the session could be cancelled only if no more than one student showed up. In the course of my student years, this happened more than once—and of all subjects, in astronomy! Dr Nechvíle (of whom much more will be said later on) sometimes found himself in this position; but even so he never stopped addressing his audience (of two) as 'ladies and gentlemen'. In how many other subjects the professor shared the same fate I cannot say; but it was to the credit of the University that any member of its teaching staff could announce any *collegium* of his own choice, and be assigned a room for it.

As to the examinations, these were also very different from those conducted in the Anglo-Saxon world. The majority of these were, in all faculties, oral (and sometimes before a board of more than one examiner). Written examinations were held only at certain stages (usually a three-hour *clausura*, in the course of which the candidate was locked up in a room all by himself, and not restored to freedom until his time was up). The oral examinations for the doctorate (called *rigorosa*) were also conducted very differently than they are in Britain or America. No external examiners were invited to assess the thesis from another university (as, with an increasing degree of specialisation, no other professor of the same subject might be available elsewhere in Czechoslovakia, and the requirement that all oral examinations be conducted in Czech precluded inviting foreign examiners). However, every professor of the respective faculty at the candidate's own university, whatever his subject or field of specialisation, had a right to be present at any oral, and to ask questions (usually of a general nature) should he so desire. The commission appointed by the faculty to adjudicate the thesis and examine the candidate (five in my case, of whom three—Dolejšek, Nušl and Záviška—have already been introduced in this chapter) was presided over by the Dean of the Faculty, who was to be present in person at each examination, and to pronounce *ex officio* the committee's verdict.

According to an old academic regulation (going back for centuries in the past) a candidate had the right to present himself for doctoral examination after four years of his matriculation. In most cases it took, of course, much longer. It may come as something of a surprise to graduates of American or British (though not Continental!) universities that not only examinations for a doctorate, but also all other university examinations, could be taken (after a given statutory interval) whenever the candidate came forward to apply for them; and, in the case of failure, repeated any number of times (on payment of an appropriate fee). No stigma was generally attached to late presentations (except that dilatory candidates might have forfeited the privilege of having their tuition paid by the state); but the economic facts of life (this was still a time of great depression) were such that very few students could afford to linger on their course longer than necessary.

No system of examinations in any institution is ideal or infallible; and all have their merits as well as drawbacks. From the student's point of view, some candidates can put up a better show orally than in writing; and for others it is the other way around. In an oral examination lasting 2–3 hours an experienced teacher can assess the quality of the student much better than he could from a written paper of any length; and in the course of a dialogue much more can come to light than could be read between the lines of a script. On the other hand, the idiosyncrasies of the examiner are also to be reckoned with: I know of professors at Charles University who (almost) never passed a candidate at the first try (claiming that the first exam is only an occasion for the student to find out what he does *not* know); and others were again almost too liberal in granting a pass—but always with the lowest grade (claiming that only professors know the subject well enough to deserve the top mark).

In the course of my student years I heard many stories about oral contests between examiners and candidates; but some (for the authenticity of which I can vouch) should be rescued from oblivion. Thus (the story came from a medical school) a professor of anatomy examined an over-age candidate, and things went none too well. When the examination had already lasted too long, the old professor remarked testily: 'I have not learned much from you so far'; but the candidate rose to the occasion with a reply which disarmed the examiner: 'This is because you have not asked me everything: and also I did not tell you all that I know'. And how often was I examined orally in the early years of my life by teachers who, when I started to answer with some alacrity, quickly brought me to a standstill: '...Stop, stop. I see you know that; so let us try something else'—as though the main aim of the examinations was to find out, not what the student knew, but what he did not!

But let me finish this rather sorry tale with an account of what can happen in a written examination, as it did to me in the third year of my university course, to prove that written examinations can be as tricky as any oral. The subject was analysis, and the field differential equations, in which I had to prove my competence in a three-hour written exam. It was in June—a beautiful morning—when I met my prof (let him remain unnamed) at 10 AM to receive my problem, and to be locked up until 1 PM to deliver the fruits of my labours. To begin with, I was in good spirits; for I thought I knew the subject well enough from the prof's own lectures. However, appearances are often deceptive. No sooner did I try a given method than it turned out not to be applicable to the circumstances on hand. And so it went as time passed: 11 AM, 12 noon; and still I did not move from the spot. My nervousness grew in inverse proportion to the time left, until shortly before 1 PM, I heard the prof's steps in the corridor, and the key rattling in the lock. The door opened; the prof (obviously after a good lunch) entered, beaming benignly at his distressed student, to ask: 'how did you do?' I confessed with a sad voice that I had not managed to solve the problem at all; and knew nothing more about it than at the outset. This did not, however, seem to bring the professor out of his good mood, or disconcert him. 'What did you try?' he asked. I went on with a sorry recital of my failures; but the longer I went on, the more cheerful the prof appeared to become, until he ended with a jubilant comment: 'Fancy that! I thought the problem would have no solution right when I gave it to you—and so it turned out to be!' The professor was obviously proud of his intuition; and to reward me for three hours of anguish he gave me the top mark for the effort—if not for the accomplishment.

The reader may have gathered from these experiences of fifty years ago that I am not inclined to overestimate the value of any examination system—oral or written, in Prague or elsewhere; and I consider them to be only a necessary evil. For the really sufficient test (and proof) of one's ability is one's life; and history teaches that its intrinsic worth is only weakly correlated with examination results in one's youth, especially at the highest level of accomplishment. For consider the (recently extinct) institution of Cambridge 'wranglers'! If we look through the lists of such champions in the sport of examinations in the nineteenth century, we see that very few of them fulfilled the hopes raised by their examination results. John Couch Adams (1819–1892), John William Strutt (Lord Rayleigh; 1842–1919) or Arthur Stanley Eddington (1882–1944) certainly did so; though Lord Kelvin did not do as well in his examinations as his chief rival (one Parkinson). Most of the senior wranglers (like many others) owed their success, however, to being good learners (which is a very different thing from having an original mind!) or to having received good training in the race-horse stables of renowned 'coaches'. Most of them no doubt became worthy citizens and holders of the right kind of prebends (whether they were senior wranglers or 'wooden spoons'); but otherwise did little to make the world of science remember them with any particular gratitude. Darwin or Rutherford (albeit Cantabrigians) were no wranglers; and Michael Faraday (1791–1867) did not attend any university at all!

When I count the blessings of my life (and we shall do so more systematically in Chapter 5), one of them should be the fact that examinations (other than those for the doctorate) played no important role in my teaching career of over four decades; and whenever I had to assist in them, I did so with mental reservations concerning my inadequacy for such a task; for if any foolproof system to grade students in their formative years free from bias has been invented, I am yet to learn about it. With PhD orals it is a different matter; there the candidate was known to the internal examiner for several years prior to the defence of his thesis; and whenever I served as an external examiner, I seldom had any need to differ from the opinion of the internal thesis supervisor.

### Student Years in Prague

In the preceding section of this chapter the academic ways of life at Charles University during my student years were briefly sketched; and now let us say a few words on how I fared in these myself. A student duly matriculated could, in principle, enrol in any number of courses; and (apart from seminars or *practica*) voluntarily; no-one would check (especially in larger classes) if all those who enrolled were actually present. The only formality was to have enrolment attested by the signature (or initials) of the professor or lecturer in question at the beginning and end of each course. In this way, the *Index Lectionum* of my father acquired several signatures of Thomas Masaryk, as mine did of Jaroslav Heyrovský and others.

In looking over this *Index*, still in my hands after more than fifty years, I note that in the first two years of my university course I attended courses on a wide variety of subjects—not all of them at the science faculty. Ever since my youth (and still today) I have been interested in history (and prehistory); and the faculty of arts of Charles University could in those days boast several great names in that field, such as Professor Josef Pekař (1870—1937), the greatest historian of the Czech nation since Palacký, or Bedřich Hrozný (1879–1952) who in his youth was the first to decipher the script and language of the Hittites. In my student years, Pekař (who served as the University's rector two years before I joined our *alma mater*) was an engaging (albeit rather irregular) lecturer; but Hrozný's performance at the rostrum was so dull that I soon gave up any further attendance.

But, in spite of such occasional experiences, in the second semester of my course I enrolled in lectures and seminars totalling 54 hours a week—not that I ever intended to follow them all, but rather to find out which ones were worth while attending. For I soon made a discovery that the profit from listening to lectures depends much more on the qualities of the lecturer as a teacher than on the subject professed: a good lecturer can make any subject interesting and informative; while those not born with that gift (and especially if their own thoughts are not on that particular subject) can douse the students' interest in it for the rest of their lives. If necessary, such subjects can be studied to greater effect and economy of effort from books, especially if the classes are large, so that it is impossible to initiate any dialogue. Even at undergraduate level, the supporting literature (if available) becomes an aid of increasing importance. A book, to be sure, cannot answer questions arising in the reader's mind in the course of its perusal; but it can be re-read any number of times, and so serve as an *aide-mémoire* to supplement the lectures.

Much of the efficacy of the printed word also depends, of course, on the quality of the book or pedagogic gift of its author, and the gifts of lecturing and writing do not always go hand in hand together: in fact, for people of outstanding intellect they are often very different. The greatest disparity of this kind which I ever had a chance to observe at close range was in Sir Arthur Eddington, perhaps the best and most engaging writer on astronomical subjects of this century, but a dull speaker (because of his excessively shy nature).

At postgraduate level, of course, no student of almost any subject can expect to learn all he needs to know to reach the frontiers of current research from live lectures. Such as may be given should serve as an introduction to this no man's land (albeit along very narrow corridors) and the space in between the student must fill in largely through his own efforts. Indeed, the main aim of lectures at that stage should be to serve as an introduction to the methods of research, by those who have already had enough experience to serve as 'mountain guides' to young climbers dedicated to following in their footsteps.

In fact, one of the lessons of my life has been that no-one can learn to become a scientist from books or lectures alone. By far the most important element of the evolution of a young scientist is the possibility of observing a living example which inspires imitation. And this is so throughout our lives, from cradle to grave. It is difficult (if not impossible) to teach young children good behaviour if the adults around them (to whom children have every right to look up) are not seen to practise what they preach. The children and young people are, in general, keen observers, who may acquire (to various degrees) the art of self-delusion only in the course of time. It is impossible to impart to them a love or respect for learning unless they see their elders actually practising it as a pursuit worth following—just as one cannot learn the experimental method in science from experiments performed on the blackboard, not in the laboratory.

Nothing can, indeed, be more detrimental to students at postgraduate stage, or inhibit further development of their minds more effectively, than finding themselves exposed to faculty members who pay only lip service to what they are supposed to profess, and who camouflage this fact by undue administrative preoccupations (which could easily be handled by others), or who spend their time on endless committee meetings as a socially acceptable way of covering up the fact that their creative days are over, and that they only go through the motions of still being intellectually alive. Good students will, of course, realise this before too long and be disappointed; but—worse still— others may notice that there are other ways than the quest for knowledge to keep their names from oblivion; and more comfortable to practise.

But trends of this kind were to emerge only later in my life to cast their shadow on an increasing fraction of academic life; and I encountered them mainly abroad in post-war years, when science was catapulted to the position of 'big business' and a worthwhile target for professional operators. In the days when I was young there was still very little evidence of this anywhere in the world; and the students' noses were still kept pretty close to the grindstone.

The lectures during term-time started generally at 8 AM; and with the language courses usually held between 7 and 9 in the evening I did not spend much time at home. Not all students, perhaps, took their courses in the same stride; and I remember that my colleagues from the same years who did not all share my catholic interest used to poke fun by referring to me as 'the last of the polyhistors'. But it was good-natured fun, and I enjoyed every bit of it. To this day I am grateful for the fact that the academic system in which I was educated allowed its alumni to pursue broader interests at a time when their minds were still open, and not too overloaded by compulsory work to take advantage of their opportunity.

Even after more than half a century I still recall with some nostalgia the nocturnal returns from various lectures or public libraries (which generally closed at 9 PM)—not by tram or bus, but on foot, across the Charles bridge of the Old Town with all its historical associations (how often Johannes Kepler must have walked the same way three centuries ago, through surroundings which have changed but little from his day), along the Vltava river through the gas-lit gardens of Kampa island; and up the Petřín hill through quiet streets on a track which brought me home after 10 PM, relaxing from the exertions of the day or making plans for the future. When asked what I was doing so late at night away from home, my usual answer was: 'I learned many new things', some of which I have not forgotten to this day!

How did I manage to avoid oversaturation? I still wonder today; but the answer probably goes back to instinctive perception. Unlike many of my fellow-students, I seldom bothered to take many class-notes, and never merely copied what I heard. Instead, I tried to follow the essential parts of the argument, and noted mainly where to find out, in case of need, more details about it. This already came to me almost subconsciously during my secondary-school years; and I still think it is the right way to learn. For while details are easy to forget (and soon likely to become obsolescent), understanding is not, and once grasped it naturally becomes a part of one's nature; while mere memorisation of class notes results (at best) in passive acquisition of other people's wisdom. To do this should, however, never become the main aim of learning. Instead of becoming passive receptacles of words or ideas presented in lectures, young scholars should respond to what they hear productively, by ideas of their own, stimulated by what they hear from their teachers (and their lectures should, whenever possible, be aimed at this end).

On the whole, the academic freedom which was a privilege of the students of science at Charles University in my time was probably more suited to benefit those of above-average ability or motivation than (say) the more closely supervised American system (in which a PhD degree has of late become a necessary prerequisite of any professional career). The current American system may indeed be more suited to produce better qualified specialists in the field of their thesis research. On the other hand, by its increasing emphasis on team-work, in which individual candidates become small cogs of a complicated academic machine, such a system fails to endow its 'finished products' with sufficient flexibility. Very few PhDs of the world are likely to be allowed by circumstances to continue working in the fields recom-

## Student years in Prague

mended to them by their academic supervisors, and their future employers may have other requirements in mind; a premature emphasis on too much specialisation may then prove to be a handicap.

How did all this begin to reflect itself in my own life? As my university terms progressed, it became increasingly obvious to my teachers as well as to myself that such aptitudes as I seemed to possess lay in theoretical rather than experimental work; and it was mathematics with theoretical astronomy which occupied my mind more than anything else. But how much did the regular curriculum offered by our academic *alma mater* to its alumni at that time meet my increasing appetite for these fields? Not enough for my liking, I am afraid; mathematics at Charles was then still almost synonymous with pure mathematics; and applied mathematics had no place in our curriculum. Self-respecting and Göttingen-trained mathematicians of Hilbert vintage still delighted to dwell in their ivory towers of 'splendid isolation' from the world of physics; and—perhaps not to spoil their minds (or soil their hands)—the teaching of such subjects as 'special functions' was resolutely left to the physicists. This the latter perforce had to do; but the time spent in physics courses on differential equations and similar subjects was sadly missed when it came to discussion of relativity or quantum mechanics.

I recall a course on differential equations I took with a professor of mathematics, of some renown in his particular field, who lectured on this subject for three hours a week the whole term without ever solving one such equation; and when I diffidently mentioned once that I hoped to learn one day to do so, he replied in some surprise: 'But why do you want to solve any equations? Is it not more satisfying to observe their general properties?' And so we did; and the one term which stuck most firmly in my mind from this course was the 'Lipschitz condition'—I hesitate to explain to the more general reader what it stands for; but actually *doing* anything with the differential equations describing natural phenomena I had to learn elsewhere. I was soon to find out that to solve most such equations was a task beyond the means of formal analysis; and that other methods must be resorted to in order to construct their solutions. But, needless to say, the subject most suitable to this end, called 'numerical analysis' (to which I contributed in 1955 my first major book) was unmentionable in the sacred precincts of our academy, from fear of losing caste. In this, as well as in many other branches of our learning, I had, of necessity, to become an autodidact—not at all a bad method, as I was to discover in the course of my life.

And what was true of mathematics at Charles University during my student years was even more true of astronomy. The senior professor of this subject was then František Nušl, at that time also Director of the National Observatory of Czechoslovakia (figure 3.1). He was an amiable old gentleman (I should really be careful in saying so, as he was then several years younger than I am now while writing these lines); and as a teacher a delight to listen to. Unfortunately, he no longer lectured very regularly; and when he did, his topics were concerned with nineteenth-century 'classical' astronomy, rather than with a more contemporary state of our science.

**Figure 3.1**  Professor Dr František Nušl (1867–1951).

Nušl was a gifted instrumentalist and applied mathematician; his initiation in observational astronomy went back to postgraduate work at the Vienna Observatory under Edmund Weiss; and later at Bamberg in the early years of Friedrich Hartwig's directorate (a destination to which he was later followed by his younger colleague Ladislav Pračka). The circumzenithal instrument developed by Nušl with his friends Josef and Jan Frič (founders of the present-day Ondřejov Observatory) in the 1890s was a direct predecessor of Danjon's *astrolabe à prisme*, and its performance received an honorable mention from Frank Schlesinger (1871–1943) in his George Darwin Lecture (cf Schlesinger 1927) before the Royal Astronomical Society.

In post-war years, Nušl was criticised (justly, no doubt) for the indolence with which he treated his position as Director of the National Observatory between 1920 and 1937; for he had no administrative talents, and (truth be told) he treated it without too much embarrassment as a sinecure. Much of what could have been accomplished in

Czechoslovakia between the wars (and was done afterwards, under much more difficult conditions, under Dr Šternberk and his successors) was merely postponed by Nušl *ad calendas graecas* during the last two decades of his life. However, I always retained a soft spot for him because of his kind nature and sincerity, for he never tried to cover up his shortcomings. He had, moreover, a real understanding of young people, and encouraged their efforts to the best of his ability (albeit with a varying degree of effectiveness). Amateur astronomers of Czechoslovakia will always remember him as the genial third President of their Society between 1926 and 1947.

Of astrophysics Nušl knew but little; but his university lectures on fundamental astronomy were interesting; and it is through what I remember of them that I can still recapture today some of the spirit of the astrometry of the nineteenth century, the patron-saint of which was Friedrich Wilhelm Bessel (1784–1846), followed by Argelander, Auwers and Küstner—a knowledge which served me in good stead in later years of my life. Nineteenth-century astronomy, comprising by and large the period between Herschel and Schwarzschild, was dominated in spirit by astrometry and celestial mechanics. Although the principal advances in the latter, from Laplace to Poincaré, continued to be forthcoming from France, Germany remained pre-eminent in astrometry at least until World War I; and it was this which Nušl brought from there during the years of his own *Lehr- und Wanderjahre*, and continued to share with his students, of whom I was to be the last.

The second professor of astronomy at Charles University during my student years (and the only *ordinarius*, Nušl's chair being an honorary one), was W W Heinrich (1884–1965); but perhaps the less said of him the better. A scion of a country family of some local distinction, he spent some time in his postdoctoral years with Wirtz in Strasbourg and Karl Schwarzschild in Potsdam. On his return to Prague he embarked upon a rapid career culminating with an appointment to full professorship at Charles University in 1923, at the unusually young age of 42 years.

So rapid a promotion soon proved, however, to be an embarrassment to the University, and an almost unmitigated disaster for astronomy in Czechoslovakia during the time between the two world wars. What Heinrich lacked to live up to the confidence expressed in him by the University in his accelerated promotion was, not brains, but character. He soon began to quarrel with everyone around him. He parted in ill will from all his assistants (from Dr Bohumil Šternberk, subsequently a distinguished director of the Astronomical Institute of the Czechoslovak Academy of Sciences at Ondřejov, to Dr Hubert Slouka with whom I travelled in 1936 to Japan) and quarrelled with most of his professorial colleagues on the faculty of science (in particular, with his senior colleague Nušl) to which he rapidly became an acute embarrassment.

In the course of time, Heinrich degenerated into a source of amusement for the students as well as a deterrent for the study of his subject. Most of his lectures consisted of arguments with absent colleagues (at home as well as abroad), whom he kept attacking assiduously with chalk if not rapier; while for the more general public he soon became a confirmation of the view (held then by many) that astronomers are oddities of whom no sensible behaviour could be expected. It is a sad fact that, in the 1930s, Heinrich educated not a single PhD in astronomy—in stark contrast with the efflorescence which after 1945 (when he was shunted onto the sidelines) produced in quick succession a series of young astronomers of the calibre of Bumba, Burša, Hruška, Kleczek, Kopecky, Kříž, Kvíz, Perek, Plavec, Sekanina, Švestka or Vanýsek (in alphabetic order) and others. A good half of these, sadly enough, emigrated after 1968 to provide, together with the professional refugees from Hungary (after 1956) and Poland (1981), an influx of astronomers to the western world—an export which, since the beginning of the twentieth century, had been very largely in Dutch hands.

But, to return to the past, by the time I joined the University in 1933, Professor Heinrich had already been relieved of the direction of the University's Astronomical Institute (a function which he vainly tried to recover for several years); and was by-passed in 1934 for the office of dean of the faculty—facts which were bound to lower the prestige of the subject, and discourage students from its study. I found his treatment of so serious a subject so frivolous and out of place as to avoid Heinrich's lectures altogether; and doubt, in fact, if he and I ever exchanged more than half a dozen words with each other.

This point is perhaps of some interest, because, ten years later, when I was already at Harvard and when changes in the constitution of the Charles University made it at last possible to rejuvenate astronomy there, I was considered as a possible successor to Heinrich, still during his lifetime. In the summer of 1947, at a time of celebration for the 175th anniversary of Princeton University, Professor Bohumil Bydžovský (then Rector of Charles University) visited me at Cambridge, Massachusetts on his return from Princeton (see figure 4.15) for preliminary discussions of my appointment. It will be explained in the next chapter why these negotiations came to nothing; and why my entire subsequent academic life was spent on both shores of the Atlantic. But this was still far in the future when I was a student in Prague.

For the present, my aim is to pay my homage, and record my gratitude, to other teachers at Charles University to whom I owe much more than to Professors Heinrich or even Nušl; and the primacy among them belongs to two distinguished scholars of past generations, one Czech, the other of German-Scottish parentage: namely Professors Vincent Nechvíle and Erwin Finlay Freundlich. The latter was by some years the older; though both were destined to depart from this world in 1964, Nechvíle at the age of 74, while Freundlich attained almost 79 years (Heinrich died a year later at almost the same age).

Both Nechvíle and Freundlich were largely responsible for setting the initial conditions of my own career. Nechvíle's influence extended from his return from Paris to Prague in 1930 (when he chose me, still a grammar-school boy, to assist him in the observations of Eros during the latter's opposition in 1930–1931 at the Štefánik Observatory in Prague), until my own departure for the wider world in 1938; and it was he—a pupil of Andoyer at the Sorbonne—to whom I owed my introduction to celestial mechanics in general, and to Clairaut's theory of the figures of equilibrium of celestial bodies in particular. Therefore, my dedication to him of my *Close Binary Systems* in 1959 represented only a belated act of gratitude (one I can never repay in full) for what I learned from him, and for the encouragement received in the formative years of my life. I may be the only one who deserves to be called his student; and have always regarded this as an obligation to accomplish that for which fate did not vouchsafe him the opportunity or the time.

**Figure 3.2** Professor Vincent Nechvíle (1890–1964)—*anima candida* among Czech astronomers—in his late forties.

## 94  Lehr- und Wanderjahre

Vincent Nechvíle was born at Prague on 10 April 1890 and, like many other astronomers of his generation, he entered our science as a mathematician. Soon after he received his doctorate in that subject from Charles University (as a pupil and, later, assistant of Professor Karel Petr—*der Altmeister Tschechischer Mathematiker*, as he was described by another of his former students, Václav Hlavatý, in a dedication to one of his books), Nechvíle's career was interrupted by almost four years of World War I, most of which he—the gentlest of men—spent in the uniform of a cavalry officer. After his demobilisation, a grant from the French Government enabled him to spend several years in the 1920s in France—first at the Observatoire de Toulouse, and then at the Observatoire National de Paris where most of his life's work was accomplished.

Under the influence of Professor Andoyer at the Sorbonne, Nechvíle turned his mathematical talents to the restricted problem of three bodies, and his treatment of its elliptic case has become basic to much of the work carried out in this field since that time; while a collaboration with George Willis Ritchey (1864–1945), the creator of the 60 inch and 100 inch reflectors of Mount Wilson Observatory (who was working in Paris at that time; see figure 3.3) led Nechvíle to study the theory of catoptric optical systems, which have since become well known under the name of Ritchey–Chrétien telescopes. As is well known, Ritchey possessed truly 'green fingers' for astronomical optics, but he was no mathematician; the theory behind his work was carried out for him largely by Nechvíle, who made fundamental contributions of his own to the subject (alas, but little known abroad; and most of this work was published only later in Czech).

However, Nechvíle's principal work at Paris concerned stellar proper motions. In the 1880s, the Henri brothers obtained more than one hundred negatives with their prototype of the 'normal astrograph' of the *Carte du Ciel*, with sufficiently long exposures to record accurate positions on them of stars down to the fourteenth apparent magnitude. At the encouragement of Professor Deslandres (then Director of Paris Observatory), Nechvíle took new plates of the same star fields with the same instrument after a time interval of more than forty years; and from the combined material deduced proper motions for more than 3800 faint stars distributed widely over the sky, representing the largest homogeneous material on apparent proper motions of faint stars available at that time.

This work, which earned Nechvíle his *doctorat ès sciences* from the Sorbonne, and the Lalande Prize of the French Academy, also paved the way for his academic career at home. On his return to Prague in 1930, Nechvíle became *Dozent* of astronomy at Charles University, and was recommended for a personal chair in 1939, three months before the

University was closed for six years during the Nazi occupation. During the difficult years which followed, Nechvíle served also (for a time) as acting director of Prague Observatory; but increasing concern for daily life left him but little time for further research. The years following his retirement from both institutions in 1960 were for him a time of increasing solitude. He never married; his family dispersed. His only brother, a post-war expatriate, found his final resting place in England; while his niece Zorka, an outstanding tennis player now settled in Canada, made a career for herself at Wimbledon as a distant forerunner of Martina Navrátilová.

**Figure 3.3** George Willis Ritchey (1864–1945), the great astronomical optician and creator of the 60 inch and 100 inch reflectors of the Mount Wilson Observatory (left) with Dr Nechvíle in the gardens of the Observatoire National de Paris in 1927.

Professor Nechvíle died in Prague on 5 July 1964, and will be remembered with affection by all who knew him as the type of man, so increasingly rare in these days of competitive life, to whom Horace's epithet *integer vitae scelerisque purus* can truly be applied. Always kind and gentle, he instinctively shied away from controversies; and it is doubtful if he ever made a single enemy in his life. Although of somewhat frail health, his strikingly youthful appearance (see figure 3.2) did not desert him until almost the end; and as such he will live in the memories of all who knew him.

In contrast with Nechvíle, whose whole life was centred on Prague (and who could almost have been regarded as a family friend of old standing), Professor Freundlich, the second great influence on my student years, appeared suddenly on the horizon in 1936, when the German University of Prague (closed in 1945) invited him to return from Istanbul to fill the chair vacant since the retirement of the late Professor Adalbert Prey. Freundlich came; and his inaugural lecture on 13 January 1937 on the internal constitution of the stars was a revelation—at least to us, the students, who had never heard anything like it from our Czech professors before. Only weeks elapsed before I became a regular visitor to Freundlich's institute; and the friendship between the young student and mature scholar, then struck, lasted until Freundlich's death.

Erwin Finlay Freundlich was born at Wiesbaden-Biebrich on 29 May 1885, of mixed German–Scottish parentage (his mother, Ellen Finlay, namesake of my faithful secretary of many years, hailed from Cheltenham), and of a large family in which a scientific bent was not unknown; the renowned physical chemist, Herbert Freundlich, was one of his older brothers. His university studies in Göttingen led him first to a doctorate in mathematics as a student of Felix Klein; and it was Klein who deflected his career to astronomy by recommending his young scholar in 1910 for an assistantship at the Royal Observatory at Berlin.

The time when Freundlich joined the ranks of astronomers was a pregnant period in the growth of our science, with relativity theory approaching the prime of its life; and Freundlich with his mathematical background was particularly well prepared to follow its implications. Somewhat to the consternation of his more conservative contemporaries (including his director, Hermann Struve), Freundlich joined with enthusiasm the early protagonists of relativity and eventually wrote one of the best expository textbooks on it at that time (*Einstein's Relativitäts-Theorie*, Berlin 1917), an achievement which earned him a lifelong friendship with Sir Arthur Eddington.

Not only did Freundlich set out to defend relativity on theoretical grounds, but he was also one of the first to undertake its experimental verification. To this end, with the moral support of the Prussian Academy of Sciences (and the financial support of Baron Krupp von Bohlen), he embarked (with W Zurhellen) in August 1914 on the first solar eclipse expedition to measure the deflection of light in the gravitational field of the Sun, in the southern Crimea. This effort unfortunately came to naught; for the outbreak of World War I a few weeks before the date of the eclipse caused all equipment to be impounded and the astronomers interned. It was some months before the luckless travellers were eventually repatriated, and the actual detection of the relativistic deflection of light had to await the outcome of the Greenwich expeditions in 1919.

The decade which followed World War I marked the high noon of Freundlich's life. It brought him in close touch with Albert Einstein, then himself at the peak of his powers; and it was largely his efforts and close contacts with German industrial circles which led in 1921 to the foundation, in Potsdam, of the Einstein Institute with its solar tower (later renamed the Institute for Solar Research and amalgamated with the observatory), of which Freundlich became the first director. The work of the distinguished group of investigators (including von Brunn, von den Pahlen, von Klüber, Grotrian, Wurm and others) which Freundlich gathered around him is too well known to need elaboration in this place. In a study of the $K$ effect, von den Pahlen and Freundlich came (by their detection of the double-wave of the $K$ term in longitude) within an ace of the discovery of galactic rotation. The eclipse expedition to Sumatra of 1929 made up for the abortive one of 1914, and as a result Freundlich, von Klüber and von Brunn thought that their measurements indicated the deflection of light in the gravitational field of the Sun to be significantly larger than that predicted by Einstein's general theory of relativity. They were not the only ones to arrive at such a conclusion—Professor A A Mikhailov and his Pulkovo colleagues seemed to confirm Freundlich et al.'s value some years later. Yet they were both wrong; for when new (radio-astronomical) methods were brought to bear on the problem, the deflections of radio sources so observed turned out to be in agreement with theoretical expectations, within the limits of observational errors; and previous measurements at optical frequencies must have been affected by systematic errors of as yet undisclosed nature.

But, to return from the Sun to the Earth, the advent of the Nazi Government to power in Germany in 1933 interrupted Freundlich's work; and because of his partly Jewish origin he left Potsdam to accept the Professorship of Astronomy in the University of Istanbul. He remained there for three years (the present Istanbul University Observatory was erected under his direction), and left in 1936 for a similar post in the German University of Prague. The Munich agreement and its aftermath uprooted Freundlich once again after little more than two years of residence in a congenial city and state. This time (declining an invitation to America) at Eddington's initiative he returned to his Scottish motherland and the University of St Andrews, where he spent the rest of his active life.

Shortly before that time, the University had received a bequest from the late Professor Gordon Lang for the establishment of a new university observatory; and Freundlich was entrusted with its erection as the observatory's first director. The outbreak of World War II interrupted this work almost immediately; and in the grim years that followed it took all of Freundlich's energy and determination to carry out his mission in the face of protracted material and manpower shortages, not to speak of perennial financial problems caused by post-war inflation. As the principal instruments for the new observatory, in keeping with

St Andrews' Gregory tradition, Freundlich wished to erect reflectors of the most modern type—the flat-field Schmidts. In collaboration with Linfoot and Waland, he succeeded in commissioning the 19 inch pilot model in regular service in 1949; but it was one of the disappointments of his life that he was unable to render the full-scale 39 inch flat-field Schmidt (the largest telescope of this type in the world) fully operational before his retirement.

On 1 January 1951, then 66 years old, Freundlich became the first Napier Professor of Astronomy at the University of St Andrews; and as my appointment to a professorship at the University of Manchester preceded his by two months, fate brought us together once more on this side of the Atlantic for the rest of our lives. At that time I still had thirty years of service ahead of me; and in the 1950s Manchester and St Andrews maintained close academic cooperation. I served repeatedly as External Examiner for the Scottish students at St Andrews (where I met, in this capacity, my subsequent student and future colleague Dr Alan Batten); and Professor Freundlich reciprocated similarly as External Examiner for my first Manchester PhDs, again including Alan, who is now senior staff member of the Canadian Dominion Astrophysical Observatory at Victoria, BC, and has been a faithful student of close binary systems ever since.

**Figure 3.4** Professor Erwin Finlay Freundlich (1885–1964), at the age of 78 years (one year before his death).

However, on reaching the age of seventy in 1955, Freundlich retired from the Napier Chair of Astronomy at St Andrews; though he retained directorship of the observatory until 1959, when he finally left Scotland for his native Wiesbaden to spend his years in 'retirement'. If his middle age was often stormy, the evening of his life was calm. In a sense, however, Freundlich never really retired; for until he died he held an honorary professorship in the University of Mainz, and his mind continued to remain active. His last book on *Celestial Mechanics* appeared in 1958. Death came gently on 25 July 1964 to claim the life of this veteran astronomer; and his passing was mourned not only by his surviving wife and children, but also by at least two generations of grateful students in Germany, Turkey, Czechoslovakia and Britain to whom Freundlich had offered generous help and encouragement at different stages of their careers. The present writer is certainly among them; and together with others will remember Freundlich as a man of forthright character and wide culture (he was, among other things, also an accomplished cellist) whose tall stature, unbent by age, and leonine head (see figure 3.4) towered in any gathering of his contemporaries. He was always forthright in expressing his views. There was nothing of the politician in him, a true liberal of old vintage, and of the type who instinctively evoked strong feelings in others. Very few people who ever met Professor Freundlich at close range remained indifferent to the challenge of his personality. What more can one say about such a man?

## First Travels Abroad

But let us bid farewell to our late teachers of bygone days, and turn to other means by which young scholars of developing countries (albeit countries sometimes of distinguished past) can become integrated into the mainstream of contemporary scientific progress more effectively than can be done by classroom education, or from printed literature. The best way to accomplish this, I soon realised, was to travel to the sources of current progress in one's chosen field, to meet the principal carriers of the sacred fire on their own home ground and in their workshops. For one born in central Europe in the early part of this century, such sources were still mostly abroad. An eagerness to reach them went, I confess, hand in hand with the lure which faraway lands can exert on young minds. For a grandson of grandfather Lelek (Chapter 1) how could it have been otherwise? Neither of my parents shared this inclination to the same extent, or only theoretically so; but to their praise let it be said that they never placed any serious obstacles in the way of my plans, then or later, so long as I could finance them from my own income (earned by tutoring or from literary proceeds).

The vacations following my *matura* in 1933 were the last spent wholly in my native country; and the summer of 1934 took me abroad for the first time. A group of students (of different fields and seniority) teamed up under the leadership of Dr Otakar Matoušek, at that time lecturer in geology at Charles University, to visit several countries of western Europe, and some of their academic institutions, for as long as their purses would permit. Naturally I was among them; and with a congenial group of about twenty colleagues we took a tour through Germany, Belgium, England and France which lasted about a month. We travelled together (by train, of course); but on arrival at each city or other destination each went largely our own way according to his own interests.

The first stop of this tour was Berlin; and, for me, this meant a visit to the University Observatory at Neubabelsberg (there was no time to visit Potsdam as well). With Professor Paul Guthnick (1879–1947), the Director of Babelsberg Observatory and one of the pioneers of photoelectric photometry of variable stars in Europe, I had already been in correspondence for some time; and although this was my first visit to his observatory, at least some of its interiors were known to me from a well known movie of about 1930 vintage called *Storm over Mont Blanc*, in which its star, Leni Riefenstahl (one-time favourite of Adolf Hitler), played the role of an astronomer (and double-star astronomer at that!); the movie had shown her working at the micrometer attached to the Observatory's 24 inch Zeiss refractor. Needless to say, when Professor Guthnick took me into the dome to see this instrument (now, I believe, in Soviet Russia), we did not find Leni Riefenstahl there, but a variable-star observer, Margarete Güssow, whom even her best friends would not have taken for a film star. But I should not write so disrespectfully of a lady who incorporated my observations of the minimum of $\varepsilon$ Aurigae of 1929 in her summarising publication on the subject (cf Güssow 1936). Only in one respect did Margarete Güssow resemble Leni Riefenstahl: namely, in her admiration for Adolf Hitler; and her role in the Nazi party debarred her after the fall of the Third Reich from the pursuit of any further professional career.

Professor Guthnick introduced me also to his colleague Professor Richard Prager (1883–1945), whose name was already a synonym for variable stars in the interval between the wars; and whom I met then for the first time, little knowing how closely fate would bring us together a few years later at Harvard College Observatory after Prager was forced to leave Germany as a non-Aryan in the spring of 1939. He too did not survive the downfall of Hitler's Germany in 1945 by many weeks; but more about that in the next chapter; while Guthnick himself survived the end of the Third Reich by another two years.

This visit to the first astronomical observatory outside Czechoslovakia was memorable for me also for another reason—as the portent of a storm which was to shake Europe in a very few years to its foundations. I started from Berlin for the Neubabelsberg (a garden suburb of the capital) quite early in the morning; and the walk from Babelsberg Station to the observatory took the better part of an hour. The streets on the way seemed, however, completely deserted, with no-one in sight to ask if I was heading in the right direction. Not far from the observatory I heard, however, two cracks—as though a car exhaust had backfired twice in rapid succession—and then all became quiet again. I did notice, however, that people at the Observatory (especially Guthnick) exhibited signs of some nervousness without apparent reason.

The reason transpired, however, after that night's train ride arrived in Brussels; and we read in the newspapers the first press reports about Germany's 'night of the long knives' on 30 June and its aftermath, as a foretaste of the shape of things to come. It was the first of many bloodbaths which Germany had to suffer as the price for having allowed its government to fall into criminal hands; and the number of their victims began to mount with frightening rapidity. As is well known, two of these were General Kurt von Schleicher (the last German Chancellor before Hitler) and his wife, who were shot dead at their door by SS thugs on the morrow of the night when Adolf Hitler elevated himself (and not for the last time) to be chief accuser, judge, and executioner of his rivals. I learned later than von Schleicher's villa was quite near the observatory at Neubabelsberg; and, therefore, it is at least possible that the two cracks I heard on my approach to that temple of Urania were shots which killed the general and his wife.

Full of forebodings about the fate of central Europe we then proceeded from Brussels to Ostende to cross the Channel for Dover—the first time I stepped on to the British soil that has become my second home in the latter part of my life. However, as soon as we were cleared by the immigration authorities and allowed to land 'on condition of not accepting any work, paid or unpaid, while in the United Kingdom', another shock awaited us. The billboards of all newspapers on sale within sight carried a message: 'ENGLAND LOST!' Full of innocence, we connected this message with recent events in Germany; and could perhaps be excused for being still completely ignorant of cricket and the test matches (not that I know much more about them today). But, perhaps the Britons, too, should learn to welcome their visitors with portents of lesser despondency. Only six short years later, a similar outcry would have sent the scaremongers to jail; and if one calls 'wolf, wolf' too often, the wolf may come—from without or within! But such a time—if it ever comes—was, in the summer of 1934, still far in the future. King George V still sat on the throne, and Mrs Simpson had not yet become a national problem.

London (which we reached after a short train ride) in summertime was a city almost given over to visitors; most Forsytes were at the seashore (or in Scotland); and a foreigner in quest of his location in the great metropolis was as often as not addressed as 'guvn'r'—in a language which bore but distant resemblance to the King's English, as taught at the English Institute in Prague by Oxbridge graduates.

We settled down in London at the university quarters (not far from the British Museum), and embarked at first on 'doing' the familiar sights of the city (some of which, like the Tower of London, I have never revisited since). One of those days we visited Cambridge, mainly to admire the architecture of its colleges largely (it was already vacation time) devoid of their inhabitants; and my pilgrimage to the observatories at Madingley Road was likewise unsuccessful: astronomers in pre-war years took their vacations more seriously than they have been doing since; and none of the staff was present to answer a shy knock at the door by a youthful visitor. Only one astronomer did I eventually find—Mr H E Green (second assistant of the University Observatory). He showed me a telescope with which Neptune should have been discovered—though it was not (because of Mrs Challis and her untimely cup of tea! Cf Smart (1947)). This failure did not, I am sure, diminish my respect for that venerable instrument, and, in later years, I had a chance to look at it more often.

My first expedition (from London) to enter the Royal Greenwich Observatory was likewise unsuccessful. It is of interest to recall that, in 1934, the price of a tube ticket from Charing Cross to Greenwich was only fourpence halfpenny (one of the most expensive trips which one could take by London Underground by that time!). Several of us invested that sum to be able to step on the Greenwich meridian; but this was as close as we got to astronomy on that occasion. For we underestimated the travel time necessary to reach Greenwich from Westminster; and when we reached the doors of the observatory, we found them already closed. Vigorous ringing produced at length an aged caretaker, who explained to us politely that office hours had ended fifteen minutes before; and that their staff were pretty punctual in leaving the establishment on time. As proof of this fact he showed us a nearby office in which all desks were already covered by oilcloth to protect papers on them from dust; and we had no reason to doubt the reality any longer. I do not know why we felt disappointed; in Prague the reception of any visitors after hours would certainly not have been any different; but I suppose one always hopes that the grass is greener everywhere else.

Perhaps to overcome some disappointment, we resolved to walk all the way back to London; and the interminable Commercial Road of East London gave us a good chance to see another facet of the capital, very different from Westminster or Mayfair. This was one of my longest walks through the streets of the British metropolis. The second (some years later) came in 1938 when, having missed the last Underground train at Mill Hill, I had to return on foot all the way to Euston Station to catch the first morning train for Cambridge; I should certainly not like to do so these days!

After a week or so in London we went to spend two weeks at St Helens on the Isle of Wight for a pleasant stay at the sea-shore; and did some sightseeing of more fashionable places of that lovely island (which I have again not revisited so far). We eventually left British shores by ship from Southampton for Dieppe in France, and on to Paris. I remember vividly the impression made on us all by some of the Cunard giants (*Aquitania* among them) then berthed in that great port, in the last spell of their glory; but nothing could exceed the treat for our eyes awaiting us in Paris after arrival at the Gare St Lazare from Dieppe. Our temporary home in the French capital was in the Cité Universitaire (south of the Porte d'Orléans), but only for an overnight stay: the days belonged to the sights of the city. The impression of the grand prospect from the Louvre through the Champs Elysées to the Arc de Triomphe kept us spellbound as we climbed slowly the Chaillot Hill, and, eventually, the Arc itself, to behold a panorama having probably no equal in the world.

The origins of Paris (like those of London) go back to Roman days; but little that meets the eye from the top of the Arc de Triomphe has survived from the Middle Ages (the cathedral of Notre Dame being the most significant exception). Its glories spread out before you from that vantage point belong to the Renaissance and subsequent times, from the Louvre of the Valois to the post-revolutionary nineteenth century of Balzac or Victor Hugo. The system of avenues diverges from the Etoile in all directions, with the old Trocadéro (then still standing, but soon to be demolished to make room for the new world exhibition to be staged in the capital) and the Eiffel Tower to the south standing on the Champs de Mars, and dominating the Paris skyline as the last sentinel of the glories of the creative spirit of the French bourgeoisie between the time of the Revolution and the First World War. All these sights were bound to make an unforgettable impression on young visitors whose cradles stood a thousand miles to the east (about as far as the French Gothic architecture reached in the Middle Ages)!

The second day of my Lutetian pilgrimage belonged, of course, to the Observatoire de Paris, one of the oldest institutions of its kind in the Western World, still occupying the historic building erected for it by Claude Perrault in 1667–68, on the orders of Louis XIV (to which the French astronomers of my time referred rather condescendingly as *la vieille maison*). An entrance past the statue of Leverrier led to the building in which my teacher Nechvíle had spent several years of his life; and his name proved to be *open sesame* to the laboratories of the Carte-du-Ciel, where Nechvíle was well remembered by his former colleagues and contemporaries. Many of the plates I was shown carried his notes; and as his student I was received with an attention far beyond my age or deserts.

**Figure 3.5** The Observatoire de Paris. (Left) The Perrault building from the time of Louis XIV. (Right) The statue of U J J Leverrier (1811–1877) by Chapu, in front of the entrance.

The next day belonged to a visit to a modern branch of the Paris Observatory at Meudon; while the third was set apart to pay respects to *tous les gloires de la France* at Versailles, and to its Sun King. By that time, however, our time was up and our purses were almost empty. What else could we do than repair to the Gare de l'Est, and return by an overnight trip on the Prague express back to the place of our origin, where we arrived around noon, a little more than one month older by age, but how much older by experience!

My summer travels of 1935 already proved astronomically much more rewarding, for their main goal was to attend the Fifth General Assembly of the International Astronomical Union, held that July in Paris. This Union had been founded a little more than ten years before as a union of the National Committees on Astronomy of the adhering countries, a character which it retains up to this time, and by 1935 the Czechoslovak National Committee agreed to include me in its delegation as its junior member.

It is easy to imagine that, in the circumstances, we—Dr Hubert Slouka (chief assistant of the Astronomical Institute of Charles University, and at that time Editor of the Říše Hvězd, the monthly journal of the Czech Astronomical Society to which I used then to make regular contributions) and I—did not travel to our destination by the shortest route. After the first night's journey, our first stop was in Constance, on the shores of the Bodensee. There we were to pay our respects to the place where Master Jan Hus (1369–1415), one of morally the greatest personalities of medieval Bohemia, professor and sometime rector of Charles University, died at the stake for his efforts to bring about the reformation of a corrupt Church (with three popes contending for recognition, each better qualified to head the Mafia than the Church of Christ); thus dying a martyr's death for the sins of the contemporary hierarchy. A year later, the same fate was shared by his younger colleague M Jeronym (Hieronymus) of Prague (see p. 78). Centuries later, when the cause of Church reform prevailed in that part of the world, a monument was erected to both martyrs, before which we now stood for the first time (though I have since revisited it more than half a dozen times).

From Constance, we proceeded (by train) via Offenburg to Strasbourg, where we stayed overnight for the next day's visit to the battlefields from Verdun to Sedan—of especial interest to Dr Slouka (who was a keen army reserve officer). Though I was not an officer at that time, no-one could have left the place, with its remaining forts (of which we visited Forts Vaux and Douaumont) without the profound impact which the carnage and destruction still evident on all sides must have made on anyone beholding it less than twenty years after the event. The battle of Verdun in 1916, like that for Stalingrad in 1942, represented the culminating monument of human suffering and folly, which was almost bound to make a pacifist of anyone who beheld the remains of the destruction, until even more durable monuments of destruction and death were erected by men at Hiroshima and Nagasaki in 1945.

From Verdun our trip continued via Mezières and Sedan (seats of decisive Franco-German fighting in 1870 and again in 1940) in the Ardennes (the forests of which have long since lost any resemblance to the charms once conjured up in this region by the imagination of William Shakespeare) for Brussels. There we visited the Royal Observatory at Uccle and spent some days at the World's Fair being held in the city (the remnant of which stands in place still to this day in the form of the 'Atomium', a monument to the atomic structure of matter) before turning southward towards Paris, which we reached on 9 July, a day before the opening of the Fifth General Assembly of the International Astronomical Union.

That opening on 10 July was an occasion not to be forgotten by someone of my age and the like of which I have never seen since—the closest to it being, perhaps, the Union's Eighth General Assembly in 1952 in Rome. The opening ceremony, attended by the French President Albert Lebrun and several members of his Government, took place in the ceremonial halls of the French Ministry of National Defence in the Rue St Dominique; and in the course of much splendour (accompanied by orchestral music) the assembled astronomers witnessed their President, Professor Frank Schlesinger of Yale University, being decorated by the French Minister of Education, M Mario Roustan, with the officer's cross of the Legion d'Honneur.

This impressive ceremony lasted the whole morning, in the course of which I beheld (and heard) many distinguished astronomers, known to me so far only by name. The dignity and splendour of the entire occasion has never been surpassed in the course of the whole history of our Union ever since, and never again has the Head of State graced the entire session with his presence, as a token of the esteem in which our science was held in his country. And yet the membership of the IAU was still very limited, not more than 500 (of whom not all, of course, were present)—a far cry from the throng of upwards of 6000 constituting our Union at the present time.

The more limited size of the membership was bound to make the General Assemblies in those days more intimate occasions—more dignified, as well as amusing. How often one would see a pair of aged astronomers clinging to each other, and attempting to read through thick glasses the names on their respective badges worn on the lapels of their coats (for everyone wore a coat in those days), until one (or both together) burst out into a joyful exclamation: 'Oh, I remember you!'

So it was at the conclusion of that memorable assembly, when astronomers of all countries and ages were filing out of the hall for the sunshine outside. Juniors like myself (there were not many younger ones around) kept their mouths closed, but their eyes wide open at this intellectual splendour. As I stood there in some corner observing all this—behold, all of a sudden I saw our President, Professor Schlesinger, heading through the crowd straight towards me, and addressing to me some kind words. He asked where I was from and about my own interests; when he heard that I hailed from Prague, he recalled his own former visits to that city, and the names of astronomers he had met there. In a few minutes of this fleeting conversation nothing was said

## Fifth General Assembly of the IAU

that was of any importance; but nothing could have brought home to me more effectively the fact that astronomers of the whole world constitute one family, motivated by the same interest—a feeling which, I hope, will never be lost in our profession.

In the evening of the same day M Ernest Esclangon, Director of the Observatoire de Paris (and President-Elect of the Union) held a reception for members of the Union in the sacred precincts of the historic building of his observatory (see figure 3.5). Soon we were lining up in corridors once trodden by the Sun King on the way to the flood-lit gardens of the observatory, now full of gentlemen in evening dress and white-toiletted ladies getting ready to dance.

On Thursday afternoon, our Union (together with members of the Congress for Scientific Photography) was received at the Hôtel de Ville de Paris, in a grand hall whose walls and ceiling were replete with historical frescoes, which were attracting the attention of the astronomers even more than the address of Monsieur Chiappe, the Mayor of Paris (and, only a few years later, minister of the Vichy government of Marshal Pétain); from whose speech I understood only that he held astronomy in great esteem and welcomed us sincerely to Paris.

A day later (on Friday, 12 July) the astronomers congregated once more at the front court of the observatory around the statue of Leverrier, to mount a fleet of excursion buses which were to transport us to visit Fontainebleau. The temperature in the early hours of the afternoon must have been as high as ever before, and air-conditioning was yet to be invented; so that after a two-hour ride we left the buses almost hard-boiled for a tour of the palace once belonging to Louis XIV, and later to Napoleon I. In these circumstances, all the eloquence of the accompanying guides was barely sufficient to evoke appropriate interest in the bedroom of Mme de Maintenon, or in a hat which Napoleon wore when inspecting his Old Guard.

After this visit to the Château, the same convoy of buses transported members of the Union to the nearby Hôtel Savoy for afternoon refreshment. The temperature must have attained a maximum, and what followed would have required the talent of a Homer for adequate description; for even the gods on Olympus must have behaved themselves with greater decorum than did the two hundred thirsty astronomers in the Hôtel Savoy at that time. In fact, in no time at all they took over the entire reception hall; and to the discreet consternation of the waiters occupied all free positions not only in front of the bar, but also behind it!

Without knowing how, I found myself among the latter, vainly struggling with a container of ice-cream to get at its contents. 'Wait; let's try it again', said a voice behind me, of a rather short man in a hat, who obviously knew the business and would not be defeated by so simple a mechanical problem. He pulled out of his pocket a knife which could have belonged to a boy-scout's outfit; and in no time at all he and I were sharing the contents of the canister to assuage our thirst. I do not think this astronomer ever looked at his young accomplice in the course of this operation (we were both sitting on the ground to hide our treasure from others); but I looked at him. This was to be my first 'collaborative project' with a man who was later to play so important a role in my life—Harlow Shapley.

On the terrace in front of the hotel, in full glare of the Sun, Professor Russell could be seen taking a promenade with umbrella in his hand. This was not mere professional habit, but a wonderful prescience of the retribution which nature had in store for thirsty astronomers; for well before we reached Paris on our return, the heavens, irked by so much frivolity among the priests of Urania, released a downpour which brought our buses to a standstill, and even after arrival persecuted the luckless astronomers on the way home according to their deserts.

Saturday afternoon was set apart for an excursion to Versailles. As the weather turned sunny again, we walked through the great park surrounding the palace, with its rondels which served as scenarios for the plays staged there in the days of Louis XIV and his successors. The white marble of these dazzled the eye in full sunlight, and the bronze statuettes of Psyche bashfully tried to cover their innocence, while chubby-faced little cupids smiled at them as sweetly as they did on the astronomers passing by.

Sunday, 14 July, in the evening: the *fête nationale*, when all of Paris abandons itself to love and jubilation, while astronomers were dining on the first floor of the Eiffel Tower. A menu (with a photograph of the Orion nebula on its cover) featured nine courses. Corks of champagne bottles were popping continuously, and even the most convinced teetotalers were seen sadly observing their empty glasses which had sparkled with wine only a minute before. The proverbial *Ville de la Lumière* sparkled with countless fireworks. Notre Dame, Place de la Concorde, Champs Elysées, Arc de Triomphe were all floodlit tonight—a grand view indeed from the Eiffel Tower!—and music, music was everywhere, while people danced in the streets.

Monday, 15 July, in the afternoon. With a more respectful feeling than in the days before we held in our hands an invitation card informing us that *Monsieur le Président et Mme Albert Lebrun* requested the pleasure of our company at a garden party in the Elysée Palace, official seat of the French President, through which so much of European history had passed in recent centuries; and which today opened its gates to other species of humanity than the politicians. We took off our wraps; the chamberlain announced our names; and the President with his wife pressed our hands.

In the garden of the Elysée was an improvised buffet for the guests; but the astronomers behaved themselves today in front of it with much more decorum; for the President was among us. I had, moreover, one more unforgettable experience on that occasion: for it was there that I was introduced (by Dr A Beer, I believe) to Sir Arthur Eddington (of whom much more will be said later in these recollections). Sir Arthur was very kind; and listened with patience to the questions which young tiros were eager to ask those who (in their opinions) were omniscient. A brief note that I wrote afterwards on the IAU Paris meetings for the *Říše Hvězd* (Kopal 1935b) allows me to refresh my memory on the subject of our conversation.

At that time, my mind was already full of eclipsing variables; and some months before I had published an article—one of my first in the English language (Kopal 1935a)—in which I attempted to deal with the physical properties and evolution of such systems. One aspect of them which already puzzled me at that time (as it has done ever since) was the curious fact that, while in Algol (and Algol-like systems in general) the cooler component was invariably the less massive of the two, for $\zeta$ Aurigae (and this alone) the converse is true, a fact which (much later) we learned to refer to as the 'evolutionary paradox'. I asked Sir Arthur if he could think of some explanation of this curious phenomenon. He said with a smile that he could not (his mind was, I later learned, already occupied with other problems); but encouraged me to do so myself. I took his words to heart, for I have never ceased to think about it ever since!

On the evening of 16 July our French hosts scheduled a *banquet suivi de bal* in the Hôtel Continental, to bid a social goodbye to their guests. It was an occasion on which the French demonstrated once more the superiority of their race in the field of gastronomy; and after dinner astronomers of different countries had an opportunity to defend their colours also on the dance floor. After midnight, as a token of *entente cordiale*, and to the sincere applause of all present, Madame Camille Flammarion appeared on the parquet with Professor Henry Norris Russell (the latter without his proverbial umbrella this time); and it was not till after 3 AM that astronomers began at last to disperse, wishing each other *au revoir à Stockholm* in three years' time.

These are, in brief, some of my recollections of the Fifth General Assembly of the International Astronomical Union held in Paris in 1935, the first in which I was privileged to participate; and at which I was also elected to membership of the Union as a member of its Commission 27 on Variable Stars—a Union to which I have belonged now for more than fifty years. I do not know what earned me this early attention at the age of 21. Possibly Professor A A Nijland (the President of Commission 27) remembered the work I had done on some of his light curves of eclipsing variables, or Professor Jean Dufay (the President-Elect of the Commission) remembered me from the French Association of Variable Star Observers.

In the past half century I have had an opportunity to participate in most of the General Assemblies of the IAU held in different parts of the world, as a member of several of its commissions and (between 1948 and 1955) as first President of Commission 42. There is no need (nor space) to recall all these from our memories in the same detail (though some of these occasions, like the reception of the IAU in the City Hall of Stockholm in 1938, or at the Campidoglio in Rome fourteen years later, will surely remain in the memories of all surviving participants as long as they live).

Yet, with advancing years, irreversible events have taken place which have largely deprived the meetings of the Union of some of their old-time intimacy and charm. The principal reason has been the growing number of participants. The First General Assembly of the IAU, held in Rome in 1922, brought together 207 participants, a number which, in Paris in 1935, had increased only to about 500; this is why, in those years, it was still possible for the inviting countries to 'roll out the red carpet' for their colleagues as the French did in Paris, with their proverbial panache. However, after the advent of the Space Age in the late 1950s, the membership of the Union began to escalate by leaps and bounds: by the time of the Thirteenth General Assembly in Prague in 1967 it already exceeded 3000; and now (in the mid-1980s) it stands somewhere above six thousand.

This increased democratisation of our profession should, of course, only be welcomed (at least in principle); but it also entailed several other consequences, not all of which have turned out to be beneficial. Thus, in the early years of the Union, and in consequence of the way in which our Union came into being after World War I as a part of the Treaty of Versailles, its presidency for a time became almost an unofficial perquisite of the Directors of the Paris and Greenwich Observatories—a restriction long since happily removed. Moreover, the locations where the General Assemblies are held have been (quite rightly) decentralised from the shores of the Atlantic to a wider theatre of global proportions—a fact which tends, however, to maximise travel costs for the majority of the participants. In the 'roaring sixties' it was still, perhaps, possible to raise these from public funds of national origin even for young astronomers who may not yet stand too high on the official ladder of their organisations. However, in the more recent past, the costs of travel have kept escalating (and the sources of national support drying up) to such an extent as to make the participation especially of young astronomers at IAU meetings increasingly difficult—a problem of serious concern not only to themselves; for without a steady influx of young blood the Union could easily wither into insignificance.

And there are, of course, other costs for the participants in Union activities besides travel. In bygone days, the participants from foreign countries could usually rely on a considerable amount of local hospitality. Due to the economic facts of contemporary life, the extent of such hospitality has, however, by now all but disappeared. On the contrary, few national organisations can now afford to invite the Union to meet on their ground without the guests bringing money in, usually (though not wholly) in the form of 'registration fees' charged to all participants and their families, fees which (like everything else) are escalating with inflation (and have to be paid in 'hard currency'). Older participants can still well remember the days when such charges were non-existent. At the Seventeenth General Assembly in Montreal in 1979, they still amounted to only $50; but at Patras (1982) they had already risen to $100; and, at New Delhi (1985) to $150; whether their future escalation will follow an arithmetic or a geometric progression is anyone's guess.

## Postgraduate Years and First Trip to the Far East

There are many more thoughts which come to mind in this connection; and we shall return to some in Chapter 7. But, for the time being, it is time for the author to return from Paris to Prague to continue his studies and take further steps preparatory to a subsequent professional career. These steps now unrolled in rapid succession. At the commencement of the 1935/36 academic session I completed examinations for the degree of a candidate of science (RNC), which should qualify me to apply for the doctorate in another two years.† Unlike at American or British universities, the candidates of science at Charles University were not officially assigned any specific supervisors.

---

† It should be borne in mind that students used to enter Charles University as undergraduates two years later than American students enter college, and one year later than their British counterparts enter their own universities (indeed, up to the middle of the nineteenth century, the last two years at the *Gymnasien* in Czechoslovakia were a part of the university; and while the Chancellor of the latter was *ex officio* the Archbishop of Prague, the last two senior years of the *Gymnasien* were under the jurisdiction of the local bishop). In my own days, the degree of RNC was essentially equivalent to that of bachelor of science (the latter title existed as such at Charles University in the Middle Ages, but disappeared since); while the modern doctorate became equivalent to the medieval title of Master.

Unofficially, however, they always grew up under the wing of some senior faculty member, usually of their own choice: in my case it was Professor Nechvíle and (since 1936) Professor Freundlich. Since the beginning of the third year of my university course I was, however, increasingly on my own; and my class attendance dwindled considerably; but I cannot by-pass the debt which I owed at that stage of my educational career to the late Professor František Záviška; not only for the encouragement which he bestowed upon me with such generosity—he had already accepted from me articles for publication in the *Czech Journal of Mathematicians and Physicists* (of which he was one of the editors) when I was still a grammar-school boy—but also for setting up a model in my eyes of what a science professor should be in the classroom.

There was, however, one event in my third year which overshadowed everything else in importance by its impact on my subsequent career: and that was a journey to the Far East to observe the total eclipse of the Sun on 19 June 1936 in Japan. It is this trip and the experiences acquired in the course of it that I wish to recall in this place.

This eclipse, whose shadow-track crossed almost the whole Eurasian continent, from Greece to northern Japan, was for many months the centre of astronomical attention, and had been under extensive discussion by Commission 13 in the course of the Fifth General Assembly of the IAU. The Czech Astronomical Society (in which I had been active since 1928) decided to sponsor two expeditions to observe this eclipse: one to the town of Sara (near Orenburg) in southern Russia; and the other, to Hokkaido, Japan, where the Moon's shadow cone crossed the northern part of that island (not far from the shores of the Okhotsk Sea) to disappear in the vast expanses of the Pacific Ocean.

The expedition to Russia consisted of Drs V Guth and F Link (accompanied by Miss B Nováková and J Vlček); while the one to Japan (organised as the 'second string', if the first were clouded out) was formed by Dr H Slouka, Mr Walter Jaschek (our guest from the Vienna Observatory, recommended by its then Director, Professor K Graff)—and myself! As we had all been born before 1916, during the reign of Emperor Francis Josef I, we were called an expedition of *Altösterreicher*; though all of us (including Jaschek, who later forsook astronomy for chemistry) were of Slavic origin.

Needless to say, I volunteered for the expedition with the most distant goal (as grandson of grandfather Lelek I could not have done otherwise!) and, as its junior member, I was entrusted with carrying its heaviest cargo: the Zeiss 8 inch photographic objective of the double-astrograph of the Society's Štefánik Observatory in Prague, of 343 cm focal length (see figure 2.6); and its collapsible mounting, which consisted of aluminium tubes. The objective I hand-carried in

## Preparing for the journey

my suitcase and the mounting in the other hand. So we were to travel together, the telescope and I, for the next four months, over more than forty thousand kilometres. It was a wonder that we also returned together in one piece, and were not separated by Customs at different frontiers, or by some other misadventure. The scientific aim of this telescope was to photograph the solar corona during totality for photometric purposes; and as I had used its optics already before in Prague, as an assistant to Professor Nechvíle during the 1930 Eros campaign, it was thought that I could handle it in the field as well. When the time came, I did not perhaps handle it so badly.

But, to return to the beginning, to volunteer for participation was not enough; much more important was to raise the necessary funds without overstraining the family exchequer. The Astronomical Society could not spare anything beyond the loan of the instruments; and as the Society was a private one, with no Government connections, the public purse remained closed for this purpose as well. What to do? The eclipse would not wait, and so I had to resort to self-help.

In the earlier part of this chapter I have already mentioned that my first acquaintance with the English language was the translation into Czech of *The Mysterious Universe* by Sir James Jeans. In the 1930s this became one of the best known popular books on astronomy, and was translated into 27 languages. The Czech translation (then still in my drawer) was to be the 28th. Instead of offering its text to a commercial publisher, my colleague (than candidate, and later doctor of science in geology) Josef Štorek, already then possessing a keen commercial sense which enabled him, in later years, to become a successful economic geologist in the Third World, and I decided to publish the translation at our own expense and risk. Cambridge University Press offered us the copyright at a nominal price (of, I believe, £5); and we had the book on the market in no time. As expected, it sold well; and my part of the profit from its sales was sufficient to solve almost all my financial problems connected with the trip to the Far East.

Not all, however; for in order to do so, it was also necessary to convert these funds into the appropriate foreign currency. This meant a round of visits to the officials of the Czech National Bank, which alone could grant the requisite permits. I recall, in particular, the call on one of these officials (let him remain unnamed) who received us with some impatience and vented his displeasure at our project with the classic words: 'I cannot really understand why you have to travel all the way to Japan to see that eclipse. Can't you wait till the day after to read about it in the newspapers?' Needless to say, we could not; and some further wire-pulling was necessary before all our affairs were set in order and we could depart on the way east.

The loss of time entailed by all these preparations ruled out travel by sea; and trans-Asiatic air travel was still a dream of the future; thus what else could we do than go by train? And so we departed from Prague on our long journey on 19 May 1936, one month before the eclipse day, through Warsaw to Moscow (across frontiers which in another ten years were to undergo such profound changes) where we changed for the Trans-Siberian Express to carry us to eastern Asia and to the shores of the Pacific Ocean, across the Volga to the Ural Mountains, the Rivers Irtysh, Ob and Jenisei, on to the Baikal Sea, which we reached on the fifth day after our departure from Moscow.

The Trans-Siberian Express, in those pre-war days, was hardly an express in today's sense of the word, as its average speed (including long stops in larger cities *en route*) was not much more than 40 km per hour. However, it was a comfortable train for a trip lasting several days. Each passenger was provided with sleeping accommodation and his place in the dining car, where one spent a good part of the travel time. No-one should fear having nothing to do, especially if one played chess! The Russian officers returning to their trans-Baikal garrisons were good players; and acquaintance which started over a chessboard developed further over glasses of vodka in the restaurant car. Several of them gave me their names; but I never heard of any one of them since. Who knows if they survived the war, then only a few years in the future? But if they did, none of them became a Marshal of the Soviet Union!

At last, after a seven-day journey aboard the Trans-Siberian Express, we arrived at the Soviet–Manchurian frontier. The train itself continued on to Vladivostok; but as the latter was a fortified port, we were not given exit visas for Japan from that locality. Instead, we had to leave the Vladivostok-bound train at Karimskaya, to reach next day the frontier station at Otpor; and to continue our journey southward, through Manchuria and Korea to Fusan, and to enter Japan (after a six-hour sea journey) at Shimonoseki.

In 1936, Manchuria and Korea were already firmly under Japanese influence; and at the border station of Manchou-li we already met exclusively Japanese personnel. Although the *laissez-passer* issued to us by the Japanese legation in Prague enabled us to navigate our way with ease through passport control, the ten suitcases containing all our equipment attracted considerable attention on the part of Japanese Customs—especially the aluminium tubes for the mounting of our refractor. 'This is to mount the lens of a telescope', we were trying to explain to the relevant officer; but only with limited success. 'Telescope?', asked the latter, not taking his eyes off the suspect tubes. 'What is it for? Shooting?' It took us some time to explain to the suspicious officer that one does not normally use telescopes for this purpose; and astronomers (when they do not fight with each other over competitive theories) are the most peaceful creatures of all mankind. Eventually, we were able to convince the Japanese inspector of these facts, and to move safely all our equipment to the train bound for Charbin.

In travelling southward aboard this train we soon saw the reason for the concern of the Japanese staff on the frontier: for we were actually passing through a zone of hostilities. The whole train was guarded by Japanese soldiers; beyond Hailar, the blinds had to be pulled down and it was forbidden to look out of the windows. Railway stations where our train came to a standstill were often surrounded with barbed wire; and heavy military transports were moving to the frontier (it was interesting, however, to observe that most of the railway personnel we saw were not Japanese, but Russians—possibly remnants of the White armies who preferred exile to a return to Soviet Russia after the war).

Our first port of call, which we reached after an overnight journey, was Charbin, where we were cordially received by the local Czech colony headed by our Consul, Mr Rudolf Hejný, and his staff. A couple of days' rest on the ground permitted us not only to recuperate somewhat after the long train journey of the preceding nine days, but also to get the first feeling of the Far East, for Charbin was a city where east met west, China with Russia: a place where you could see Orthodox churches side by side with Buddhist temples, in the streets of which *invoschčiks* competed for passengers with rickshaws; and where inscriptions could be seen in the Russian alphabet as well as Chinese characters.

But time did not permit us to linger too long in this gateway to the Far East; and in the evening of 30 May we boarded the train again to travel through Hsinking (then the capital of the puppet state of Manchu-kuo) and Mukden to the Korean frontier, and through Korea to its southernmost point of Fusan. The train did not even hesitate in crossing the 38th parallel north of Seoul (then called Keijo); and on the eleventh day after we left Prague we arrived at last on the shores of the Pacific. From here we could still return home on foot! But none of us changed his mind to do so; and two hours after we reached Fusan, a Japanese ship carrying the name of *Kaifuku Maru* left the port with our expedition for the island Empire of the Rising Sun.

We reached the shores of Japan after a six-hour trip by sunset; and after landing received our second lesson in oriental patience and politeness in baggage and passport control. The *laissez-passer* issued by the Japanese Government did not prevent the passport control officer from asking us all kinds of questions about our family trees or our *Weltanschauung*. I was, I remember, asked for my opinion about the revolt of young army officers the previous February† and possibly disappointed the examining officer by a confession that I did not know there had been any. At any rate, it was only a matter of time before we were politely released, and permitted to board an express train from Shimonoseki through Kobe to Kyoto, which we reached on the morning of 2 June. Only one fleeting recollection remains in my mind from that night journey: namely, of a short stop which our train made at a station bearing the name of Hiroshima. No one could as yet dream of the gruesome fate in store for that unfortunate city in less than ten years' time!

The arrival at Kyoto marked the first stage of our visit to Japan; for there we were to meet the friends who had invited us to come. Professor Yssei Yamamoto (1889–1959), Director of the Kwasan Observatory, who awaited us with his staff at the station, was primarily responsible for our earlier decision to observe the eclipse at Hokkaido. Most of the equipment needed for this purpose we brought with us from Czechoslovakia; but one of the most important pieces we did not have: namely, a coelostat for mirroring the image of the Sun in our horizontally-mounted telescope. This represents a heavy piece of equipment, not easy to transport across Asia in the time at our disposal. It so happened, however, that of the two expeditions which Kwasan Observatory planned to send to observe the eclipse, one to the Soviet Union and the other to Hokkaido, the latter was equipped with a coelostat of 40 cm aperture which was too large for the use of the Kyoto astronomers alone; and Professor Yamamoto kindly allowed us to mount our horizontal telescope alongside one of their own, with enough light for both. Yamamoto himself planned to go to Omsk in Siberia, with a major part of his staff; while we should be accompanied to a Hokkaido village called Nakatombetsu by his American-educated wife (whose English was better than my own) and his younger colleague Professor Shinichiro Takeda (1901–1939), of whom much more will be said in Chapter 6.

In Kyoto itself we found ourselves in the very midst of classical Japan, which for centuries had been the seat of its Emperors; and of its culture (especially since the advent of Buddhism in the eighth century AD). Kyoto and its neighbourhood is still the seat of more than 3000 shrines and temples, apart from many institutions of higher learning, including one of the foremost of Japan's imperial (now

† On 26 February 1936, a group of young officers of the Japanese army revolted against their Government, and murdered some of its leading statesmen. After a few days the revolt collapsed, but in so far as it endeavoured to influence subsequent Japanese policy, it largely attained its aims; though, at that time, the fact that certain army officers failed to obey the Emperor's rescript exerted a traumatic effect on the Japanese mind.

national) universities of international renown (which it has been my privilege to visit many times since).

But, in the first week of June 1936, we had as yet no time fully to enjoy the lure of this ancient city; for our distant goal of Nakatombetsu was still several days' train journey ahead of us; and the eclipse day was approaching. Thus, only three days later we had to leave Kyoto for Tokyo, and further north cross by sea (aboard *Matsumae Maru*) from Aomori to Hakodate; and on through Sapporo, Asahigawa and Otoineppu to Nakatombetsu; which we reached on the evening of 9 June, ten days before the eclipse and twenty-one days after we left Prague, after possibly one of the longest train journeys one could make on Earth.

When we had left Tokyo two days before in the late afternoon, we got our first glimpse in the distance of the beautiful cone of Fujiyama—a good sign, according to our Japanese companions; and our arrival at Nakatombetsu two days later was no less joyful. We arrived late in the evening; but the whole village was illuminated and most of its inhabitants were at the station, old, young, schoolchildren with Japanese as well as Czechoslovak flags in their hands, and from everywhere we heard the greeting '*Banzai*' as the village-master, Mr Tomotaro Sato, and his councillors led us through the crowd-filled streets to a Japanese inn for a ceremonial dinner, at which we had a new opportunity to extend our acquaintance with Japanese cooking and put to good use our but recently acquired skill with chopsticks. Many toasts were drunk on that occasion (with *sake* as well as good Sapporo beer) to the Sun and the dragon which was soon to devour it (albeit only for 116 seconds!), as well as to the visiting astronomers. It was not long before daybreak before we could at last find some rest on *tatami*, in the guest quarters of the local school building, where all the astronomers and their equipment were housed.

In June 1936, Nakatombetsu became the site of no less than three expeditions to observe the eclipse of the Sun: those from the Kwasan Observatory in Kyoto, Mitaka Observatory in Tokyo, and ourselves (who had come from the greatest distance). It is with pleasure and affection that I recall today, almost half a century later, our hosts as well as colleagues whom the eclipse of the Sun brought together at that time, and whom I was not to see again till 47 years later. When I visited Nakatombetsu in October 1983, of all the participants in that event, only two (as far as I know) are still alive, the second survivor being my dear friend Sigemaro Kibe of Kyoto, my contemporary in age. It was Kibe with whom I was to share the large coelostat, which can be seen in figure 3.6; we became close friends at that time, and have remained so ever since.

Kibe was a very talented amateur astronomer with veritable 'green

## 118  *Lehr- und Wanderjahre*

fingers' for astronomical optics. I was, therefore, sorry to learn from him that he could not contemplate the career of professional astronomer, as fate had another future in store for him. Little did I realise at first that Sigemaro Kibe was the descendant of an old *daimyo* family, whose father was one of the four most important Buddhist dignitaries in Japan, and hereditary abbot of one of the oldest monasteries in the Kyoto region. When an abbot died, he was succeeded by his son for several hundred years; and young Sigemaro had no other future. And so it came to pass. The last letter I received (when already at Harvard) from Kibe, before Japan entered the war in 1941, informed me of his forthcoming ecclesiastical elevation; but otherwise he was full of dark forebodings. 'The clouds are gathering on the horizon', he wrote, 'and I do not know how long we may be able to communicate in the future. But whatever happens, I want you to know that my feelings towards you will always remain the same as at Hokkaido in 1936.'

**Figure 3.6** The optical system used to take photographs of the total eclipse of the Sun (figure 3.9) on 19 June 1936 at Nakatombetsu. The 21 cm objective (lower right corner of the configuration) had travelled with me from Prague in my suitcase; but it was the Kyoto coelostat which had brought us to Japan.

I am not sure if my reply, animated by the same spirit, reached him before the disaster at Pearl Harbor. But when I visited Japan again in 1962, Kibe was already installed as abbot of the Kinshoku-ji monastery founded in 1285 (in the difficult post-war years, he combined the large facilities of his temple with his talents as an optician for commercial optics work to keep his monastery financially out of the red); and I shall never forget the scene when Kibe came to see me off at Kyoto Railway Station: he in his abbot's full regalia, accompanied by his charming young daughter in a miniskirt! Since that time we met each time I revisited Kyoto, and at one of these occasions I met him also in the office of the Chief Education Officer of Shiga Prefecture, a function Kibe held with great distinction in the 1970s. Figure 3.7 shows us in 1983 on the way to visit the Hiei Buddhist Monastery on the outskirts of Kyoto.

**Figure 3.7** Abbot Sigemaro Kibe (left) and the author, old friends from the 1936 eclipse camp, *en route* to the Hiei Monastery above Kyoto in late October 1983.

But let us return to Nakatombetsu in June 1936 where Kibe and I met for the first time. In the first few days after our arrival, the local weather was fine, and the installation of our instruments could proceed at a satisfactory pace. Later, however, it suddenly took a turn for the worse, and mud produced by heavy rain made it almost impossible for us to reach the eclipse camp. At last, the rain stopped; and winds from the west rapidly dried up the land, but they also became strong enough one night to threaten the stability of our equipment. We had to stay up the whole night to safeguard what we could, for the storm lasted till morning; but when it finally subsided, the weather was fine. Our hopes were rising again! In the dining room, where all the astronomers took their food, Professor Hashimoto of Tokyo Observatory listed daily the weather forecasts (in Roman as well as Japanese characters) which were increasingly optimistic: the day before the eclipse there was no longer a cloud in the sky and, what was even more important, the forecast for the next day was equally optimistic.

And, indeed, when I opened my eyes on the day of the eclipse at about 5 AM, a bright Sun was rising above the horizon and the eastern sky was clear; but a bank of cirrus clouds was rapidly approaching from the west. However, these soon passed; and for most of the morning the sky remained completely clear. But nature can play queer tricks with astronomers. Just before noon another bank of clouds, this time cumuli, reappeared from the west and were soon overhead; but they passed again, and were followed by isolated formations. The situation began to get exciting: if any one of these was to cover the Sun for us at the critical time for only two minutes, our journey from Prague to Japan would have been in vain!

The partial eclipse at Nakatombetsu was to commence shortly after 2 PM local time; and two seconds after the predicted time of first contact the lunar limb made its appearance in the focus of my telescope. However, the success of our expedition was still far from assured; for during the partial phase of the eclipse clouds kept covering the Sun, sometimes for several minutes at a time (see figure 3.8). The more the eclipse progressed, the greater the tension became in the eclipse camp. But, at last, the merciful heavens took pity on travelling astronomers: the last visible clouds passed the Sun (or, rather, the remaining crescent of it); and at least half an hour would elapse before the next cloud field (just visible near the horizon) would pass overhead; while, in the meantime, we should be able to observe the total phase of the eclipse under near-ideal conditions.

**Figure 3.8** Japanese amateur observers at Nakatombetsu (outside the enclosure set apart for the professionals). (Left) The eclipse advancing towards totality can become rather a gruesome sight (Mrs Yssei Yamamoto at left). (Right) All's well that ends well, as the Sun reappears from behind the Moon.

## The 1936 solar eclipse expedition

And so it came to pass. A total eclipse of the Sun represents a phenomenon of overwhelming and incomparable beauty, without equal in the whole world of nature. Just before the last ray of the solar photosphere disappeared before the western limb of the Moon, the whole disc of the latter became suddenly encircled with the glistening ring of light of the solar chromosphere, in which three large prominences stood out by their pink colour, enveloped in an extremely fine cobweb of the solar corona reaching in beautiful streamers far from the Sun to the fathomless depths of the dark-blue sky in which the entire apparition (how small it appeared to be, and so high!) seemed to be floating (figure 3.9). Not far from the Sun the planet Venus shone dazzlingly in the sky like a diamond; below it one could see the principal stars of the constellation of Orion (down to about second apparent magnitude); while the horizon was girdled in all directions with a blue–green aureola of light from distant regions outside the belt of totality—a heavenly wonder almost frightening in its esoteric beauty, like a miraculous vision from the Apocalypse!

**Figure 3.9** The total eclipse of the Sun on 19 June 1936, photographed by the author at Nakatombetsu with an objective of 21 cm free aperture and 343 cm focal length. Exposure times (left to right): 15, 30 and 60 seconds.

But it was no time merely to behold this heavenly wonder in platonic admiration; my task was to photograph the solar corona in white light with our refractor of $f/16$ focal ratio, with exposure times of 15, 30 and 60 seconds (5 seconds being required for each change of casette). Once totality commenced, Dr Kenzo Inamura counted seconds for us all: *ichi, ni, san shi, go,* ... etc. At pre-set times, one could hear an abrupt click of casette exchanges, until, 116 seconds later, with the appearance of the first ray of the Sun at the opposite limb of the disc, a cry of '*Banzai!*' reverberated from all the telescopes, and was taken up by hundreds of lay spectators from Nakatombetsu and the neighbouring villages. These had congregated in the meantime around our camp (perhaps feeling safer there in the proximity of our astronomical guns, which had come under such suspicion by Customs at the Manchurian frontier), but were kept by the police at a safe distance from the telescopes, in order not to interfere with our astronomical work. Only afterwards could they visit the camp, where our Japanese colleagues provided a guided tour for them and an explanation of what they had witnessed with us that memorable afternoon. And that the celebrations continued even after the end of the partial eclipse is further attested by figure 3.10.

**Figure 3.10**  *Kampai* to the Sun! The afternoon of 19 June 1936, upon the conclusion of successful observations. Third to seventh from the left (sitting) are: the village-master, Mr Tomotaro Sato; the author; Dr Takahashi, a local physician; and the station-masters of Ottoineppu and Nakatombetsu.

This was the first total eclipse of the Sun which I was privileged to witness in my lifetime; for the second I had to wait almost another half century, until 11 June 1983 at Magelang in Central Java, when the Sun was totally eclipsed by the Moon for more than five minutes—over twice as long as it had been on 19 June 1936 at Hokkaido. Since that time, I have had the opportunity of revisiting different parts of Japan no less than eight times—but not Hokkaido. It was not till the autumn of 1983 that, thanks to the kindness of my friend Professor M Kitamura of Tokyo University Observatory, I had another opportunity to revisit our old eclipse camp at Nakatombetsu after an interval of more than 47 years.

Return to Japan                                                          123

When I did so on 10 October 1983, it was a brilliant day, and the skies were as blue as they had been on the eclipse day of 1936. Unfortunately, none of our old friends were there to meet us any more (the kind village-master Sato had died only a short time before). But the eclipse of 1936 was not forgotten in the region; and it was almost nostalgic to see our photographs as we were young on exhibit in the local museum (to which we were conducted by the new mayor). The school in which we were housed in 1936 had been replaced by a much larger modern building; but a memorial tablet erected in one corner of the school's playing field proclaimed for posterity that, at that spot, the total solar eclipse was observed by an expedition from faraway Czechoslovakia. These are not the exact words of the inscription (which the reader may decipher from figure 3.11!), but approximately so.

**Figure 3.11** The tablet (of post-war origin) marking the place on the school playing field at Nakatombetsu where the Czechoslovak Solar Eclipse Expedition of 1936 had mounted its telescope to observe the total eclipse (see figure 3.6).

In fact, not only the school itself, but everything else at Nakatombetsu has changed in the meantime almost beyond recognition. The introduction of modern industries has increased the town's population more than ten-fold. The old railway station, at which we were received with such enthusiasm in 1936, has all but disappeared in the meantime; and has been replaced by filling stations on the roadside offering petrol to cars, and 'Coca-cola' (a name completely unknown in those parts in

## 124  *Lehr- und Wanderjahre*

1936) to thirsty tourists. Only the hills around the horizon and their forests have remained without change.

The astronomical results obtained by us during the 1936 eclipse (cf figure 3.9) had a long way to travel before we could properly evaluate them; first, we had to return home to Prague, and this still took some time. Three days after the eclipse we left the friendly village where we had had such good luck with the weather (much better than our colleagues in many other localities, both in Japan and abroad); but while at Hokkaido we visited some of its striking national parks (Daisetsusan, Akan, Noboribetsu) and reservations of the Ainu, the aboriginal inhabitants of the island. After once more crossing the sea separating us from Honshu we headed south to visit the International Latitude Observatory at Mizusawa, then directed by Professor H Kimura, the islands of Matsushima, and the magical Nikko (according to an old Japanese proverb, 'do not say "gekko", before you have seen Nikko'); and on to Tokyo, which we reached still before the end of the rainy season, and where we remained for almost seven weeks.

We returned in time to attend a party given by the Japanese Imperial Academy at the University of Tokyo for astronomers who had come to observe the eclipse in different parts of Japan (see figure 3.12). It was there that I first met Professor F J M Stratton and Dr R O Redman from Cambridge, England and Dr T Royds from the Kodaikanal Observatory in India, who were clouded out in Kamishari at Hokkaido (some 50 km from our observing site); and some of the senior scientists of Japan (including both Professors Hirayama). Of the persons recorded on the group photograph of figure 3.12, I am probably the last survivor; some (like Redman or Stratton) I often met in the years to come in England, but others I never met again.

It was also our good luck to meet in Tokyo the great architect Antonín Raymond, born and educated in Prague before World War I, an American citizen by adoption, but resident in Japan for many years. Raymond enriched Tokyo with many new buildings (such as the American Embassy, and others) which survived the great earthquake in Tokyo in 1923—a feat which founded his renown in Japan, and caused him to become acquainted with everyone of importance (including members of the Imperial family).

In addition, and to our good luck, Raymond also discharged with rare devotion the duties of Czechoslovak honorary consul in the Japanese capital. He and his Canadian wife made us feel as if we were at home, not only at their residence in Tokyo, but also at their weekend villas at Karuizawa in the Japanese Alps, or Hayama by the sea-shore in Kamakura Bay (when we arrived there for the first time, we were promptly visited by the Japanese secret police, guarding the Emperor's residence in close proximity to Raymond's villa, to establish our identity). He introduced us to many important dignitaries of the realm. We were quite surprised that the name of Czechoslovakia seemed to be quite well known in Japan since World War I, when the Czech legions had held for a time a part of eastern Siberia as Japan's allies. Many years of residence in Japan had made Antonín Raymond almost an honorary Japanese: he spoke the language fluently; and although he had to go back to the US in 1941 (during the war he lived in New Hope, Pennsylvania), he returned after the war to Japan to find there his last resting place.

**Figure 3.12** A session at the Tokyo Imperial University in July 1936 to commemorate the results of the observations of the total eclipse of the Sun at Hokkaido on 19 June of that year. Front row (left to right): W Jaschek (Austria), the author (CSR), Professor R Sekiguchi (Director of the Tokyo University Observatory at Mitaka), Professor F J M Stratton (Cambridge, UK), President Nagayo of the Tokyo Imperial University, Professor H Nagaoka of Tokyo University, Dr T Olczak (Poland) and Dr H Slouka (CSR). Professors of Tokyo University (standing; from left to right): K Hirayama (4th), Dean Shibata (6th), S Hirayama (8th), S Katayama (11th), M Hashimoto (14th) and N Fukumi (16th; extreme right).

The Tokyo we visited in the summer of 1936 exists no more; and we caught a glimpse of it in transit between two events which profoundly changed its face: the 1923 great earthquake (and fire which followed that disaster), and the even greater disasters which were in store in 1944–45 in the last stages of World War II. At the time when I saw Tokyo for the first time (I have visited it many times since), it was already a great metropolis of a thousand faces: the white palaces of the Marunouchi business quarter could have been transported there from America; and its main thoroughfare, Ginza, would put in the shade (literally) anything that could be seen in Europe (Avenue des Champs Elysées or Piccadilly). In the immediate proximity of large department stores, in the whirling lights of street advertisements, and in the shadow of the willow trees (which disappeared during the war), one could see countless businessmen offering wares of all kinds: apparel (European as well as Japanese), books, toys, musical instruments, goldfish as well as fossils! And not far away, at Kojimachi, one could see against a reddish sky the black silhouettes of willows or cypresses of the gardens of the Imperial palace, surrounded by a moat of greenish waters where goldfish could be seen between lotus leaves—a real metropolis of Eastern Asia, a second *Ville de la Lumière*!

Six weeks we spent in that remarkable city among good friends, and we visited not only all the principal astronomical establishments within its confines (the old observatory at Azabu had only recently moved to Mitaka, then a suburb, and now (1985) an integral part of the city), but also many other sites in central Japan, the most remarkable of which was an ascent of Fujiyama which we accomplished on 8–9 August of 1936. It was indeed a memorable occasion; and with the aid of some notes I used after our return to describe it to the listeners of Prague Radio Station I shall attempt to recapture some of our experiences and adventures of that expedition.

Everyone has no doubt heard the name of that sacred mountain of the Land of the Rising Sun, which, apart from *kimono* or *geisha*, became one of the first Japanese words to be found in the dictionaries of most European languages. And no wonder; for the fame preceding it is fully borne out by the observed facts. On Isu peninsula, in the midst of virgin countryside which the Japanese have adopted as one of their National Parks, the snow-covered cone of this highest of Japanese mountains rises almost from sea-level to an altitude of 3800 metres as the veritable queen of the realm (figure 3.13). There are no less than thirteen prefectures in Japan from which Fuji-san can be seen on the horizon; and if a Japanese can spot it at sunrise, he believes that this will be his lucky day.

The name of Fuji itself is not, strictly speaking, Japanese; and in the language of the Ainu (the aboriginal inhabitants of the region, whose communities still survive locally at Hokkaido) it signifies a goddess of the fire. It is easy to see why; for Fuji is a volcano; and, while its activity has largely subsided since the Tertiary period when it was formed as a part of the volcanic chain girdling the Pacific Ocean, still less than two centuries ago it managed to cover Tokyo with ashes, almost 100 km away. Today, the volcano is dormant, and its crater, several hundred metres deep, is bottled up by debris; only at isolated points does an escape of hot gases testify to the fact that it is not yet completely extinct.

Although Fuji is located at the latitude of North Africa, its summit is for most of the year covered by snow which does not melt except for a few weeks of high summer. During that time, when the ascent is easiest, Fuji becomes the goal of countless pilgrims from all parts of Japan: for just as every Muslim should visit Mecca at least once in his lifetime, so every Japanese should climb Fuji-san to pay his respects to the rising Sun from its summit.

**Figure 3.13** Mount Fuji, queen of Japanese mountains: a semi-extinct volcano rising to an altitude of 3778 m above sea-level.

This summit can be approached along seven different trails; and we chose the one which leads most directly to the goal. We left Tokyo's Shinjuku Station by train for Subashiri, at the foot of the trail we chose to follow. We were not alone, for the first half of August is the time most suitable for an ascent, and dozens of Japanese detrained with us, bound for the same destination. Foreigners who wanted to climb Fuji attracted their interest; and if we found the Japanese to be always pleasant and kind, on this occasion we were doubly the centre of their attention.

By the time we left Subashiri Railway Station, dusk was already descending on the whole region. The dark silhouette of the great mountain stood in front of us with its top high in the clouds. The way towards it goes at first through a deep forest, by that time almost completely submerged in darkness. We passed through the shrine of the goddess of the Mountain; and the mute sound of our steps was barely audible in the soft forest soil. But some distance behind the temple a surprise was in store for us: our eyes, accustomed to the forest darkness, were suddenly dazzled by the glare of flashlights; and from the forest depths suddenly emerged... a ghost?—no; a little man in peaked cap, opening an official bag: 'tickets please'. Yes; even for the ascent of Mt Fuji one must have a ticket!

Afterwards we were loaded into an old Ford car (probably older than Fuji itself) and transported to the place from which the real climb begins. For the first two hours or so the trail still goes through a deep forest. It was a beautiful quiet night. The waning Moon sailed peacefully between the clouds, the trees rustled gently overhead; and the coolness of the night offered a pleasant respite from the daytime heat of the lowlands. Moreover, the trail was not deserted; here and there we passed white-clad native pilgrims, equipped with a long staff, and wide straw hats to offer protection against the Sun. Occasionally a small dwelling emerged from the darkness, and through its half-open doors we saw the flickering glare of an open fire. These are the pilgrim inns, where one can take refuge in bad weather, or refreshment by hot tea. On the way to the summit we encountered altogether nine such inns; and in each you can get its name impaled on your staff as a testimony that you safely passed through it.

We passed the first, then the second and third; the forested areas gradually remained below us while we proceeded higher and higher. Midnight approached; occasionally we sat down on the ubiquitous lava boulders for a brief rest, and looked around the landscape illuminated by pale moonlight. A faint glare along a stretch of the northeastern horizon—that was Tokyo. But no rest could last too long; for time passes, and the summit continued to loom high above us. Time and again we arose to continue our upward journey—one o'clock, two o'clock in the morning; the landscape below was getting lost in great depths; and a cold wind which started at about that time kept reminding us all the more unfeelingly that we were already more than 3000 metres above sea-level. Also the trail which accompanied us to these altitudes gradually disappeared between the countless boulders of broken lava; and steel cables fastened at more difficult points remained the only markers of the further upward climb. Three o'clock in the morning—300 metres higher, but still well below the summit! The sky was deep indigo; and the boulders of the broken volcanic ground cast black shadows in the pale moonlight: scenery worthy of Dante's inferno!

This last part of the climb was also the most difficult. The lava no longer offered a firm grip to hands or feet. Repeatedly we had also to negotiate our way across snow patches which the night frost had converted to an ice film: and both breath and heart-beat became irregular from sustained physical effort.

However, the eastern sky by that time had begun to show the first signs of the dawn. And behold, as though by a miracle, the road ahead of us came to life: like a magic glow-worm it became alive with quietly moving white creatures with long staffs, carrying little lanterns. These were the pilgrims who had taken a night's rest at the last of the nine inns closest to the summit, and were now *en route* to 'make' the summit on time. I lost my Czech friends, and joined one of the glow-worms heading for the summit. The entire sky was visibly brightening up, and a glance ahead revealed that we could not be very far from the goal. At long last, a second rim appeared in front of us across the crater. We were on the top (see figure 3.14)!

It was the first time in my life that I had ascended so high a mountain; and although, in later years, I stood on peaks higher than Fuji, I never bettered an altitude difference of almost 3500 metres in less

**Figure 3.14** The summit (with crater) of Mount Fuji, photographed early in the morning of 9 August 1936, after our nocturnal ascent of the mountain.

than eight hours. Inexperienced as I then was, I reached my goal in a very queer state: while in Tokyo we were accustomed to daily temperatures upwards of 40°C, on the summit of Fuji there was a keen frost and strong wind, against which my 'climbing gear', consisting of tennis shoes, shorts, and open shirt with only a light sweater, offered altogether insufficient protection against the frosty wind which penetrated to the bones. However, the good Japanese pilgrims took pity on me in their inn on the summit's rim, lent me a warm kimono and gave me hot tea; they smiled at my broken Japanese and told me many things which I did not understand.

A sunrise at the summit of Fuji-san belongs among the most fascinating as well as dazzling phenomena that one could experience. A splendid drapery of clouds which piled up in the east ran through the colours of the rainbow, as the horizon was brightening up fast— until a sudden glare forced us to close our eyes: the Sun was rising! It was not as the red disc to which we are accustomed in the lowlands; but as an intensely white furnace which (at that latitude) rose steeply in the sky as though flying on wings! And below, at almost fathomless depths in the surrounding landscape, it was daybreak wherever we looked, in the interminable forests still covered by bluish haze; to the north was the wild chain of the Japanese Alps; while towards the east one beheld the infinite plains of the Pacific Ocean glistening at sunrise.

In the meantime, the Sun continued to climb rapidly above the horizon; and soon it became unsafe to linger for too long at our exposed post. In no time at all, small flocks of cumulus clouds, formed far below, ascended to cover the peak of Fuji with ourselves on it; and cases were not unknown when, under such circumstances, whole groups of pilgrims had frozen to death in sudden snowstorms. Therefore, a rapid descent was the order of the hour, for which we chose another trail (towards Gotemba), and which lasted till the afternoon. In the course of it we got drenched to the skin by rain several times, and tired (also sunburnt by exposure to the Sun on the top of Fuji for about half an hour after sunrise), but otherwise satisfied, we returned safely to Tokyo that evening.

Only once more during our descent, almost at the foot of the mountain, did Fuji appear to us again in the sky in its full grandeur for a few moments, before low clouds covered it from view again; and this was the last glimpse of it we had during the rest of the summer. When we left Japan two weeks later by sea, and our ship rounded the Isu peninsula after its departure from Yokohama, it was a beautiful day; but the queen of the Japanese mountains hid its face completely behind clouds; and it was not till 1962 that I next had occasion to get another glimpse of it.

And now, fifty years later, as I still pass sometimes within the sight

## The observatories of Shanghai

of Fuji's summit glistening in sunrise, or in a pink glow long after sunset, and can no longer run to the top as in the days of old, I have to be content in repeating with the French medieval poet (and so many others) his nostalgic words, '*où sont les neiges d'antan?*'

Why do I recount our adventures on Fuji in 1936 in such detail? Because the story now also belongs to the past. Tourists today can approach the summit of Fuji much more comfortably, by cable-car! When, many years later, my Manchester colleague and former student, Dr Edwin Budding, came to spend the years 1973–75 at Tokyo Observatory, our mutual friend Professor Kitamura (himself a 'Mancunian-in-exile') saw to it that Ed would not return back home before scaling (albeit with the aid of the *téléférique*) Fuji as well, claiming that every visitor should do so once in his lifetime (though, Kitamura added wryly, only fools would do it twice). But, to return to the past, were we leaving for home at that time? Not yet; for the *Asama Maru*, one of the largest ships sailing the Pacific Ocean, carried us first south to the main Asiatic land mass which we were to re-enter at Shanghai in China. At the request of the Czechoslovak legation, the Chinese Embassy in Tokyo provided us with another *laissez-passer* permit which facilitated the crossing of all frontiers; and our principal goal at Shanghai was to visit the astronomical observatories at Zi-ka-wei and Zô-se, then managed (as in past years) by the French Jesuits.

The observatory at Zi-ka-wei was, at that time, already located in the midst of the great Chinese commercial metropolis (since to become the largest city of the Eurasian continent). However, Zô-se, located some 30 km west of the city (and renamed in 1962 the Sheshan Station of Shanghai Observatory of the Academia Sinica) was like a French island in the Chinese countryside, where one could easily imagine being somewhere in central France. A large church, proudly raising its two spires to the heavens, would certainly have been in place at Clermont, or some other locality in the Massif Central.

Father de Villemarqué, director of the station, always extended a sincere welcome to guests from abroad with whom he could converse in his native tongue, and we were no exception. Under his direction, the principal activities of the stations were the photographic observations of asteroids with a long-focus refractor (not too unlike the one which, decades later (see Chapter 5), Manchester astronomers were using in France at Pic-du-Midi), the measurements of their positions, and computations of orbits. All this work was being carried out by twenty or thirty gifted young Chinese assistants, trained for the purpose by Father de Villemarqué, with all computations carried out on the abacus; it was a genuine surprise for us to see this ancient aid to computation being used with such skill and effectiveness by so many capable young hands. Moreover, Father de Villemarqué taught his young Chinese staff not only to carry out competent astronomical work, but also to speak French—probably better than ourselves. When, towards the evening, we were to be rowed back to the city (there was as yet no surface road connecting Zô-se with Zi-ka-wei at that time) and, at the time of the Angelus we failed to make the appropriate responses, our young rowers asked with disarming frankness: 'Messieurs, êtes-vous chrétiens ou payens?'

The observatory in Zi-ka-wei, founded in 1873, was dedicated primarily to solar work; and sunspot studies there by Father S Chevallier are still well remembered to this day. Its director, Father P Lejay, was away in Europe at the time of our visit; but his resident colleagues (one of whom was Father Javorek, a native of Slovakia) received us more than kindly, and so did the Czechoslovak diplomatic and business representatives in Shanghai whose guests we were at the time of our visit. What a pleasant surprise it was for me to learn that the Councillor of the Legation, Dr Max, had been born in their family's mill ('Max Mill') about half-way between Hřmenín and Važice (see pp 3–5) in the native corner of the Kopal family in Bohemia!

But it was not only astronomy which kept us fully occupied during our visit to Shanghai. On the social side (as I have already mentioned in the Preface) we were allowed to enter that European Country Club where 'dogs and Chinese were not permitted', and take tea with the fossilised membership. But much more pleasant for us was the entertainment extended by the local Czech community (and, especially for me) in a tavern by the docks maintained by Mr Mareček from Cerekvice (a village of about an hour on foot from my native town of Litomyšl) as a temporary haven of refuge for sailors of all races and nationalities. Mr Mareček had been a sailor himself—once, before he deserted from the Austrian Navy of the Emperor Francis Josef to help subdue the Boxer Rebellion in 1900. Mr Mareček had no taste for quelling any rebellions, so he walked out of his ship one day and decided to remain in China, where he had now been settled for 36 years. He married there (a White Russian refugee); and his grown-up daughters no longer knew a word of Czech, so adding to the loneliness of their father in his advancing years.

Mr Mareček insisted on entertaining us in his tavern, and we readily accepted; for we had never seen a real sailors' den anywhere except in the movies; and how could we return home from Shanghai without visiting at least one of them? Mr Mareček's tavern lived up to our expectations, with its smoke and noise and brawl of drunken sailors. When I looked in one of its corners, however, my heart almost skipped a beat; for, barely visible through the smoke-filled room, I saw hanging on the walls the portraits of Master Jan Hus, the religious reformer, and Jan Žižka, the terrible Hussite warrior (their English equivalents would be John Knox and Oliver Cromwell), looking down with obvious disapproval on what they saw below, and no doubt thinking: was it for such people that we risked our lives to bring about their eternal salvation?

Driving back late at night to our hotel, we could not fail to notice flocks of native rickshaws pulling home their white *taipans* dead drunk in their fragile vehicles to be delivered to their doorsteps. Even to us it was evident that the end of the old times was near, though we did not realise how close we were to it!

From Shanghai, our return journey by the Shanghai express went through Nanking, where our train had to cross the Yang-tse River on a raft; and after one and a half days via Tientsin we reached Peking, where we spent two weeks as guests of Fu-jen Catholic University (figure 3.15). In the pleasant company of one of its teaching brethren (Brother Brückner, originally from Vienna) we were able to visit the most important antiquities of Peking and its surroundings (including the Great Wall, some 100 km to the west). The 'Forbidden City' is the largest as well as oldest royal enclosure of the world (its present dimensions go back to the twelfth century AD). Since the advent of the republic it has opened its gates to globe-trotters from all parts of the world; and if its old courts and temples have been kept in some kind of repair, it is only to enable the government to collect entrance fees at its gates. What a difference I encountered there 47 years later!

**Figure 3.15** The Catholic University of Peking, whose guests we were in September 1936 *en route* from Shanghai to Dairen.

And the same was true of the old astronomical observatory, established in Peking in the second half of the sixteenth century by the Jesuits—still in the pre-telescopic era. Its beautiful instruments (contemporaries of those of the era of Tycho Brahe) could still be seen on their original site (cf figure 3.16) as an echo of ancient times. I later learned that, on the eve of the Japanese invasion of the Peking region the following year, these instruments were removed to Nanking, where they survived World War II; they can be seen again at the Purple Mountain Observatory, in the midst of more modern instruments measuring the passage of time.

**Figure 3.16** The old observatory of Peking, equipped with instruments of the pre-telescopic era which were brought over to China by the Jesuit Fathers in the second half of the sixteenth century. During World War II most of these instruments were removed south, to the Purple Mountain Observatory, where they can be seen at the present time.

After two weeks in Peking, the road ahead of us went straight home. First from Peking back to Tientsin; from which after two nights and a day at sea we reached Dairen, another western outpost in the Far East, the possession of which, since the beginning of this century, has changed hands not less than three times. In company with the Czechoslovak Minister to China (who was vacationing in Dairen at that time) we also visited the old fortress of Port Arthur; and a day later, having cast our last glimpse at the Pacific Ocean (which I was not to see again till twelve years later from its opposite shores), we boarded the Japanese Asia Express to deliver us back to Charbin in the hospitable hands of our Consul Hejný. After a further ride of almost ten days on the Trans-Siberian Express (which we picked up again at Karimskaya) we reached Moscow. On our outward journey in May the region around the Baikal Sea had been still largely snowbound; now we saw it from the train in the full glory of an Indian summer.

When we reached Moscow, however, it was already autumn. As we were changing stations, the streets in the capital were lined with rows of soldiers—not to guard Moscow against any invasion, but to offer a hero's welcome to the polar explorer Sigismund Levanevsky and his crew, who were to reach Moscow that day after a successful flight to the North Pole. We did not, however, wait to see them; and thus lost that chance for ever: since only two years later (like Amundsen in 1928) Levanevsky disappeared in the northern mists somewhere near the Pole, in a flight from the US West Coast to Moscow—a trip which, not many years later, became a scheduled flight for more than one airline. I myself flew over the North Pole at Easter 1963, aboard an SAS DC8 airliner from Copenhagen to Tokyo; and to this day have not forgotten a magnificent view of the sunrise over Mount McKinley (6240 m) and other Alaskan giants (some of which were climbed by one of my sons-in-law in the years to come).

But this was still 27 years in the future; and a trip from Moscow to Prague, the last leg of our long journey, which now takes only two hours by air, in 1936 still needed two days by train. In Warsaw, on the penultimate day of our return from the East, we found the funds remaining in our pockets too low to allow us to spend the night in a hotel: instead, we checked in all our baggage at the station from which we were to depart next morning to Prague, and spent the time walking till dawn through the streets of Warsaw, largely deserted after midnight. We walked through the modern city and Old Town (at one time, I remember taking a nap on the pedestal of Thorwaldsen's statue of Copernicus at Nowy Swiat), without any premonition of the gruesome fate which was in store for that beautiful city only three years later; or that when on my next visit to Warsaw in December 1960, I should see many of its buildings completely re-built from the ruins caused by the war.

But in September 1936 this was all still in the future (albeit not too distant); and in dozing under the statue of Copernicus I was consoling myself with the thought that this was the last night of our long journey to be spent abroad, and next day I should be home again. We indeed returned to Prague on 15 September, in the evening, four months after our departure for the Far East in May; having covered in between (mostly by train) a distance just about equal to the equatorial circumference of the Earth.

This was my first long trip to destinations outside Europe; and (as we shall see later) it was not to be the last. But only once again, 42 years later, did I return from Japan to Europe in 1978 via Siberia; though this time not by surface, but by air. A Boeing 707 jet made that journey, however, in no more hours than the train took days in 1936! For, leaving Tokyo's Narita Airport at noon in October 1978, by 7 PM (local time) of the same day we were already in London; and two hours later I could be home in Wilmslow for dinner (when it was already morning in Japan). We have certainly learned to move about much faster since 1936; but whether to better ends is a somewhat moot point.

## 1936/38 : The Doctorate and the First Postdoctoral Year

Our return to Prague in September 1936 coincided with the start of the last academic year of my university studies, at the end of which I would become eligible to apply for the doctorate. By that time my thesis research was already quite well under way. It concerned the empirical information of the internal structure of stars which one can deduce from the observations of close binary stars which, on account of the chance inclination of their orbital planes, happen to be eclipsing variables.

That variables of this type attracted my attention from almost the beginning of my astronomical career has already been related in the preceding chapter; and earlier in the present one we referred to a paper (Kopal 1935a) from my undergraduate years which marked my more serious entrance into this field. By that time, the subject still had not made much progress since 1914 when Shapley's PhD thesis appeared in Princeton. Following him, most investigators regarded the components of such systems as spherical; or (if the variations of light exhibited between minima made this impossible) as similar ellipsoids, regardless of their fractional dimensions or mass ratio. That this was unlikely to be true was first emphasised by the German astronomer K Walter (1931) of Königsberg, who pointed out that, in general—and in systems of the Algol type in particular (Walter called these the *spezielle Algolsterne*)—the components should be expected to be of very different shape. Walter still regarded, to be sure, the components as rigid bodies which might differ from each other in shape to an arbitrary extent. But to me this already appeared at that time to be impossible; for the physical conditions obtaining in their interiors were bound to make them behave as fluid bodies and, consequently, their actual shape should depend on their masses, absolute dimensions and internal structure, in the manner specified by the Clairaut theory of the figures of equilibrium of self-gravitating fluids (the importance of which had been emphasised to me by my teacher Nechvíle in my undergraduate years).

Stimulated by Walter's work, I set out to apply Clairaut's theory of the figures of equilibrium of stars of arbitrary structure to a number of actual eclipsing systems, with the aim of establishing what their observed properties might disclose about their internal structure. This work soon led to concrete results, which I wrote up briefly for publication; and with some daring I submitted the text to Sir Arthur Eddington, then the foremost authority on the subject, with a request that, should he find the method and results of some interest, he might recommend them for publication in the *Monthly Notices* of the Royal Astronomical Society in London (of which I became a Fellow in 1936). Eddington did so; and informed me of the acceptance of my paper by the Society's Council with a few flattering lines, from which I gathered that he must have forgotten the young man who had been introduced to him in 1935 in the gardens of the Elysée Palace in Paris; for his brief note was addressed to 'Professor Kopal', a title so which I could not raise a legitimate claim for another ten years.

However, an anticipation of this kind from Eddington's pen could not but favourably impress my Prague teachers. At any rate, the paper did appear soon thereafter (Kopal 1936); and attracted enough attention on the part of the Editor of *Nature* to deserve a brief report on my result in his renowned journal (cf *Nature* 1937); and the same was true of the *Observatory* magazine, whose reporter was nobody less than S Chandrasekhar (1936), my senior by four years. At the encouragement of Professors Nušl and Záviška in Prague, I therefore decided to submit an expanded version of this work as a thesis in partial fulfilment of the requirements for the doctor's degree at Charles University at the end of the academic session 1936/37, the earliest time then permitted under existing regulations.

Before it came to the defence of my thesis, another event occurred which may be mentioned in these recollections. In the autumn of 1936 I returned from my first trip to the Far East not only with reasonable eclipse plates which awaited photographic photometry, but also with a sizeable debt which had still to be repaid; and with which I did not want to burden the family exchequer. For part of it I could rely on my lecture fees, or on honoraria for articles of a more popular nature; but the bulk of it had to be earned in a more effective way.

Encouraged by the financial success with which my friend Štorek and I had marketed my Czech translation of *The Mysterious Universe* by Sir James Jeans, for our next joint venture I selected *The Stars and Atoms*, a series of truly delightful popular lectures by Sir Arthur Eddington, which were brought out in book form by Cambridge University Press in 1930. Their Czech translation took only a few weeks (by that time, my knowledge of the English language was already reasonable); but since the text of the book was already somewhat behind the times, I ventured to ask Sir Arthur to bring it up to date. He very kindly agreed; and although the text of his additions reached us too late to be incorporated in the book, it was published at almost the same time in the January 1937 issue of Říše Hvězd, the journal of the Czech Astronomical Society, read by most Czech astronomers at that time. I need hardly add that Eddington's book enjoyed no less success among a Czech readership than had Jeans' a year before. The proceeds from our second venture in book publishing solved all my outstanding financial obligations, so that the next goal was to qualify for the doctorate.

This I did in June 1937; and as all my examinations (beginning with the first year of the *Gymnasium*) had been passed 'with distinction', I was qualified to receive the degree *summa cum laude*, in a ceremony at which the Head of State (or his representative) traditionally presented the successful candidate with a souvenir of the occasion (usually a gold watch). Such an occasion could, however, take place only once each year; and it so happened that another student (a girl) had earned the right to a similar distinction only one month before. The University put before me, therefore, an alternative of waiting for another year; or of being granted the degree immediately, but at a ceremony not involving the Head of State.

On the advice of my father—and what good advice it turned out to be!—I opted for the second alternative; and thus received my degree on 30 June 1937 at the ceremony shown in figure 3.17. The Rector of that year (centre), Professor Karel Weigner, a distinguished anatomist of the Faculty of Medicine, who then had less than a year to live (he already suffered from cancer at that time), addressed some kind words to me expressing hopes for my future; and Professor Nušl (on the right), who was retiring that year at the age of seventy, read (in Latin) a promise which I had to make to receive the degree. I no longer recall its full text; but remember that I was called upon to cultivate science '... not for mere vainglory, but so that the light of the truth should radiate all the brighter', to which I was expected to say 'Spondeo ac polliceor.' The readers of subsequent parts of this book can judge for themselves the extent to which I have lived up to that promise.

Then I had to deliver my own oration (likewise in Latin), and received the degree in the midst of the congratulations of teachers and friends alike. It was the first doctorate I received in the course of my life (the others were honorary); but it gave me great satisfaction; for I was the first of my class-mates in our *Gymnasium* to attain this academic rank; and in doing so *suo anno* I only lost the chance of receiving a gold wrist-watch from President Beneš at that time; for a year later, in June 1938, the President had already too many other worries to be able to give much thought to academic candidates. In fact, I had to wait for a golden wrist-watch until my seventieth birthday in 1984, when I received one from Mrs Ellen Carling, my faithful secretary at Manchester for almost thirty years; and although this watch carries no engraving of presidential arms, it is much more accurate than any watch could have been half a century ago, to keep reminding me of the passage of time (now sadly running out).

**Figure 3.17** The author's doctoral graduation ceremony on 30 June 1937 at Charles University in Prague. In the Large Hall of the Law School (the medieval Carolinum then being under repair) on the rostrum reserved for academic dignitaries are (left to right): the Dean of the Faculty of Science (Professor Josef Kratochvíl), the Rector (Professor Karel Weigner of the Medical School) and the Promotor (Professor František Nušl). The newly-promoted doctor (extreme right) closes the 30-minute ceremony by giving his thanks (in Latin) to his teachers, and affirms his promise to abide for the rest of his life by the University's regulations.

After the doctorate, what next? By that stage of my life it had already become clear to my teachers as well as to myself and my family that I was destined for an academic career; and that all my future steps should be directed to that end. The long vacations which followed my graduation (the last ones I was to spend with my family for many years) were interrupted only by two weeks in Paris where I went to visit the 1937 World Exposition; and in September I volunteered for the post of a junior librarian in the University Library of Prague, an unpaid position (as no paid one was available at that time), but one usually kept in reserve for future academic teachers; and which kept me in touch with the books through one of the largest libraries in Central Europe (going back to the Middle Ages). I may add that (since

1918) the University Library was amalgamated with the National Library of Czechoslovakia; and as such it had the statutory right to receive every book published in the country (and priority for acquisition of other books of historic value that might become available from any other source).

Whole libraries of old monasteries (antedating the foundation of the University in 1348) were deposited there, and serviced by a permanent staff of devoted bibliophiles who had grown up (and aged) among its treasures, housed in the historic buildings of the Clementinum going back to the sixteenth century; and whose walls have no doubt seen both Tycho Brahe and Johannes Kepler among the library users. In fact, the Týn church where Tycho is buried is no more than a few hundred metres from the Library's entrance; and it was from its windows that I was able to watch on 21 September 1937 the impressive and moving funeral of President Masaryk, who had died a week before at the age of 87 years, and whose cortège was also taken around the buildings of the University Library, in which he (as Professor Masaryk) had spent so much time in his active life. To many of us who remember that occasion, Masaryk's funeral in 1937 (like that of his son Jan Masaryk in March 1948) represented subsequent stages of the sunset of Czechoslovak independence—awaiting resurrection to this day.

My temporary position in the library was not all that I had in mind at that time. Professor Nušl, then on the point of retirement also as Director of the National Observatory, by one of his last acts in office recommended me to the Czechoslovak Ministry of Education for the post of the Observatory's junior astronomer—a post which was likely to become vacant in the near future; and my honorary position in the library (the only one, as it turned out, which I ever actually held in Czechoslovakia) would also sooner or later have been converted to a regular one.

Such was, at least, the hope of my parents; father probably saw a career opening up before me, and leading to a higher position in the course of time. Although fate willed it otherwise, after so many years I still recall my short spell of library service with pleasure tinged with nostalgia. Our director, Dr Jan Emler, was the son of Professor Josef Emler (1836–1899), distinguished paleographer (born in Libáň, the same place as my mother); and his deputy was Dr Antonín Hrozný, brother of the famous linguist who deciphered the script and language of the Hittites. My immediate superior was Dr Anna Dvořáková, widow of Professor Max Dvořák (1874–1921) of Vienna University, the distinguished historian of art, whose (posthumous) work *Kunstgeschichte als Geistesgeschichte* is still well remembered today after more than half a century. As I was at that time the youngest member of the staff, I doubt if any of my colleagues senior to me are still alive

## Postdoctoral year abroad

today; but the cultural atmosphere pervading the halls of our ancient library did much to make me appreciate the arts side of human culture; and I thoroughly enjoyed my brief stay within its walls.

That this stay was brief was partly of my own doing; for I was unwilling to settle for the time-scale which a library career would have entailed. The aim uppermost in my mind was, instead, to pursue my postdoctoral *Lehr- und Wanderjahre* abroad, at the primary sources of contemporary learning, more systematically than only in the course of occasional visits; and to this end I now turned all my attention.

In 1928, as part of the celebrations of the tenth anniversary of the recovery of Czechoslovak independence, the Government of the Republic created a number of postdoctoral fellowships for its citizens to study abroad for longer periods of time, and named in honour of Ernest Denis (1849–1921), the distinguished French historian of the Czech nation. I decided to apply for one of them in the autumn of 1937; and, if successful, to spend it at Cambridge, England, in work under Sir Arthur Eddington (who by that time already knew that I was not yet a professor).

In response to an inquiry about such a possibility, Eddington promptly agreed to accept me. Moreover, he backed his promise with so strong a letter of recommendation that the Ministry agreed to award me one of the Denis Fellowhips, tenable at Cambridge, and vacated then by the retirement of Dr František Wolf (a young Czech mathematician who was then studying at Cambridge with A S Besicovich, and who eventually ended his academic career as Professor of Mathematics at the University of California in Berkeley). The official award was made before Christmas 1937; and in January 1938 I departed thus for Cambridge, unaware that I was leaving my native country for good, not to return in future except for occasional visits (which became shorter and shorter with the passage of time). The journey to England in the winter of 1938 through Germany and Belgium was uneventful (by then I was—or at least I thought I was—an experienced traveller); and after a brief stay in London (where I had to report my arrival to the Czechoslovak Legation) I reached Cambridge.

Before going up to the Observatory to report my arrival to Professor Eddington, I had one experience which, albeit minor, I have remembered ever since. On arrival, I had to report (as an alien) my presence to the local police; and did do by depositing my passport at the Cambridge constabulary. A day or so later the passport was duly returned to me 'With the Compliments of the Chief of Police'! To an Englishman, this may sound nothing out of the ordinary. However, any visitor from the Continent was bound to be impressed by so polite a gesture as to remember it for the rest of his life. For, on the Continent, the police of all states have traditionally been accustomed to deal with any individual who came within their purview with suspicion appropriate to the criminal element of society. My arrival at Cambridge in 1938 was indeed my first experience with the British 'Civil Service' in the true sense of the word; and it has not been the last.

Professor Eddington, whom I now had an opportunity to meet at close range, has been generally (and justifiably) regarded as the greatest astronomer of the first half of the twentieth century (see figure 3.18). So much has already been written about his remarkable personality (cf, for instance, his biography by Vibert Douglas, 1956) that whatever I can add to it in this place can, at best, only supplement the picture already known. In the late 1930s Eddington was, however, gradually retreating from the position of premier 'astrophysical engineer' that he had occupied in the preceding two decades when, almost single-handed, he had created the theory of the internal structure of the stars. In the 1930s, Eddington's main interest began to turn to more fundamental problems of physics and cosmology, to the structure and evolution of the Universe, as witnessed by his books on the *Relativity Theory of Protons and Electrons* (1936) or *The Fundamental Theory* (1946), in which he went so far ahead of his contemporaries that almost no-one could follow him in his reasoning, except, perhaps, E T Whittaker (and he too, in 1956, took its secrets with him to the grave).

It was far from me even to attempt to follow Eddington's cosmological reasoning at that time; my main aim was to profit from his reasoning on astronomical matters which were within the limits of my comprehension; and I count as one of the blessings of my life that I had an opportunity to do so, albeit for only a short time. It was not easy; for Eddington was a very shy man, not given much to lecturing, let alone to small talk! Many stories exemplifying his shyness were well known to his contemporaries; but the following one (which I heard from the late Dr Harold Babcock) is perhaps worth recording here. Babcock was one of the veterans from the early days of Mt Wilson Observatory in California; and the following story of Eddington's visit to Mt Wilson in the early years of the Observatory's existence he related to us on a convivial occasion at the Athenaeum in Pasadena some 25 years ago.

In those days, trips up the mountain used to be far less easy than they have since become; and the visitor had to be accompanied *en route* by a local member of the staff. One of the technicians was detailed to accompany Eddington; and, since the trip lasted several hours, tried to engage the visitor in some talk. This was, it seemed, mostly unilateral, the guide pointing out the place where there had recently been some forest fire, where some of the Observatory's people had seen

## Sir Arthur Stanley Eddington

a mountain lion, etc—to all of which Eddington apparently responded only with non-committal grunts.

Some days later, the same technician met in Pasadena a colleague who, in the meantime, had been detailed to return Eddington back to civilisation; and the two happened to exchange some comments. 'Wasn't that English professor you took down a stuffshirt?', asked the first one. 'What do you mean?' retorted the other incomprehendingly. 'Well, when I took him up the mountain, he said hardly a word for the whole trip.' 'Impossible', said the other; 'when we came down he was a very pleasant companion.' 'What did he tell you?' 'Well', replied the second guide, 'as we were coming down, he pointed out to me where we had recently had a forest fire, or where one of our chaps had seen a mountain lion—and he had it all right'—not realising that what he had listened to was only a replay of what Eddington had learned going uphill.

**Figure 3.18** Sir Arthur Stanley Eddington (1882–1944).

To me, when I hear it, the story brought home both Eddington's effort to be friendly with people whom he happened to meet, and his habitual difficulty in accomplishing this. In a more serious vein, his office as well as his renown brought him invitations to deliver formal lectures, many of which also appeared in print. When one re-reads them today, one is greatly impressed by their felicity of expression and penetrating power, which can keep the reader spellbound. The latter can, however, also lose sight of the fact that, on the actual occasion, these were mostly read from a prepared script in a way which made it difficult for the audience fully to appreciate their contents. It is the readers rather than listeners who have been their principal beneficiaries. When in 1938 Eddington was elected President of the International Astronomical Union in Stockholm, he left the meeting before the end of the Sixth General Assembly, in order to avoid having to give the final address customary on such occasions—and he never gave one; for before the IAU met again in Zürich ten years later, he was dead.

Eddington was indeed as superb a writer as he was shy as a speaker, on public rostrums as well as in the classroom. In the Lent term of 1938, when I arrived, he read a course on the internal constitution of the stars, which attracted three listeners: Wen San Tai from China, H C Corben from Australia—and one Czech! In reading this course, based almost entirely on his famous book under the same title (Eddington 1926; second edition in 1930), he never looked at his audience (though he must have sensed the arrival of the last student; for it was not till afterwards that he closed the door of the lecture hall at the Cavendish Laboratory); and was easily upset by any noise from the benches. It was enough for anyone from there (usually Corben) to whisper 'it's all wrong, Sir' for the Professor to erase all he had written on the blackboard so far, and start over again!

No; the principal place to gain from the proximity of this truly wonderful man was Eddington's study at the Observatory (which was also the place of residence where Eddington, a confirmed bachelor, lived with his sister). There he was available to his students almost any time of day (or night)! The walls of this study were lined with books, about half being professional literature and the other half detective stories which Eddington (like Rutherford) obviously used to reach for to 'unwind' after hours of concentrated work.†

When a student (or postdoctoral fellow, like myself) came to see him, the door of Eddington's study was never closed to him; and the

---

† It may be added that while Eddington no doubt bought himself such literature for this purpose, Rutherford (as a good Scot) used only to borrow these from a lending library at the corner of Wilmslow Road on Fallowfield in Manchester, which, as we learn from his biographer A S Eve (1939), could not keep enough new ones in store to satisfy Rutherford's appetite!

There must be something in the Cambridge atmosphere that calls for such a kind of relaxation. Ask today my friend Raymond Lyttleton about (say) the first names of the 'Three Garridebs' from one of the Sherlock Holmes stories by Conan Doyle, and he will rattle them off for you more easily than the history of mountain-building on Earth on which he is supposed to be an expert.

liberal side); and so was Dick Woolley's brother who came occasionally to see them. But the Woolleys did not remain in Cambridge long; and (I believe) left for Australia the next year for Richard Woolley to become Director of the Canberra Observatory—the last Englishman and Cantabrigian to hold that post.

In 1955, Woolley eventually returned to England to succeed Sir Harold Spencer Jones as the thirteenth Astronomer Royal and Director of the Royal Greenwich Observatory, the last one while that institution was still (as for the previous 280 years) administratively under the Admiralty. That choice was perhaps not unexpected; for (I believe) Woolley was an admiral's son; but (as is referred to in Chapter 5) his performance in the latter job demonstrated perhaps the disadvantages of spending sixteen years far away from more active centres of current research.

With Dr R O Redman (1905–1975) then Assistant Director of the Solar Physics Observatory, we knew each other already from the 1936 solar eclipse in Japan; and he too was to leave Cambridge and serve his term as Director of the Radcliffe Observatory in Pretoria, South Africa, before his return to Cambridge to succeed Professor Stratton. Redman's inborn pessimism was already increasingly evident at that time; and his proverbial saying 'it won't work, my boy' became only too well known to his students. He told me once (much later, of course) that he considered his main task in British astronomy was to play the role of a 'wet blanket', to keep everyone's feet on the ground and not let things get too far above it; and in this role, it must be admitted, he proved perhaps more successful than in anything else.

The third staff member of the Solar Physics Observatory with whom I became very friendly was the late Andrew David Thackeray (1910–1978) with whom I spent perhaps more time over the chessboard than was good for the advance of science. Always gentle and kind (there was nothing in him of the superciliousness or arrogance which sometimes marred the image of young Britons of the Eton–Oxbridge class), during the war to come he (a conscientious objector) served his country by driving an ambulance in Africa for the Eighth Army. After the war, he too went to South Africa (to succeed Redman at Pretoria), but never came back for good; for a car accident caused by a rare quirk of nature cut short his life on 21 February 1978 while he was still not far from the prime of his life.

The fourth astronomer of my age with whom I became acquainted at that time in Cambridge was Hermann A Brück, likewise of the Solar Physics Observatory, who had reached England only recently from Germany via the Vatican Observatory. The scion of a noble Prussian family (his uncle used to be a German Ambassador under the Weimar Republic), he had to leave Nazi Germany on account of religious intolerance (his first wife was of Jewish descent). He made his home in the British Isles, first as Director of the Dunsink Observatory near Dublin; and later as Director of the Royal Observatory of Edinburgh and Astronomer Royal for Scotland, a position from which he retired in 1975. Brück's recent continental origin brought us together more closely than with other young scientists in Cambridge of our age, partly because of apprehension about the shape of things to come†, and we have remained good friends ever since (hopefully still for a long time to come).

But it was not only astronomy that was in my mind at Cambridge in the spring of 1938; for the political events of the day began to claim our attention with increasing urgency. As is well known, on 12 March 1938, two months after my arrival at Cambridge, the troops of Adolf Hitler marched into Austria to '*Anschluss*' that country; and all Czechs suddenly felt in their bones that their country might be next in line for Nazi aggression. And what was worse was to observe the bewilderment, if not indifference, with which such a course of events was contemplated at different levels of British society. The source of it was largely a lack of information about a storm so suddenly and so menacingly rising above the horizon. It was not only that Miss Eddington (Sir Arthur's sister and housekeeper) asked me at one of the teas in their home, 'Prague is in Hungary, isn't it?'; but also Mr Neville Chamberlain, the Conservative Prime Minister of that time, who expressed (somewhat later in the season) regrets that his country should get involved in disputes concerning nations 'of whom we know nothing'. There were people in Britain who knew all that was to be known, and what it would mean for Britain in the very near future; but they were not listened to in time. Experience is indeed a hard taskmaster, especially if one has to learn in a hurry; and in less than two years the tragic folly of old men in 1938 had to be paid for by lives of the young pilots of the RAF.

We—the Czechs at Cambridge in the spring of 1938—did not need to be warned twice in advance. Two other Denis Fellows from Czechoslovakia (nuclear physicists both working at the Cavendish Laboratory) were in residence in Cambridge at the same time. Dr (later Professor) Václav Petržílka of Charles University in Prague, then a partner of the trio Curran, Dee and Petržílka, and (years later)

---

† I recall that, in September 1940, after the fall of France to the (then victorious) Nazi armies, and with an invasion of Britain expected at any time, Professor Eddington wrote to Dr Harlow Shapley to find a place for Brück (then interned in Britain as a technical 'enemy alien') at Harvard; for should the Nazi invasion have succeeded, this would have meant curtains for Brück (who would have been treated by the Nazis as a deserter). Shapley indeed did so; and we were expecting Brück to join us at Harvard at any time; but (fortunately for Britain) this did not turn out to be necessary.

Deputy Director of the Dubna Laboratory for Nuclear Physics near Moscow, was one; Dr V Viktorin from Brno (who died during the war under unknown conditions) was the other. We had not known much of each other in the old country, but the stress of the times brought us together in almost daily contact at Cambridge. Petržílka was the most apprehensive of us three, Viktorin the most optimistic. As for myself, it was at that time that my plans to leave Britain for the United States (see the next chapter) assumed their final form.

I remained in residence at Cambridge till the end of the summer term, and then returned to Prague, though no longer for carefree vacations, nor to the University Library any more. With some time in England behind me, the Czechoslovak Foreign Office encouraged me to maintain personal contacts with foreign guests and certain junior diplomats or journalists accredited to Prague at that time. The journalists were sharp, intelligent and surprisingly well informed; they mostly knew the score, and asked very specific questions. But the contacts with diplomats were discouraging. One junior attaché of the British Legation (a casual acquaintance, whose name I have forgotten) listened to me with only half an interest to what further Nazi triumphs could mean for Britain; then lit up a cigarette and blowing off its smoke with a nonchalant gesture he asked, 'what can they do to us?' He learned only too soon; for (I later heard) within less than two years his dead body was lying on the beaches of Dunkirk, with a bullet through his head fired by the German strafing planes. Such is the cost of freedom; and the price to pay for not being wise in time! Will the lesson at least be remembered by posterity? Sometimes I doubt it.

## References

Chandrasekhar S 1936 *Observatory* **59** 189
Eddington A S 1926 *The Internal Constitution of the Stars* (Cambridge: Cambridge University Press) (second edition 1930)
—— 1927 *Stars and Atoms* (Oxford: Oxford University Press) (Czech translation by Z Kopal appeared in 1936)
—— 1936 *Relativity Theory of Protons and Electrons* (Cambridge: Cambridge University Press)
—— 1946 *The Fundamental Theory* (Cambridge: Cambridge University Press)
Eve A S 1939 *Rutherford* (Cambridge: Cambridge University Press)
Fletcher A 1957 *Math. Gazette* **41** 232
Freundlich E F 1917 *Einstein's Relativitäts-Theorie* (Berlin: Springer)
—— 1958 *Celestial Mechanics* (London: Pergamon)
Güssow 1936 *Veröff. Univ. Sternwarte Berlin-Babelsberg* Bd **11** Heft 3

Jeans J H 1929 *The Universe Around Us* (Cambridge: Cambridge University Press) (Czech translation by B Mašek appeared in 1931)
—— 1929 *The Mysterious Universe* (Cambridge: Cambridge University Press) (Czech translation by Z Kopal appeared in 1936)
Kopal Z 1935a *Z. Astrophys.* **9** 239–257
—— 1935b *Říše Hvězd* **16** 134–144
—— 1936 *Mon. Not. R. Astron. Soc.* **96** 854–861
—— 1959 *Close Binary Systems* (London: Chapman and Hall and New York: Wiley)
Kopal Z and Vand V 1933 *Atlas d'Etoiles Variables* (Prague: Czech Astron. Soc.)
*Nature* 1937 **139** 293
Schlesinger F 1927 *Mon. Not. R. Astron. Soc.* **87** 506–523
Smart W M 1947 *Occasional Notes R. Astron. Soc.* **2** (no. 11) pp. 63–64
Vibert Douglas A 1956 *The Life of A. S. Eddington* (London: Nelson)
Walter K 1931 *Veröff. Königsberg Obs.* no. **2**

# Chapter 4

## The American Years

*Fulgens, sequar*

In order to open this most significant chapter of the first part of my life, we have to return to the story a year before, after I received my degree and entered the service of the University Library of Prague. After my return from the Far East, the urge to see the wider world would not leave me; and my next destination outside Europe was to be the United States of America—between the wars the undisputed leader in astronomical progress in the world. Therefore, even before I applied for the Denis Fellowship which was to take me to England, I had also applied for possible postdoctoral fellowships that would be available for young scholars in America for the next academic year.

The Czechoslovak Ministry of Education forwarded my application (duly endorsed) to the Institute of International Education in New York, which then served as a clearing house for European applications of this nature. The responses to my application did not reach me till I was already in Cambridge next spring, but they exceeded all expectations; for I received offers of postdoctoral fellowships from no less than three leading American institutions: Harvard, Mt Wilson, and Berkeley.

Dr Walter S Adams (1876–1956), then Director of Mt Wilson Observatory, offered me one to be tenable at Pasadena; Professor Armin Leuschner (1868–1953) of Berkeley recommended me for the Martin Kellogg Fellowship at the Lick Observatory; and Dr Harlow Shapley offered me one at Harvard College Observatory. The offers were ample to take care of my financial needs for the academic year 1938/39; but—and this turned out to be the problem—none of these institutions was in a position to contribute to my travel expenses from

Prague (where I had returned from Cambridge in the meantime) to the respective destination in the US—and the American West Coast was twice as far from Europe than the Harvard Observatory in Cambridge, Massachusetts (a suburb of Boston), making the travel costs almost twice as expensive. Dr Adams (figure 4.1) went out of his way to intervene with the Rockefeller Foundation on my behalf to raise an additional grant to defray my travel costs—but, unfortunately, to no avail; for some weeks later the Secretary of the Foundation had to inform me that, to his regret, no such grants were available for young scholars of my age at that time. Those were still the years of the Depression, when any such support was hard to come by; and since the Czechoslovak Government was also not in a position to help, I was left on my own.

**Figure 4.1** Dr Walter S Adams (1876–1956); Director of Mount Wilson Observatory 1923–1946.

The situation became all the more urgent for me as, in the summer of 1938, I intended to get married, so that two fares would have to be paid for; and since, moreover, my wife-to-be was to accompany me to the US on a visitor's visa, her return fare had to be prepaid in advance. We eventually managed to raise the fare, but the money at our disposal turned out to be sufficient only to transport us to the East Coast; and so our final choice was limited to Harvard. Not that we ever had any reason to regret it!

Before we crossed the Atlantic, one more pleasant occasion awaited us in Sweden: the Sixth General Assembly of the International Union, held at Stockholm between 3 and 10 August 1938. Like the preceding assembly in Paris, it was a memorable occasion. Few of its participants will ever forget the opening reception in the City Hall in Stockholm, the architecture of which is worthy of Flaubert's *Salammbo*! It was an occasion—the last, it turned out—when formality was still much in evidence among astronomers, not only in speech, but also in dress. Evening dress (with decorations) was still *de rigueur* for certain occasions (no great inconvenience, as participants still travelled mainly by surface, and excess baggage was scarcely a consideration).

I renewed many personal acquaintances from the Fifth General Assembly, and made new ones. Of particular significance for my future was to meet—now formally—Dr and Mrs Shapley (about whom so much will be said in the rest of this chapter) and other members of the Harvard family of which I was to become a member in the autumn. It was also in Stockholm that I spoke for the first time with Professor Henry Norris Russell, with whom I was to have so much to do in the next twenty years. And among European astronomers, two new friendships stand out in my memory which have outlasted decades: namely, with Dr and Mrs Hannes Alfvén, and with Miss Wilhelmina Ivanowska, a charming young Polish astronomer who was then spending several months in Sweden. But, on the sad side, it was also in Stockholm that I saw Sir Arthur Eddington for the last time, little knowing that when the IAU reconvened again after the storm which already then loomed threateningly on the horizon, he (then President-Elect of the IAU) would no longer be with us in this world.

Soon after I returned to Prague after the Stockholm meeting, an event took place of even greater significance for my future. On 7 September 1938, I entered holy matrimony in St Wenceslas' chapel of St Vitus' Cathedral in Prague (where, in the fullness of years, our second daughter Zdenka was to undergo the same ceremony at the end of the Thirteenth General Assembly of the IAU 29 years later) with Professors Nechvíle and Freundlich acting as witnesses of our union. Soon thereafter my wife and I were to depart from our native country for a brief honeymoon in Italy (and, eventually, for the United States).

Who of us still living can forget that September of 1938, when the Indian summer clothed the land we were leaving in a beautiful dress of fiery red leaves, and ordered the Sun to transform its hills into heaps of pure gold, its rivers into glittering streaks of silver? It dispersed fogs, and bid merciful fairies spread azure-blue skies over our homes in daytime, while at night it lit up the fathomless depths of the glacial lake of the stars glistening like diamond-studded coronation glory. All this we were leaving behind as we crossed the Alps by air towards the Mediterranean, not to see it again for the next nineteen years.

The first leg of our long journey from Prague to Trieste (with stop-overs in Bratislava and Klagenfurth) was over before noon, and this was just as well; for planes were not pressurised in those days; and the crossing of the Alps (we flew over Triglav) gave us something to remember. Fortunately, the rest of our journey aboard the Italian liner *Vulcania* from Trieste to New York was to remain strictly at sea-level; and its sailing dates gave us some time for a brief honeymoon in Venice, Florence and Rome.

It was a glorious time of late summer in Italy as well, still unmarred by much news from north of the Alps about the political situation in central Europe. In Italy, we were largely unaware of what was going on, for we could scarcely read their newspapers; and as the Italians are generally peaceful people (fighting, if need be, with their tongues rather than weapons), any idea of a possible war conflagration was obviously distasteful to them. In spite of some resolute slogans like '*Credere, obbedire, combattere*' over the gates of military barracks in Italy, martial spirit was quite conspicuous by its absence in the country at that time; and at some stages of our honeymoon we even encountered situations which did not lack a touch of humour.

Thus when we arrived at Florence *en route* to Rome, and settled in a small *pension*, the dining-room table was occupied by another young couple (American, I believe) who had arrived earlier and probably on a similar mission. The first evening the *patrone* asked the American couple (just back from sightseeing): 'How did you like Botticelli?' 'Thank you', responded the young husband with some hesitation; 'we prefer Chianti'. The *patrone* smiled, but the young wife did not; and kicking her husband none too gently under the table, she whispered (loudly enough for us to overhear): 'You fool: it is not wine; it's cheese!'

Yet indications that something serious was brewing north of the Alps were also not absent. When we returned to Trieste to embark on our transatlantic journey, we found that our hold baggage had failed to arrive from Prague by train—probably (as we were told by the ship's purser) because of the 'political situation'. What this situation was, however, the Italians kept in the dark; and no news about the Munich 'agreement' of 30 September was broadcast on the ship till we were already half way across the Atlantic. Our long journey took us first to Patras, which I saw then for the first time, little knowing of the ties which would connect us with this city and its university (yet to be founded in 1967) in the future; then via Naples, Palermo, Algiers, Lisbon, and the Azores to New York. The journey was slowed by the hurricane which lashed the north Atlantic Ocean at that time from coast to coast, reputedly the worst hurricane to hit the East Coast of the United States before '*Gloria*' of September 1985, so that we did not reach New York till 13 October. It happened to be Columbus Day; and

## To the New World

another unforgettable experience as the Statue of Liberty emerged from the morning mist, and later the skyline of Manhattan, to welcome us to the land which was to become our second home. It was for the first, as well as last, time, that we entered the US (and on such a beautiful autumn day) by sea up the Hudson river; for the days of sailing ships were approaching their end, and travellers in and out of the country would soon be taking to the air.

At the pier where our ship eventually landed we were met by the representative of the Institute of International Education in New York; and it was there that we learned of the full extent of the disaster which had befallen our native Czechoslovakia at Munich while we were in transit at sea.

In retrospect, the event itself should not have come to us completely unexpected. In August 1936 (while I was in Japan) the Nazi armies reoccupied the Rhineland. Its prompt fortification cut off France from all its eastern allies (of which Czechoslovakia was one), and thus reduced all her mutual-defence treaties with them to 'scraps of paper' *de facto* if not *de jure*, without any effective protest on the part of the signatories of the Versailles Treaty. Already at that time Britain and France wrote off, in effect, their claim to remain in the 'first league' of the world powers; and they left the timetable for subsequent events wholly in Hitler's hands. What happened at Munich that 30 September was only a belated recognition of that situation, by another 'scrap of paper' which should have brought (but did not) 'peace in our time'; and if the overt betrayal of Czechoslovakia in 1938 must legally be laid at the door of France, Britain did the same to Poland only a year later (as well as at the end of World War II in 1945). The only comment one could add today in defence of the leading statesmen of that time is the fact that none of them were of sufficient calibre to be aware of the consequences of their actions; and neither were their people (until history spelled out the price).

But a full realisation of this was still far in the future as we disembarked to take our first tentative steps in the New World. We stayed in New York only for a few days at that time, to report at the Czechoslovak Consulate and to see at least some of the principal sights of this truly incredible city, and then we left by train from Grand Central Terminal for Boston and Harvard Observatory. But before I come to the work on which I was soon to embark, let me begin with a few words of recollection on some of the principal personalities who greatly influenced my life. The foremost of these was Dr Harlow Shapley, at that time Director of Harvard College Observatory, who was principally responsible for the fact that we forsook Europe and opted for the New World which we were not to leave for the next thirteen years.

## 156  The American years

What kind of man was he? The life story of Harlow Shapley (figure 4.2) reads almost like a fairy tale, not often encountered these days in the history of our science. Born on 2 November 1885 on a farm near Nashville, Missouri, in the rural mid-west of the United States, he had no thought at first of a scientific career. Being gifted as a writer, he wanted to become a journalist (and he actually edited, for a time, a local paper in his home town). It was to improve his qualifications as a journalist that he went in 1906 to the University of Missouri, in Kansas City, to enter a course on journalism. But having arrived too late in the year for admission, he was not accepted in the department of his choice. What next? As Shapley used to tell his friends with his inimitable Irish humour (in which *Dichtung und Wahrheit* were often freely intermixed), he naturally did not want to go back empty-handed. He consulted, therefore, the University calendar in which subjects of instruction were listed in alphabetic order. The first was archaeology; he claimed he did not know what it was at that time. The next, astronomy, he thought he *did* know; so he presented himself at the

Figure 4.2  Dr Harlow Shapley (1885–1972), Director of the Harvard College Observatory between 1921 and 1952, at his office in front of his well-known rotating desk.

department, and was duly accepted. This was to be the beginning of a great career which took Shapley far beyond the confines of the Universe known by the turn of the twentieth century.

Shapley's first professor, Frederick H Seares (1873–1964), was an outstanding man of his time, who later became a senior staff member and assistant director of Mt Wilson Observatory. Under his guidance Harlow Shapley, 21 years of age at the time of his entrance, rapidly advanced in academic rank (AB, 1910; AM, 1911), and his name made an early appearance in the astronomical literature, to remain inseparably connected with it for the next sixty years.

After obtaining his Master's degree in 1911, Shapley decided to go east to further his academic career; and no university appeared more attractive to this end than Princeton, then being rejuvenated under the energetic administration of Woodrow Wilson, and with the young Professor Henry Norris Russell (1877–1957) at the head of its astronomy department. Let us allow Russell himself to tell us what happened: 'Not long before, we received at Princeton a business-like application for our one fellowship in astronomy, accompanied with a letter of recommendation expressing a very high opinion of the applicant's abilities. It was so enthusiastic, in fact, that though I knew its writer (Seares) to be a man of excellent judgement, I pondered a moment and asked myself: what sort of chap can this young man Harlow Shapley be? I soon learned!' (Russell 1948).

The time when Shapley came to Princeton could not have been more opportune. Shortly before, Russell had embarked on a new approach to the problem of the analysis of the light curves of eclipsing variables for the characteristics of the constituent stars—a problem directed at the very heart of stellar astrophysics, and one which also remained closest to Russell's own heart till the evening of his life. The arrival of a research student of Shapley's calibre at this juncture was a godsend to Russell as well; and although Russell had still more than a third of a century to teach, Harlow Shapley, his junior by only eight years, remained without doubt his most distinguished pupil (see figure 4.3).

Thousands of published observations of eclipsing variables were available in 1912 for analysis by the new methods. Aided by his neverabsent slide-rule, and unimpeded by any excess of mathematical punctilio, Russell was the trail-blazer; but without the energy with which Shapley applied the new methods to practical cases, the new methods would not have got off the ground. Shapley's PhD thesis in 1914 (*Princeton Observatory Contribution* number 3) on the orbits of 90 eclipsing binaries virtually created at a stroke a new branch of doublestar astronomy.

'There is no need to go into technical details of these calculations...' went on Russell, in the lecture from which we have already quoted; but the present writer would like to do so, in order to enable him to introduce the third silent partner of the great work of 1912–1914, whose self-effacing nature led that partner to remain consistently in the background. Those of us who knew that the main talents of both Russell and Shapley were not in the domain of mathematics, and that neither was enamoured of numerical computations, often wondered how the computations at the basis of Shapley's memoir in *Princeton Contribution* number 3 were ever performed.

**Figure 4.3**  Henry Norris Russell (left) and Harlow Shapley (right) at Harvard Observatory in the mid 1940s.

Recent information has lifted the veil on what happened: how, often, young Harlow would take a train from Princeton junction to Broad Street Station in Philadelphia, where he would meet Miss Martha Betz, his school sweetheart from the University of Missouri, then an alumna at Bryn Mawr, who would take over new computations from her fiancé (then, as always, we suppose, in a hurry) and deliver those which she had already finished. For Martha Betz (who in 1914 became Mrs Harlow Shapley) was an excellent mathematician and outstanding computer as can be testified by anyone who, like the present writer, has had the good fortune to work with her for a long time.

The spring of the year 1914 was memorable for Harlow Shapley for much more important reasons. Having received his doctorate from Princeton University, he married Martha Betz, and obtained a position on the staff of Mount Wilson Observatory. Fortune smiled on

Shapley (and astronomy) when George Ellery Hale offered him a research post at one of the few institutions in the United States where such positions were available at that time; and Shapley made magnificent use of his opportunity. His scientific interests shifted from eclipsing variables—a field in which he retained lasting interest, but to which he never really returned—to globular clusters; and in this field he made what was to become the discovery of his lifetime: that of the real dimensions of our Galaxy and of the location of its centre.

The tool Shapley used for this purpose was the globular clusters. It was, to be sure, known before his time that the distribution of such clusters in galactic latitude was highly asymmetric—in point of fact, almost all are situated in one hemisphere of the sky centred on the constellation of Sagittarius. Moreover, it was known that many of them contain large numbers of short-period ('cluster-type') cepheids whose absolute brightness could be estimated with the aid of the period–luminosity relation established for such variables from evidence nearer to hand. Armed with such evidence, Shapley embarked on a systematic determination of the distance of globular clusters in space, from the measured periods and apparent brightness of thousands of cluster-type cepheids embedded in them.

Moreover, when Shapley constructed a spatial model of the system of globular clusters, he found it to be situated in the constellation of Sagittarius at a galactic longitude $l = 325°$, some 50 000 light years from the Sun. This point Shapley boldly identified with the centre of the entire galactic system: and he was indeed right. As a result of this work, the Galaxy began to emerge in our minds for what it actually is: at least ten times larger in linear dimensions than all earlier estimates, from Herschel to Kapteyn, had made it out to be; with our Sun located eccentrically some 15 000 parsecs from its centre.

This latter distance was, to be sure, somewhat overestimated; for in 1918 (when these results were published) Shapley was not yet aware of the absorption of light which a concentration of gas and dust close to the galactic plane exerts on distant objects at low galactic latitudes. The role of this 'interstellar fog' was not fully realised until the 1930s: and it diminished our actual distance from the galactic centre from 15 000 to about 8500 parsecs—still, however, of the same order of magnitude. Incidentally, Shapley did not seem to have easily forgiven interstellar absorption for having played a trick of this kind on him; and when I came to join Harvard in 1938, I was cautioned by a well-meaning colleague (called Bart Bok) not to mention interstellar absorption more than necessary in Shapley's presence if I wanted to stay on the right side of the Director.

The years 1914–1921 mark, in retrospect, a high noon in the life of Harlow Shapley as an individual investigator: certainly he never published papers of greater originality and importance than during those years full of achievement. And these achievements were not limited only to astronomy, for Shapley's catholic interests also embraced some other departments of science; and one of them was biology. Shapley had, in particular, a deep interest in the life-cycle of the ant, which he had ample opportunity to observe in California; and by careful observation he established that the mean velocity with which ants run to and fro is proportional to the square root of the ambient temperature ('Shapley's law').

His interest in ants did not subside with the years; and, in 1945, it almost provoked an international incident. In the course of a visit to Moscow on the occasion of the 220th anniversary of the foundation of the USSR Academy of Sciences, an elaborate dinner was laid on for the guests by Joseph V Stalin. In the course of this convivial occasion (lasting no doubt a long time) Shapley noticed a rare species of ant crawl out of a basket of fruit with which the table was heavily laden. Shapley promptly caught the uninvited specimen, imprisoned it in a vial ever present in his pocket; and as a conserving liquid used the vodka that was being liberally served out (no doubt 100% proof). Needless to say, this somewhat unusual behaviour promptly caught the eye of those who had to watch for anything unusual in Stalin's presence; and some discrete explanations had to take place before things returned to normal. But (as Shapley noted) if it not had been for that ant, Stalin would never have given him a glance. At the time when the present writer learned of all this on Shapley's return, the ant was still partaking of the Kremlin vodka in the vial.

Another story illustrating Shapley's sense of humour comes to my mind in connection with another of his trips a year later. In 1946, Shapley was invited (with a group of other scientists) to visit India, then getting ready for independence; and, at some stage of that visit, the guests were received by the penultimate British Viceroy, General Lord Wavell. As they were waiting in line for a viceregal handshake, Wavell happened to drop his monocle lens, but caught it with great skill before it hit the ground. Shapley was greatly impressed, and whispered to his neighbour in line '...if he [Wavell] did that trick in Hollywood, he could earn by it a pretty penny'—loudly enough to be overheard by the Viceroy; and to receive a stern look from the old soldier (who obviously 'was not amused' by such frivolity). Of course, Shapley managed to smooth out the awkward moment with the schoolboyish smile he put on on such occasions, and was favoured with a viceregal handshake without a word of reprimand. This was not a recollection which Shapley sometimes made up to improve a story; for it was witnessed by my future Manchester colleague Patrick Blackett who happened to be there as well (he was not amused by it, either).

But the 1946 visit to India had another and more pleasant corollary for Shapley: namely, a friendship then struck with Pandit Jawaharlal Nehru (soon to become India's first Prime Minister), with whom Shapley obviously got along better than with Wavell. And so it happened that when, in the autumn of 1950, Nehru came to visit the United States, he found time also to visit Harvard Observatory together with his sister Madame Pandit; and he attended one of the Observatory's popular 'hollow-squares'. While Nehru himself remained a bit on the distant side during that occasion, Mme Pandit (then Indian High Commissioner, and later Ambassador to the Government of the United States in Washington) at once became popular with the staff, and everybody enjoyed her presence.

It became my privilege to meet Mme Pandit twice more in her life, first, while still at MIT in Cambridge (where she came to address a rally of the students); and, for the second time, in spring 1957, aboard an Indian Airlines flight from London to Prague, when I was *en route* for my first post-war visit to Czechoslovakia after an absence of nineteen years. Mme Pandit and her retinue flew all the way to India from western Europe; and there were few other people aboard. We recalled briefly our previous meetings almost ten years before; and when she disembarked for the transit lounge, I was amused to be taken by the Airport authorities for a member of her entourage, and given VIP treatment.

But to return to the beginning of our story and Shapley's early life in California, during his Mount Wilson years Harlow Shapley did not distinguish himself by research only. Soon the whole country—and not only astronomers—came to know him as an outstanding lecturer, brilliant and witty, as well as a successful writer. In addition, the example of George Ellery Hale helped to develop Shapley's innate talents as an administrator. By 1920 Shapley was rapidly acquiring national stature. Small wonder that in spite of his youth he was offered, and accepted, the Directorship of the Harvard College Observatory when the latter became vacant on the death of E C Pickering; and this observatory was to remain Shapley's academic home for the rest of his life. When Shapley came to Harvard in 1921, after a brief interregnum, to take up the reins of office, he initiated another uninterrupted era of 31 years (1921–1952) when the observatory's Directorship remained in the same hands. Thus for a full three-quarters of a century, between 1877 and 1952, Harvard College Observatory was administered by only two (albeit outstanding) Directors; and the renown which it enjoyed during most of this time was due in no small measure to that fact.

E C Pickering died in office at the age of 72; and the last decade of his stewardship was, perhaps, imbued by a somewhat autumnal air.

When, therefore, Harlow Shapley took over in 1921, much remained to be done; and Shapley addressed himself to his new task with energy and determination. The observatory's telescopic equipment was rapidly modernised; the Arequipa Southern Station was transferred to a better location in South Africa, to become the Boyden Station near Bloemfontein. In addition, the domestic observational activities were delegated to the new Oak Ridge (later Agassiz) Station and were provided with new 60 in reflectors; and, in the last decade of Shapley's Directorship, with a modern flat-field (Armagh–Dunsink–Harvard) Schmidt designed by James G Baker—a Harvard graduate and one of the brightest young astronomers who ever walked through the door of Harvard Observatory.

But perhaps the greatest and most fruitful accomplishment of Harlow Shapley as Director was the establishment of a graduate school of astronomy as part of the educational structure of Harvard University. Until the end of E C Pickering's regime the observatory was a pure research institution sharing only the name with Harvard College (and, later, University), but without any direct academic ties. One of the first reforms which Shapley introduced (with the full support of President Lowell) provided for a course of regular instruction in astronomy, at both undergraduate and postgraduate level, from which future astronomers could be recruited. This school rapidly became a great success; and a list of its graduates or research fellows from the 1930s reads like a *Who's Who in Astronomy* today.

It is easy to see, in retrospect, why this happened. By 1930, Harlow Shapley was around 45 years of age (see figure 4.4), full of enthusiasm tempered by extensive experience, and without equal on the contemporary American scene. Secondly, those were the years of economic depression; and the last age when the income from the observatory's endowment could be stretched to support a whole spectrum of pioneer activities. The Observatory's total staff had risen to about a hundred, making it (then) one of the largest astronomical institutions in the world, and new buildings had to be erected to accommodate them.

In his fifties (at the time when I learned to know him well) Harlow Shapley continued to be a paragon of hard-working man, constantly in touch with his numerous staff by means of his famous 'yellow notes' (messages dictated to his secretaries, which the addressee found on his desk next morning); but personally accessible mainly after normal hours between dinner and midnight in his office. He stopped every so often at my desk in Building C on his way from the office in Building D to his home at the Residence after midnight, when he was no longer in a hurry. These were my best hours with him, as they had been in late-afternoon with Eddington; while with Russell they used to be in the morning (for later in the day, according to my recollections, Russell was apt to get tired).

But to come back to Shapley, the executive duties in daytime he discharged with consummate skill; but their increasing load did not cause him to abandon active work in science at any time. He was a busy administrator in daytime; but at night—if at no other time—he returned to his research, conducted with the help of a devoted band of assistants; and this work continued to produce important results in the field of extragalactic research. After he left Mt Wilson for Harvard, Shapley no longer had at his disposal the large reflectors on the West Coast, with which Baade and Hubble continued to penetrate ever-increasing depths of space. The telescopes at his disposal at Harvard (the 16 inch Metcalf at Oak Ridge, or the 24 inch Bruce of the Boyden Station) possessed smaller apertures, but wider angular fields. It was mainly with these that Shapley and his collaborators continued to explore the Magellanic Clouds, and the spatial distribution of galaxies. Perhaps the most important result of this work was the discovery, in 1938, of southern dwarf galaxies in Sculptor and Fornax—discoveries which augmented in an unexpected manner the known population of the inner metagalaxy.

**Figure 4.4** The British-born theoretical astronomer (a native of Hull), Ernest William Brown (1866–1938) of Yale University, master of celestial mechanics and of the motion of the Moon (left), with Harlow Shapley at Harvard Observatory (in front of the Director's residence) in October 1929.

Then came World War II; and with it a chain of changes the end of which is nowhere yet in sight. Many staff members left the observatory during those years, not to return, attracted by other pursuits opening up in the post-war years. Other successful postgraduate schools of astronomy sprang up in different parts of the United States. Wartime technical developments rendered much of the astronomical equipment of earlier vintage obsolescent; and modernisation called for expenditures which were difficult to defray from the endowment income hard hit by post-war inflation. Moreover, the financial problems arising in this connection could no longer be resolved by the traditional method of soliciting contributions from rich men—an art which Shapley, following Hale's example, knew how to practise with consummate skill.

The post-war rich were no longer rich enough for the job—income tax saw to that—but the source of funds merely changed location: instead of waiting on men of wealth in their antechambers, one had to lobby for Federal or State funds in the 'corridors of power' in Washington and elsewhere. To practise this art with success called for somewhat different talents than those which Shapley possessed in abundant measure; and the fact that he rather fell out (over principles) with the Washington bureaucracy of the post-Roosevelt era did not facilitate matters. With his reputation as a liberal free thinker, it was almost inevitable that Shapley would come to the attention of certain reactionary elements in the US Congress—in particular, of the infamous junior senator from Wisconsin named Joe (not Eugene!) McCarthy, who subpoenaed Harlow Shapley to an inquisition in his 'Star Chamber' on Capitol Hill. After this session (held in private) both these men were reported to have emerged from McCarthy's quarters red as lobsters; and both refused to make any statement to the press, so that the public did not know who got the better of whom; but Shapley's friends thought they knew! To me (and many others) these and other attempts to cast Shapley in the role of a communist fellow-traveller made little or no sense. By all his instincts as well as his upbringing on a mid-western farm, Shapley was a 'rugged individualist' in the Herbert Hoover sense; and his employment practices (strengthened under the tutelage of George Ellery Hale) would have been approved by Andrew Carnegie, but not by the public authorities of any socialist country!

But be that as it may, by the end of World War II Shapley was already sixty years of age and unlikely to change colour, and by seniority had reached the stature of an 'elder statesman' whose services were much in demand for many purposes. Thus he was called upon to serve as President of the American Astronomical Society, of the American Association for the Advancement of Science, and of the National Society of Sigma Xi, almost in succession. All this was bound to take more and more of his time away from the observatory and its work. As had been the case with Pickering, the last seven years of Shapley's term of office as Director of Harvard Observatory were not marked by the same *élan* as the preceding decades. Shapley's retirement from the Directorship in 1952 (he continued to serve as Research Professor until 1956) signified the close of an era which had kept Harvard Observatory in the forefront of astronomical progress for 75 years—a record unequalled in the whole annals of our science.

Unlike his predecessor Pickering (who died in office), Shapley survived his retirement for twenty years; and during this time saw the directorship of his observatory change hands not less than three times. He saw many other changes and developments, such as the landing of man on the Moon (an event he predicted in 1945 as likely to take place within his lifetime); and when I last visited Shapley at Boulder at Christmas 1971, he wanted to know what we were making of the Apollo results. He passed away on 20 October 1972 in Boulder, a few weeks before his 87th birthday, as one of the senior and most distinguished members of our profession, the like of whom has not risen in our generation; and with whom the world reputation of Harvard Observatory has also come to an end.

Harlow Shapley was a man of many parts—astronomer, teacher, author, orator as well as man of affairs. Some of these gifts displayed prominently in the course of his life may gradually fall into oblivion as those of us who knew him in his prime may no longer be here to remember; and dust may settle on some of his work, or on the many honours bestowed upon him during his lifetime. But one title to fame will never tarnish, and will not be divorced from Harlow Shapley's name as long as astronomy is cherished on this Earth: namely, that of the discoverer of the centre of our Galaxy, and of our position within it.

In counting the many blessings of Shapley's long life we should list as foremost the fact that his wife, Martha Betz Shapley (figure 4.5), never left his side, and survived him, in fact, by nine years. She was, however, much more then the First Lady of Harvard Observatory during the 32 years of her husband's directorate, and deserves to be remembered as an astronomer in her own right in these recollections.

Martha Betz was born on 3 August 1890, in Kansas City, Missouri, of German parentage. The family had emigrated to the United States by the middle of the nineteenth century from Hanover, and (as Mrs Shapley once told me), when her grandfather was a young boy there, he had seen Caroline Herschel (who in 1781 was at her famous brother's side at the discovery of Uranus, and later discovered not less than eight comets (including Encke's!) in her own right, and attained the ripe old age of 98) more than once being driven in a coach through the streets of that city; and grandfather Betz remembered the old lady well enough to describe her to his granddaughter half a century later.

But astronomy did not become at home in the Betz family from that time. Carl Betz, father of the future Mrs Shapley, was an educator in

Kansas City and director of music in the City schools. Music was a part of the future Mrs Shapley's childhood; and remained her favourite avocation throughout her lifetime. The present writer will never forget her exquisite rendering of Chopin's *Ballades* at the piano of the Harvard Observatory Residence (when she thought that nobody was listening).

**Figure 4.5** Martha Betz Shapley (Mrs Harlow Shapley), 1890–1981; as she was in the 1930s.

In due course, Martha Betz graduated with a Master's degree from the University of Missouri which, by great good chance, happened also to be the *alma mater* of her future husband. Years later, as the latter recorded (cf Shapley 1968), 'There, in my third year, I met a brunette named Martha Betz, from Kansas City, and never got loose—or wanted to. We first met in a mathematics class—she sat in the front row and knew all the answers. She was a clever lady in those days' (added Harlow, with a touch of facetiousness which characterised some of his latter-day writing). 'She took five full courses and got top marks in them all. Eventually, it became astronomy; and she published quite a few astronomical papers.'

This latter somewhat off-hand remark did not, perhaps, quite do justice to her whom Harlow referred to as a 'clever lady'. That she was by far the better mathematician of the two was borne out not only by their school records, but by the subsequent careers of them both. What, however, may not be quite so generally known is the extent to which Martha Betz, then doing postgraduate work at Bryn Mawr and Harlow Shapley's fiancée, participated in her future husband's PhD thesis at Princeton under the benign eye of Henry Norris Russell, as previously described on p. 158.

While the resulting *Princeton Observatory Contribution* number 3 marked the end of Harlow's active interest in eclipsing variables, for his wife Martha this was only the beginning, as witnessed by her papers published in the *Astrophysical Journal* (1916,1917) as well as in the *Publications of the Astronomical Society of the Pacific* (1919). When Harlow Shapley was called upon in 1921 to accept the Directorship of Harvard Observatory and moved east to Cambridge, Martha Shapley continued for a time her independent work in the same field (see Shapley 1924 or 1927), but the number of her contributions was gradually subsiding for several reasons. In 1921 the Shapleys returned from California to Cambridge with a family of three children (Willis, Alan, and daughter Mildred, who all made a name for themselves in the world of science in due course), and in subsequent years the family was augmented by two proper Bostonians (Lloyd and Carl) who by their talents did not stay behind their Californian brethren.

The cares of a growing family exacted much of Mrs Shapley's time, but it was not only the children who claimed their mother's attention. In a very true sense, on coming to Harvard the Shapleys adopted a much wider family of budding astronomers, residents as well as birds of passage, who will remember them with affection all their lives. For at Harvard Mrs Shapley's career as a warm and gracious hostess blossomed out to the full, entertaining with equal grace the rich and the poor, famous and humble (sometimes, as with Albert Einstein or Igor Stravinsky, both in one person), the geniuses and amateurs, or scared undergraduates. The Shapleys loved to entertain; and any occasion for a party was always welcomed. On each Friday nearest the full moon, when astronomical photography (and most astronomical work in those days was photographic) was curtailed by bright moonlight, Dr Shapley was the host to a Full-Moon Club at his home, for graduate students as well as staff.

The highlight of the season used always to be the Christmas parties at the Residence, at which praises to the Heavens were raised by carols sung by many astronomers of all ages. One particular number which sticks vividly in my memory was the solo parts of 'We Three Kings', the role of Melchior being sung by the good Christian Bart J Bok (I sang Caspar), and that of Balthazar by Henry Norris Russell—a guest performance which would have penetrated the Gates of Heaven even more readily if that distinguished astronomer had been better able to carry the tune! Miss Sibyl Chubb used to act as honorary conductor of most such performances, with Dr Shapley as a genial master of ceremonies, seeing to it that the festive spirit prevailed at all times; though it was his wife who, inconspicuously, but with unfailing charm, made everyone feel truly a member of the same family.

In the meantime, years went by leaving no-one unmarked; the Shapley children left the paternal roof one by one to start families of their own. Then 'the tides in the affairs of men' brought about another Great War. This is, perhaps, not the place to tell how well Mrs Shapley served her country during those years. When, however, it came to its end, she caught her 'second wind' and returned to the science of her youth to make the major contribution of her lifetime, as co-author of the *Catalogue of the Elements of Eclipsing Binary Systems* (Manchester, 1956), which in more than one sense can be regarded as a descendant of Harlow Shapley's *Princeton Contribution* number 3, 42 years before.

The work on the new catalogue, based on systematic re-discussion of observational data on eclipsing variables, accumulated in the first half of this century, by more modern methods developed by Piotrowski and the present writer at Harvard in the 1940s, commenced in 1948 under the auspices of the American Philosophical Society of Philadelphia; and it was only the departure of its junior author from Cambridge to Manchester in 1951 that delayed the completion and publication of this work till 1956.

The preparation of this *Catalogue* (which, although now out of date as regards the material included, still remains unsurpassed in method of presentation) was the last major contribution of Martha Shapley to the subject to which she remained faithful for more than forty years. After the retirement of her husband from the Directorship of Harvard Observatory in 1952 she had still almost thirty years of quiet life ahead of her—the last of which she spent in Tucson, Arizona (see figure 4.6), with her daughter Mildred Shapley Matthews, an astronomer in her own right (who by now may have edited more books than her father ever wrote), watching the third generation of Shapleys grow to maturity. Death came gently in her sleep to Martha Shapley on 24 January 1981, to take her soul among the stars to which she gave so much work during her lifetime.

The last Director of Harvard Observatory (and successor to Harlow Shapley) whom I knew well in his younger years and had a chance to observe at close range was Donald H Menzel (1901–1976); and although he was not yet Director when I left Harvard for Manchester, it may not be amiss to add some observations in connection with the events which saw the American scientific atmosphere change from its grass-roots stage to the 'big science' which it has become—for better or worse—today.

By his roots Menzel, like Shapley, was a Mid-Westerner, born on 11 April 1901 in Florence, Colorado; and his educational career led him likewise to Princeton, where he received his doctorate in 1924 as a theoretical astrophysicist (of whom there were not many in the States at that time), bent on application of atomic physics to astrophysical situations. At first, his interests centred on physical processes in gaseous nebulae, a field to which he made many contributions of importance at that time, and some of which are still remembered today. After several junior appointments of short duration, in 1932 he joined the staff of Harvard University which remained his academic home until his retirement in 1971; and he was associate professor when I met him in 1938. It was also about that time that he developed the second main interest which he maintained to the end of his life: our Sun.

**Figure 4.6** Mrs Martha Betz Shapley, on 6 January 1978 (at 88 years of age) in Tucson, Arizona (right); with the author (left) and her daughter Mrs Mildred Shapley Matthews, a prolific editor of astronomical books.

Many old-timers will still remember the excitement which the French astronomer Bernard Lyot (1897–1952) created in the late 1930s with his observations, at Pic-du-Midi, of the 'white' solar corona outside total eclipses, and with the films of the activity of solar chromospheric phenomena shown in August 1938 at the Sixth General Assembly of the IAU in Stockholm. Menzel was the first to transplant Lyot's techniques across the Atlantic, and to develop them further at an even more favourable site than the summit of the Pic-du-Midi. His efforts led to the foundation in 1939 of the High Altitude Observatory in the Colorado Rockies, near the town of Climax (3450 m above sea-level), first as an offshoot of Harvard, and later as an independent institution which only recently closed its doors because of the increasing pollution of its environment.

Unlike Lyot, who was a genius at improvisation and preferred to perform all the crucial steps of his experiments with his own hands, Menzel was by nature a scientist of managerial type, preferring to work through other people's hands; and the principal credit for the work performed at Climax during the many years of the observatory's existence should go to Walter Orr Roberts—a student of Menzel's at Harvard, who retired only recently from Directorship of the National Center for Atmospheric Research (NCAR) at Boulder (of which the High-Altitude Observatory eventually became a part).

Then war came; and with it the changes in academic life which were fully reflected also in Menzel's subsequent career. Like so many others, after 1945 Menzel never resumed the career of an original or creative scholar; and all his activities veered increasingly to those of an administrator and promoter. The abilities of a skilled operator—at a premium in the inflationary post-war years—eventually led to his appointment (in 1954) as Director of the (then still) Harvard College Observatory, in succession to Harlow Shapley (who had retired two years before), a position which Menzel held for the next twelve years.

His illustrious predecessor had possessed, among his many talents, the uncanny gift of extracting money from rich individuals to augment income from the Observatory's endowment. In the post-war years inflation made this goal doubly desirable, but it was no longer attainable in the old ways. There were too few Carnegies left around. As a result, Shapley's successor had to raise additional support for his institution not by waiting on the rich, but by lobbying in Washington for grants or contracts from the many governmental agencies set up for this purpose, and responsive to the political powers of the day.

This trend of events was ominous for the further development of science (and not only in America); for while private individuals (or corporations) were still (sometimes) willing to make gifts for a good cause with few strings attached (and requiring very little paperwork), the new dispensers of Government largesse knew exactly what they wanted, and saw to it that they got it. As a result of such outside pressures, Harvard Observatory in the post-war years became increasingly involved in new fields of research or development without sufficient tradition or expertise—as long as such ventures brought in sufficient revenue. In fact, at times the observatory seemed to differ only in location, but not in spirit, from the many laboratories which sprang up during the 'roaring sixties' along Route 128 around Boston as offshoots of Harvard and MIT, to serve customers rather than science, at a time when men were heading for the Moon, and nothing seemed impossible on Earth.

In fairness to Harvard it must be said that, in the post-war years, the observatory faced a predicament that was common to its sister institutions all over the country. It was probably better off than most; but what history must hold against Menzel's administration will be the eagerness with which it sold its inheritance for a dish of porridge (in the form of overheads). And worse: what cannot be excused by any reason known to this writer was the mean and shabby way in which

## Arrival at Harvard

the new incumbent set out after 1954 to push the Director Emeritus (and his wife) out of the observatory at which they had served science so faithfully and so long, for no apparent reason—unless the source of this indignity (which cut across the whole American astronomical community) was an inferiority complex, sometimes felt by lesser men towards those to whom they owed so much in the past.

Needless to say, such a life has its drawbacks for the institutions as well as the individuals directing them; and Menzel was no exception. His health suffered a serious setback in 1965, from which he never fully recovered. He retired from the Directorship at Harvard in 1966; and although he continued to do some teaching until 1971 death came to claim him (after a prolonged illness) in Boston on 14 December 1976, aged 75 years—four years after Shapley.

The sixth Director of Harvard Observatory was Dr Leo Goldberg, a Harvard graduate whom I still remember from his student years; and who for many years had been Menzel's *alter ego* in most respects. He did not, however, hold that office for long; for shortly after the installation of a new President of Harvard University in 1971, Goldberg resigned his Harvard post to withdraw to the Kitt Peak National Observatory in Arizona, and was eventually succeeded by Dr George B Field of the University of California at Berkeley, another Harvard graduate whom I remember as a student (I believe he was my summer assistant at MIT one year). Even George, however, did not hold the job as long as Pickering or Shapley, and he resigned in 1982 to be succeeded by Dr Irwin Shapiro, a young physicist who (like Pickering a century before) came likewise from MIT. However, by that time my contacts with an institution at which I spent several years in the prime of my life have already become so tenuous that there is nothing of any significance I could add to this more recent past; and the time has come to return again to my beginnings there before World War II.

By the time of our arrival at Harvard in October 1938, Dr Shapley was not in residence: he was in South Africa to inspect the Boyden Station of Harvard Observatory near Bloemfontein; and was not to return to the northern hemisphere till several weeks later. However, Miss Jenka Mohr, his secretary and faithful assistant, was sent to meet us at Boston South Station on our arrival from New York, and to take us across the Charles River to Cambridge, which was to be our home for the next thirteen years.

We were received by all staff present at the observatory with extreme kindness—not, I am sure, because they knew us, but because the country we came from was then (and remained for a long time) the centre of sympathetic attention. Chance had it that the American Association of Variable Star Observers (AAVSO), of which I had been a member (albeit not too effective) during my student years, held their

Annual Meeting at Harvard Observatory; and its final dinner happened to coincide with the day of our arrival. Mr Leon Campbell, its Recording Secretary of many years (see figure 4.7) invited us to attend this dinner as guests of the Association; and he asked me to make an after-dinner speech. I did not quite know what was expected of me; but when I shared my perplexities with Dr Bart J Bok, the latter did his best to set my mind at ease. 'It does not really matter what you tell those Americans' (he was not one either at that time); 'the important thing is to start each talk with a good joke.' Needless to say, for me this was easier said than done; and I recall to this day the embarrassment with which I was groping for one; but if my performance did not quite live up to their expectations the astronomers were too kind to let me know.

**Figure 4.7** Mr Leon Campbell (1881–1951), veteran Harvard astronomer and, for many years, Recorder of the American Association of Variable Star Observers.

## Harvard College Observatory and Eclipsing Variables

Having introduced to the reader some of the principal *dramatis personae* of this chapter, let us return to its young writer and to some of his first steps on the stage of Harvard College Observatory (figure 4.8) in the autumn and winter of 1938. My first office (which I shared with Lawrence Aller and Martin Schwarzschild) on the ground floor of the

Observatory's Building C was next door to that occupied by the Gaposchkins, whom I also got to know personally for the first time (see figure 4.9). It was an interesting neighbourhood; for I soon noticed that in a state of some excitation, the lady next door did not watch her language (easily penetrating through the door in between) any more closely than (on occasions) did Professor Higgins of Shaw's *Pygmalion*. As my knowledge of four-letter words in the English language still left something to be desired at that time, it was Lawrence who did the translating; and he usually chuckled at their effect on my pious soul.

**Figure 4.8** Building D of the Harvard College Observatory, erected in the 1930s.

However, my wife and I soon became friendly with the Gaposchkins, outside the Observatory as well, and we visited them more than once in their home at Lexington. None of these visits turned out to be dull; and on occasion I had to do some translating myself. Thus, when Sergei (in some exasperation at some minor domestic difficulty) invited his wife: 'Idi k chortu!' I was too embarrassed to answer Cecilia's request for translation; and, on being pressed for it, pretended instead that *chort* in Russian signifies a darling. 'Darling?' Cecilia pursued her inquiry with some suspicion; 'doesn't it have something to do with the devil?' What else could I do but plead linguistic inadequacy—an excuse which was not very far from the truth (except for the similarity of the incriminating term to its equivalent in Czech!).

**Figure 4.9** A group of foreign scholars at Harvard Observatory in spring 1939, in front of the Observatory's Building D. Standing (left to right): F Shirley Patterson (Canada), Donald B MacRae (Canada), George Z Dimitroff (Bulgaria), Paris Pişmiş (Turkey), Zdeněk Kopal (Czechoslovakia), Cecilia H Payne-Gaposchkin (England), Odon Godard (Belgium), Richard Prager (Germany), Luigi G Jacchia (Italy) and Sergei I Gaposchkin (USSR). Sitting (left to right): Bart J Bok (Holland), Jaako Tuominen (Finland), Masaaki Huruhata (Japan) and Luis Enrique Erro (Mexico). Six of this group of fourteen have already departed from this world.

However, our relations with the Gaposchkins never became very close, in spite (or, rather, because) of the fact that Sergei too worked in the field of eclipsing variables; and Cecilia complained more than once (in her often blunt language) that I did not give her husband sufficient credit for his work. Now Sergei was a likeable man of many parts, who had brought from his native Crimea to the New World the merry spirit of his homeland; but when it came to science, one just had to draw the line. When, however, war came to Mother Russia in 1941, Cecilia and Sergei organised political discussion evenings at Harvard Observatory (under the name of the Forum for International Problems); and invited me to join its steering committee, which I did (together with my friend Luigi Jacchia and some other astronomers of the Observatory's staff)—at some embarrassment (as it turned out) later.

For when war engulfed the United States, and I was required to obtain a US security clearance (for purposes described in more detail in the next section), our infrequent visits to the Gaposchkins in the past became of considerable interest to the officers in charge of such investigations; and they quizzed me about my relations with the Gaposchkins in great detail, without disclosing any reasons. These transpired only in the 1950s (during the McCarthy era), when the political affiliations of Cecilia and Sergei Gaposchkin became a matter of public knowledge.

By that time, we had already been living in England for several years; and as I never met Cecilia (who died in 1979) since, I could not ask her whether she had brought her political affiliation already with her from Cambridge (where between 1919 and 1923 she was an undergraduate at Newnham College prior to her re-settlement to Cambridge, Massachusetts), or embraced them only later in the United States. The latter is perhaps unlikely (because of restricted opportunities); while Cambridge (England) was then a more fertile ground in this respect. To be sure, Anthony Blunt *et consortes* about whom we have heard so much in recent years were Cecilia's juniors by several years (she was born in 1900, Blunt only in 1907); but they may of course have had predecessors. Since, moreover, Cecilia's husband Sergei Illarionovich (born in the Crimea in 1878, and married to Cecilia in New York in 1934) has also departed from this world (in the autumn of 1984), their full story now rests with them in their graves. *Requiescant in pace!*

It should be added that our places of work at Harvard Observatory did not long remain in close proximity; for from the beginning of 1939 I was 'kicked upstairs' to the first floor of the same building, to occupy a corner of the room where (as proclaimed by a memorial tablet on the wall above my desk) Miss Henrietta Swan Leavitt (1868–1921) discovered in 1908 the existence of a period–luminosity relation for cepheid variables in the Magellanic Clouds—a fundamental property of such variables which has since become our main tool for distance measurements on the galactic (and extragalactic) scale. Eventually, I received an office for myself with a secretary (Miss Doris Beaudoin, who years later served in the same capacity my editorial colleague Professor A G W Cameron) in Building A (the Observatory's original site, adjacent to the Director's Residence); and I used this till my departure for England in 1951.

But to return to the variable stars, the Gaposchkins were not their only students at Harvard of recent European provenance. In the spring of 1939 two more—Dr Richard Prager (1883–1945) and Dr Luigi Jacchia (1910– ), of whom we have already heard—reached the American shores and Harvard Observatory (as victims of anti-Semitic persecutions in Germany and Italy) to reinforce our ranks (see figures 4.9 and 4.10). I was already acquainted with both in Europe, and we shall meet them again in the next section.

But of greater historical significance for these recollections is perhaps the fact that I was still able to make the acquaintance (albeit brief) of Miss Annie J Cannon (1863–1941) and Miss Antonia C Maury (1866–1952), surviving veterans of the Pickering era of spectral classification. Miss Cannon (by that time the senior member of the Observatory's staff, and already rather hard of hearing) contributed a delightful talk on 'The Story of Starlight' for a series of broadcasts I was then called upon to organise for the short-wave station WRUL (see the next section); and Miss Maury's interest in the spectrum of $\beta$ Lyrae remained active till the end of her life. After I published my own first paper on this subject (Kopal 1941b) Miss Maury often dropped in during her occasional visits to Harvard, to discuss the physical properties of this perplexing system; though who can blame us if we never arrived at any credible conclusions?

**Figure 4.10**  Professor Richard Prager (1883–1945), devoted life-long student of variable stars.

But life at Harvard Observatory was not necessarily all shop-talk! As a legacy from the Pickering era, the majority of the staff were still ladies—some of them the kindest and most charming I ever met in my life. Who could use other terms to describe Miss Henrietta H Swope, a devoted friend of variable stars and of all who were interested in them? Henrietta was particularly kind to my wife and children; and all photographs of them from a tender age were taken with her camera (Henrietta was an excellent photographer). Or who could have been a kinder auntie to our children than Miss Madeleine Harvey, curator (after Miss Hodgden) of the Harvard plate collection; or Miss Dorrit Hoffleit (never a year went by without Dorrit coming to visit us on Christmas Eve in the role of Santa Claus for our children)? There was nothing but fun and laughter whenever the vivacious Rebecca Jones (the future Mrs Boris Karpov of Aberdeen Proving Ground) dropped in to see her friends, or when Frances Wright sat down at her piano to play songs by Stephen Foster, often accompanying voices which made up in sincerity for what they lacked in training. And there were many others. These reminiscences would have no end if I were to record all that I remember from those days. Most of these ladies have now gone to their reward; but they will not be forgotten as long as the last one of us lives.

Coming back to shop, it may perhaps surprise the reader that my early work at Harvard had nothing to do with eclipsing variables, but with cepheids in globular star clusters. During the last months of our stay in Prague, Professor Freundlich had urged me repeatedly to try to investigate the actual dimensions of such clusters from the distribution of cluster-type variables contained in the parent formation: at greater distances from their centres, the ordinary cluster stars merge imperceptibly with the foreground stars (especially at low galactic latitudes), but cluster-type variables belonging to each cluster can be fished out from the substrate by their apparent magnitudes.

When I mentioned such a plan to Dr Shapley, he was immediately interested; and put at my disposal to this end the 8 inch IR camera at Oak Ridge—to no avail, however; the cluster stars were too faint (and most clusters too far south) to make much headway with the search. However, the 24 inch Bruce telescope plates from South Africa, properly blinked, enabled me to discover several new cluster-type variables in the neighbourhood of $\omega$ Centauri, which are not yet recorded in the latest edition of the Sawyer–Hogg catalogue, for the simple reason that they have never been published. Shortly after I embarked on this work, however, Shapley sent to me (through Miss Frances Wright) a gentle hint to the effect that he wanted to undertake this work himself. Of course I gave it up—perhaps to the detriment of a good cause, for Shapley never got around to doing it after all; and so these variables are yet to be discovered. There is, however, no doubt that the number of cluster-type variables associated with the globular cluster $\omega$ Centauri is larger than known so far, offering future investigators ample scope for further work.

The readers who may have heard of my more recent work may perhaps be surprised that, in my younger years, I used also to be an observer, at Ondřejov Observatory near Prague (where, during my student years, I did some photographic work with the observatory's

8 inch astrograph); and later at Cambridge (England), where I assisted the late Mr H E Green with photoelectric observations at the Sheepshanks telescope. During my early years at Harvard, it was customary for postdoctoral fellows or junior staff members to look after the night work of graduate students at the Observatory's Oak Ridge (later renamed Agassiz) Station, located some thirty miles west of Cambridge; and between 1938 and 1940 I was called upon to do so once (and sometimes twice) a week. At that time, Dr Shapley put at my disposal the 8 inch IR (having nothing to do with infrared!) telescope, and later the 24 inch H reflector for work on globular clusters, work for which these instruments (located so far north) were of only marginal usefulness. The 24 inch Jewett Schmidt went into service much later; but before it did, theoretical work on eclipsing variables began to monopolise all my time, and with the exception of brief spells at Pic-du-Midi in France and Kottamia (Egypt) in the 1960s concerned with work on the Moon, I never returned to systematic observations again.

But it is perhaps just as well that I did not continue with the observational work at Harvard at that time. For in the summer of 1939 a series of conferences was held between 5 July and 15 August at Harvard Observatory, in which I was invited to participate by giving some lectures on proximity effects in close binary systems. Although the extent of my participation was quite modest (I do not think I lectured for more than one week), the subject attracted considerable attention; and it is probably true to say that their preparation (as well as subsequent discussions of the subject with many astronomers who attended the lectures) marked the beginning of my involvement in this subject which was to last almost without interruption for the next twenty years.

When these conferences ended, my wife and I went for a brief vacation to New York in the second half of August to visit the World's Fair then being held in that city; and it was there that, in the last weeks of that month, we followed (on Times Square, and elsewhere) the final phase of the dramatic events which preceded the outbreak of World War II, which was to last for the next six years and change the face of the world almost beyond recognition.

The consequences of the events which erupted in eastern Europe during the month of September influenced life in the United States very little, at least as far as astronomy was concerned. On 20–21 October 1939, a two-day conference was held at the New York Academy of Sciences, at which R E Marshak, H N Russell, H Shapley, J Tuominen and myself presented addresses eventually to be published by the Academy (cf Kopal *et al.* 1941a); and my own contribution was

*Eclipsing variables* 179

to elaborate the topics of my course in the Harvard Summer Conferences two months before. For the academic session of 1939/40 Harvard Observatory awarded me their Agassiz Fellowship to support further research in the field of eclipsing variables; and in pursuit of certain of its objectives (unconnected with any proximity effects) I stumbled upon the first new alternative way by which the problem posed by the need to extract the elements of eclipsing binary systems from an analysis of their observed light curves could be approached—a problem which has not left my mind ever since, and which has still not been properly solved to this day.

As is well known (see also Chapter 6 of these reminiscences), this problem, attacked first by E C Pickering (1880) almost one hundred years after the first stars of this type were discovered in the sky, was first systematically investigated by H N Russell (1912), who, in collaboration with Harlow Shapley (cf Russell and Shapley 1912), developed procedures by which the desired elements could be obtained (under certain assumptions) by replacing the actual observations by smooth curves drawn free-hand to follow the course of individual observations. The relations between the actual brightness observed at any particular phase of the eclipse are, of course, highly non-linear (in fact, transcendental), a fact making any *direct* solution of the problem impossible. Russell and Shapley attempted to by-pass the difficulties arising from this source by the adoption of certain 'fixed points' on the light curves drawn free-hand, the positions of which could be read off such curves without any error. The adoption of the error-free locations of such pivotal points is, in effect, equivalent to postulating a knowledge of certain exact relations between the elements of the system to be obtained by the analysis, which arbitrarily reduces the number of 'degrees of freedom' of the problem.

If we do this (and go far enough with such a strategy) *some* solution of our problem will always result, in the sense that the elements so obtained can yield a theoretical light curve matching the observations with seeming accuracy. However, even so we are still left to confront the question: is such a solution *unique* (i.e., are these the only elements which can do so within the limits of observational errors)? The answer is, unfortunately, in the negative; for *no* method which replaces the actual data by smooth light curves drawn free-hand to interpolate between individual observations can enable us to ascertain whether the results obtained *à la* Russell are the actual elements of the system, or merely 'a right combination of wrong elements', whose only merit is that they happen to reproduce the observations within the limits of observational errors. The latter condition is, to be sure, necessary, but *not sufficient* to verify the correctness of the results. This fact was overlooked by Russell and Shapley in their 1912 work, as well as by their successors who followed their basic line of reasoning, and caused no end of confusion in the subsequent literature on the subject.

In fact, owing to the manifest non-linearity of the problem at issue, the only way to *approach* the solution of our problem, and obtain the *most probable* values of the elements of the eclipsing system would seem to be by *iteration*. Such thoughts indeed went through my mind as I set out, in 1940, to work out the first iterative process suitable to this end (Kopal 1941c). This brought me, for the first (though not the last) time on a collision course with Russell, the recognised doyen of American astronomy, for whom too the study of eclipsing variables remained a subject close to his heart for more than half a century, from almost the beginning of his academic career (cf Russell 1899) to the end of his life (Russell 1956).

I have already mentioned that Russell's 1912 work on the light curves of eclipsing variables was known to me before I entered university; and when I made his personal acquaintance in 1938 Russell was already over sixty years old. Although I could not, therefore, regard myself as his student (nor did I ever become as closely acquainted with him as I did with Shapley), these reminiscences would be incomplete without at least some recollections of Russell as a person, now that almost thirty years after his death, his life and times are gradually retreating into the past.

Henry Norris Russell (figure 4.11) was born on 25 October 1877 in Oyster Bay, New York, the son of Alexander Russell, a Presbyterian minister and (like Simon Newcomb before him) a Scottish–Canadian immigrant to the United States. There is some parallel in the early years of the future distinguished astronomer and his future President (at Princeton University and in the White House) Thomas Woodrow Wilson; himself a first-generation American born in the manse of the same religion.

No wonder that the only son born in such distinguished surroundings had to be a model boy; and while, in Woodrow Wilson, a repressed reaction against such a station eventually almost resulted in a tragedy of world dimensions, there is no evidence that, for Russell, it would have been as serious an obstacle (unless, perchance, his nervous breakdown in 1900–1901 after graduation and occasional difficulties later in life could be ascribed to the same cause).

At any rate, Russell received his first degree in Princeton in 1897 (at the age of twenty) *insigne cum laude*—a grade higher than *summa*—and his doctorate from the same institution in 1900 on the basis of a thesis concerned with a study of the perturbations of the semi-major axis of the asteroid Eros by the action of Mars (cf Russell 1900). Moreover, after having recovered good health, Russell spent the next four years

# Henry Norris Russell

(1901–1905) as a postdoctoral fellow in England (at Cambridge and Greenwich); but in 1905 he returned to Princeton as instructor (and since 1911 Professor) of astronomy, a position which he retained until his retirement in 1947. He survived (likewise in Princeton) for another ten years.

**Figure 4.11** Professor Henry Norris Russell (1877–1957) of Princeton University.

Russell's first work on the subject of eclipsing variables (Russell 1899) antedated, in fact, his doctoral thesis; and his most important work in this field (carried out in collaboration with Shapley) has already been mentioned in this chapter. Russell's subsequent work in the field of theoretical spectroscopy ('Russell–Saunders coupling') and other subjects is, of course, outside the scope of my reminiscences. However, some recollections of possible interest, based on a personal acquaintance of almost twenty years in the last quarter of Russell's life (and backed by extensive correspondence now on deposit with the Manchester University Library) may be added in this place.

First, let it be stressed that—notwithstanding any critical remarks which I have felt impelled to make here and there in the text for the sake of the record—Russell was a gentleman of the old school; and the influence which he acquired in the United States in the second half of his life did not reduce him to the role of an 'operator', in the modern sense of the word (perhaps because administration was not his *forte*). Of somewhat frail health (which did not prevent his attaining the age of eighty) he lacked the energy and drive of a Shapley or other of his contemporaries. His Princeton colleague Raymond Smith Dugan

(1878–1940) used to say that if there was a piece of work which would have taken him (Dugan) half an hour to accomplish, Russell would spend half a day thinking how to do it more easily; and should he succeed in doing so in a way 'which does not take any time', he would regard this as an intellectual triumph!

As a mathematician, Russell was certainly better than Shapley—though, of course, he would have been no match for a Chandrasekhar. Indeed, his mathematical work but seldom left the protective umbrella of a slide-rule. With problems which could be handled with its aid, Russell was on safe ground (during the years of his Directorship, Princeton Observatory never acquired a computer more elaborate than that); but anything beyond its limits Russell instinctively distrusted.

As an original scientist, Russell lacked the vision of an Eddington; but his mind was quick (which was not the case with Eddington). During the best years of his life, Russell managed to remain a step or two ahead of most of his contemporaries in the United States (though, perhaps, not many steps); and this ability earned him a well-deserved national renown. But this implies also that, when advancing age relegates one to the sidelines, it will not be long before one is overtaken by those who are younger. When, years later (and not being conditioned to Russell-worship since a tender age) I mentioned this now and then to Dr Shapley, he always used to say: 'You should have known him [i.e. Russell] twenty years ago'—which, of course, I did not.

It was, however, the transparent sincerity with which Russell accepted all accolades as his rightful due which retained for him the affection of many who were glad to accept that self-esteem (even if a bit inflated) is no serious failing, least of all in one who throughout his life was thoroughly spoiled by a doting family as well as friends and pupils. But Russell was never a hypocrite, or an operator scheming to take advantage of others for selfish purposes. As a matter of fact, the opposite may well be the case. The following story may serve to illustrate this. When E C Pickering died in February 1919, the Directorship of Harvard Observatory was apparently first offered to Russell, who indeed considered it, as disclosed from extant correspondence (cf Gingerich 1975). Thus on 13 June 1920 Russell penned the following missive to George Ellery Hale in Pasadena:

> If they [presumably the Harvard University administration] accept this plan, I will then propose Shapley for second in command ... consider what Shapley and I could do at Harvard! Between us, we cover the field of sidereal astrophysics rather fully. We can both do some theory—and I might keep Shapley from too riotous an imagination—in print!
> 
> Shapley couldn't swing the thing alone ... But he would make a bully second, and would be sure to grow—I mean in knowledge of the world and of affairs; if he grew intellectually he would be a prodigy!

In reading over this passage I am surprised to see how little Russell really knew his Shapley to think that the latter would be satisfied to work in Russell's shadow while his senior (by only eight years) would collect the kudos! Shapley too had a good opinion of himself; and in his energy and administrative talents outstripped Russell by a very wide margin. The reason why Russell's plan did not work was the fact that it would not have been to Shapley's advantage. His lecturing and adminstrative talents were already then attracting national attention; and while President Lowell of Harvard was contemplating the situation, Shapley received (I had this from his wife) an offer of the Presidency of a large Mid-Western University which swung the scales; and Shapley became full Director at Harvard from October 1921, after a sedis-vacancy that lasted a little more than two and a half years. This does not mean that Shapley was not genuinely fond of Russell (though not uncritically so); but whenever it suited his purpose, Shapley 'outoperated' his senior hands down; and perhaps his crowning skill was to remain on terms of warm friendship with Russell for the rest of his life (how many of our own contemporaries would manage to do the same in our own times?).

One more story (which I have from Russell himself) should perhaps be mentioned in this connection. When the Directorship of Yale Observatory became vacant in 1920 the University offered the position to Russell; and only on his turning it down did it go to Frank Schlesinger (1871-1943), then Director of Allegheny Observatory, who held it with great distinction for the next 21 years. Some time after Schlesinger's appointment became official, Russell asked the then President of Yale University why they had bothered with him if they could give it to Schlesinger? 'Because we did not think we could get him', was the answer.

On the personal side, since the days of his youth (and certainly since his postdoctoral years) Russell was a dedicated anglophile and admirer of Edwardian England, of old-fashioned formality (down to his rolled-up umbrella, carried in rain or shine), but not, however, in his speech (generally lively and fast) which to the end of his life carried unmistakable traces of the Brooklyn accent acquired in his youth—it would never have passed for the King's English! Conservative to the bone, throughout his life he never weakened in his conviction that the American Revolution of 1775-1783 was the last and the only legitimate one in history (a view surprisingly at variance with his general anglophilia); all other revolutions were (according to Russell) the work of the devil bent on disturbing the established order of things. In line with his wholehearted acceptance of nineteenth-century British imperialism in other parts of the world, his favourite poet was Rudyard Kipling, whose poetry he could recite with relish till the end of his days;

and his particular *bête noire* in post-war England was Nye Bevan (whom Russell did not understand at all). In other aspects, however, Russell could exhibit again a degree of tolerance remarkable in the son of a Presbyterian minister. His wife, a Greek and a very charming spouse, belonged to the Greek Orthodox Church, which observes a liturgy very different from that of the Presbyterians. However, this incongruity was resolved in a very satisfactory way (no doubt pleasing God and the Russell family alike): each Sunday the husband and wife went to divine service alternately in each other's church.

Precocious from childhood, and pampered by his family (Russell never learned to drive a car), how could all this attention he received avoid spoiling rather completely his trusting soul? Certainly, in the second half of his life, Russell was disposed to accept the honorific epithet of 'astronomer laureate' of the United States as only his natural due. Had he, moreover, lived long enough to read the introduction to Fritz Zwicky's *Catalogue of Selected Compact Galaxies...* (1971), in which that cantankerous near-genius (of mind no doubt more original than Russell's, to whom we owe the concept of supernovae and many others) referred to Russell as the 'Pope of American Astronomy', it is doubtful whether Russell would have sensed the irony; or recalled the old Roman Catholic adage: 'He who enters the Conclave a Pope, will leave it a Cardinal'. Incidentally, one reason for the esteem in which Russell (or Hubble) was held by his contemporaries in the United States (he was much less well known abroad) was the fact that (like Hubble) he was a thoroughgoing WASP—White, Anglo-Saxon *and* Protestant. On this score, Zwicky failed on one point: he was not Anglo-Saxon but Swiss, and with a Czech mother into the bargain— what could you expect?

Domineering Russell could sometimes be, but never arrogant; and his old-fashioned courtesy endeared him to many (young or old alike), including the present writer who after so many years remembers him only with affection. The only feature of his character which (to put it mildly) puzzled me at times was the fact that, on occasion, Russell would say different things to one face-to-face, and behind one's back—especially when he may have thought that his own scientific interests (or, more often, those of his Princeton students) were threatened.†

As I write these lines, I have in front of me a handwritten note scribbled by Shapley on one of his returns from Princeton, saying (in connection with the iterative methods then under development) that

---

† At one time, I thought that I was the only victim of this idiosyncrasy, and only later learned that he acted the same way against Cecilia Payne (cf Haramundanis 1984) and S Chandrasekhar (1980)—and possibly others as well.

'... you have become so accustomed to H.N.R. that you might as well see this in the raw ... please remember that he [i.e. Russell] means it well, but that he has a complex about you and eclipsing binaries'; and from more than this particular instance I can only conclude that this was indeed the case.†

To young astronomers today Russell's name is perhaps most often recalled in connection with the Hertzsprung–Russell diagram. As is well known, the empirical form of this diagram (in which division of the stars between 'giants' and 'dwarfs' first became apparent) was constructed by Ejnar Hertzsprung (1873–1967), a great Danish astronomer of bygone days, in a paper published under the title 'Zur Strahlung der Sterne' in *Zeitschrift für Wissenschaftliche Photographie* (1906); but because that journal was not read by many astronomers, its contents remained largely unknown to them until Russell (1913) attempted its theoretical interpretation. It was this interpretation which caught the interest of the scientific public—so much so that the underlying diagram was referred to only under Russell's name (at least, when I came to Harvard, Miss Gladys Wickson, the Observatory's publications editor, left no room for doubt in my mind that this should be the case). In 1952, the American Astronomical Society convoked a Colloquium at which this diagram would be suitably commemorated (and the proceedings of which appeared in *The Astronomical Journal* volume 57).

Both Russell and Hertzsprung were invited to attend the occasion. Russell came; and his brief paper appeared at the head of the proceedings. Hertzsprung (Russell's senior by four years) did not; and (as I heard from the Secretary of the Society at that time) in a rather caustic message expressed the view that the only useful task the Colloquium could accomplish was to remove the names of both Hertzsprung and Russell from any connection with the diagram altogether. For, wrote Hertzsprung, by the beginning of this century the only characteristics of the stars known to us then were their luminosities and spectra; and to plot one against the other called for no particular ingenuity! He could have added that the physical interpretation of the diagram, proposed by Russell in 1913, was exactly the opposite of what it eventually turned out to be.‡ It was indeed a pity that, in 1913, it was not Shapley who kept Russell 'from too riotous an imagination— in print'. But if Hertzsprung thought so, he made no public comment

† *Si parva licet comparare magnis*, the same could perhaps be said of Eddington and young Chandrasekhar in the matter of the relativistic degeneracy formula in 1935 (cf Chandrasekhar 1980).

‡ In 1913 Russell thought that giant stars were contracting towards the main sequence; while since the advent of nuclear astrophysics in the 1940s we now know that the converse is the case.

to this effect; as an observer who kept his feet on the ground all his life, he probably had no high opinion of theoreticians.

At any rate, by the time I came to the United States in 1938, Russell was already past sixty years of age; and it soon became apparent to me that (at least as far as eclipsing variables were concerned) he was already living off his past; and concentrating on efforts to leave a record of this work for posterity in an organised form. This was true of his book on *The Masses of the Stars* (Russell and Moore 1940) which was, in effect, a monograph on dynamical parallaxes of visual binary systems (see Eddington 1940); or of the updated version of the Russell–Shapley work of 1912 on the light curves of eclipsing variables, issued belatedly in 1952 under the names of Russell and Merrill (and commented on in Chapter 6 of this book).

In these circumstances, it was only to be expected that such a disturber of the peace as myself, stumbling upon the realisation that Russell's work on eclipsing variables of 1912 vintage, far from being the end of the road, might be only the beginning of further intellectual adventures, was not very welcome at Princeton (to say the least); and my first paper on iterative methods for an analysis of the light curves of eclipsing binaries (Kopal 1941c) brought promptly a rejoinder by Russell (1942)—in which he could not, however, make any objection to the new approach beyond that my new methods were 'laborious'. This was, however, only a matter of definition. It was true that the application of new methods could scarcely be practised with a slide-rule—Russell's favourite computing device till the end of his life; but with a Marchant electrical calculator which could mutiply two ten-digit numbers in ten seconds of real time (my favourite computer at that time) there was no difficulty, even if the work per star might take several hours.

Unlike Russell, I saw no reason why the analyst should begrudge the need to spend at least as many hours on extracting the information contained in the observations as the observer had to spend at the telescope to obtain the underlying data. But in that idle discussion both Russell and I would have been astounded if someone told us in 1942 that the day was just around the corner when automatic computers would enable us to compute $10^8$ times more quickly and $10^8$ times more accurately than could Russell with his slide-rule; and that this vastly increased capability would be bound to revolutionise also the astronomy of close binary systems. And the day indeed came when the iterative methods were programmed for use on automatic computers (cf Huffer and Collins 1962, Jurkevich 1970, Linnell and Proctor 1970, 1971); but these all belong to the post-war years, some 20–30 years later.

It was not only work along these lines which occupied me before the war. On being appointed to the Harvard faculty in June 1940, I embarked on systematic investigation of the proximity effects in close binary systems (Kopal 1941a and 1942a,b), not only between minima, but also within eclipses. The last of these—containing as it did the first evaluation of the explicit forms of the associated $\alpha_n^m$ functions for odd values of $n$ in terms of elliptic functions—represented a *tour-de-force* which one can embark on only when one is young, and is probably my first permanent contribution to the subject which has not yet been superseded. All this work, I may add, met with Russell's enthusiastic approval and support; for in this field Russell had no axe to grind; it was only the *raison d'être* of Merrill's out-of-date tables needed in the elementary ('circular') case (which eventually appeared in 1952) that he felt impelled to defend (by means both fair and foul) with the tenacity worthy of a better cause (see pp 423–25).

By now this is all in the distant past; but in 1941–42 my entry into the field looked almost like a contest between David and Goliath, and it eventually ended up that way. Many years elapsed, however, before this happened; and when the problem at issue was translated from the time- to the frequency-domain (i.e., when the object of analysis ceased to be the light curve itself, but its Fourier transform) Russell was no longer with us to witness the sequel.

But we are getting far ahead of the subject matter to be covered in this chapter. Let us, therefore, finish this section with a story which concerns not the stars, but a future President of the United States. While at Cambridge in 1938, I followed (in the papers) the arrival in the United Kingdom of a new ambassador from the United States, Joseph P Kennedy by name, with a large family of boys and girls; and much was made of the fact that these were allowed to enjoy the blessings of British secondary education. The events of 1938–39 soon extinguished all traces of the Kennedys from my mind, until I heard (I believe) in 1941 that their son John (since some time back in the States, and then an undergraduate at Harvard) desired to meet me. John was already then breathing politics; and, partly as a result of experiences in pre-war London, was then preparing a book entitled *Why England Slept* (Kennedy 1940), describing the way in which Great Britain had been sliding into World War II during the Baldwin and Chamberlain eras. A part of it concerned, of course, the Munich crisis of 1938 (which young Kennedy observed from London). On his return to Harvard, it somehow came to his notice that there was a postdoc at the Observatory who had recently arrived from Czechoslovakia and could possibly supply him with some additional information.

At any rate, we arranged to meet in the Widener Library one day (I have no record of the date), and spent a couple of hours in discussion on this subject. How useful my information could have been to him I do not know; and we never saw each other again. What stuck in my mind, however, was the fact that this young and obviously gifted student kept addressing me (only a few years his senior) as 'Sir'. I hope this was not all that he learned during his school-years in England!

In 1949, when our youngest daughter Eva was born in Boston, we received a brief congratulatory note from 'Representative John F Kennedy' (sent no doubt by his office); but afterwards we could only watch at a distance John's meteoric rise in the political sky to the White House . . . and Dallas! In thinking about him—and I do so often—I cannot forget also the old Greek prophecy, 'whom the Gods love, dies young'. How true this turned out to be of him!

## The War Years: Broadcaster to Occupied Europe

As the work described in the preceding section began to unroll, the Second World War broke out in Europe in September 1939; and my involvement in the events of that time commenced almost immediately. Since March of that year Czechoslovakia had ceased to be an independent state. However, the Government of the United States never recognised the Nazi occupation; and diplomatic representatives appointed prior to March 1939 continued to be recognised as such. Since, however, they no longer received financial support from their home country (and the legal status of funds deposited abroad prior to Nazi occupation was not quite unequivocal), any activity which they could develop abroad was limited.

One of the main tasks of the Czechoslovak authorities in exile was, therefore, to establish connections with the home front, and inform people at home of political developments outside Nazi-controlled territories (in so far as they were not covered already by the information services of belligerent countries). By great good chance, at this juncture I was in a position to render some assistance to this cause through Dr Shapley who (in addition to many other duties) was also one of the trustees of the World-Wide Broadcasting Foundation in Boston, operating a short-wave radio transmitter W1XAL (shortly to be reclassified as WRUL) at frequencies of 6.04 and 11.9 MHz, with which I had already had a previous connection as organiser of weekly astronomical radio talks by Harvard Observatory staff.

The Czechoslovak Embassy in Washington (headed by Colonel Vladimír S Hurban, son of a well-known Slovak poet, Svetozar Hurban-Vajanský; whose wife was a cousin of the late General Milan Rastislav Štefánik, the Slovak astronomer-statesman of World War I), as well as Mr Karel Hudec, Czechoslovak Consul-General in New York (and, incidentally, uncle of my friend Vladimír Vand, who was mentioned on p. 67), were supporting our project, but could not buy any broadcasting time from the meagre funds then at their disposal. However, thanks to the backing which Dr Shapley gave us in the station's advisory board, we were able to obtain broadcasting time (almost an hour per week) on credit, provided that programmes and services were supplied free of charge.

And these could, of course, be supplied at no cost: the news to be read was compiled at the Consulate General in New York (mainly by refugee journalists from Prague who had found a haven in the United States), supplemented on occasions by more scholarly contributions from Professors J L Hromádka (Princeton) and O Odložilík (Columbia), which were read before the microphone in Boston by a trio of broadcasters consisting of Mr Josef Hanč (formerly Czechoslovak Consul in New York, now teaching at Fletcher School of Law and Diplomacy at Tufts College in the Boston area), Dr Ivan Getting (a young physicist with a fresh DPhil from Oxford, and now Junior Fellow at Harvard)—and myself. Hanč and I broadcast in Czech, while Ivan Getting (the youngest son of the former Czechoslovak Consul in Pittsburgh who, though American-born, spent some years after World War I in the land of his ancestors) read the news in Slovak.

And so it went for the better part of the next year, until the fall of France in the spring of 1940, when accelerating events provided sufficient funds for the Czechoslovak Government in Exile to place the broadcasting service on a more professional footing. Both Ivan and I were soon needed for other types of service; and so was Mr Hanč who returned to the Consulate in New York, eventually to become the Czechoslovak Minister in Washington. Retired some time after 1948, he returned home where by now he must have found his final resting place. Ivan Getting was to render distinguished service in the field of fire-control in the Radiation Laboratory at the Massachusetts Institute of Technology (where I was soon to join him); and after a short academic career in postwar years he embraced industrial research, first in the Raytheon Corporation in Boston, which he joined in 1951 as Vice-President for Research and Development. He left for California in 1958 to become President of the Aero-space Corporation in El Segundo, Los Angeles, a position from which he retired a few years ago to enjoy a well-earned rest.

My own fortunes since 1940 will be detailed in subsequent parts of this book. But before we pursue their story, the question can be raised: what was the role of our broadcasting activities between 1939 and 1940, and did anyone listen to them? During my infrequent post-war visits to Czechoslovakia between 1957 and 1975 I did not come across anyone there who did. This does not mean that our broadcasts were not listened to at that time; but the casualties on the home front during the war years were heavy; and the identity of broadcasters remained anonymous. It is true that, in the autumn of 1939, our home in Prague was visited twice by the German Gestapo in quest of incriminating evidence (and they got away with a considerable part of my pre-war correspondence in English, including the letters from Eddington, which they failed to return!). However, I do not think that these visits were connected with my broadcasting activities; for had they been aware of them, the results would have been more drastic. It is true that what we broadcast about the Germans across the Atlantic they would not have framed and put on the wall; but we were probably too small fry to merit more detailed investigation.

One more fact should, however, be mentioned in this connection and in this place: namely, that (at Dr Shapley's prompting and because of the world situation) my wife and I initiated steps at that time which eventually could lead to American citizenship. In 1938, I was admitted to the United States on a non-quota visa as an academic visitor (my wife as a visitor); and these had to be periodically renewed. Their change to regular immigration visas had to be done abroad; and in 1941 we underwent the necessary transformation at the American Consulate in Montreal, Canada. However, this could be done only on submission of the appropriate supporting documents: some were provided by the Czechoslovak Consulate in New York; but others had to come from the local authorities of the place of our American residence; and we were required to prove that we had no criminal record.

I applied for this from the Cambridge police, and sent the necessary fee; but no reponse was forthcoming for several days. When time was running out, I decided to call on the Cambridge Constabulary myself, and was promptly ushered to the office of the Chief—a short but tremendously heavy man, sitting on a swivel chair behind a battery of telephones and one typewriter, all of which he could reach if he turned his chair to the appropriate angle (I think it would have been difficult for him to get up). He asked me in a thick Irish brogue (virtually the entire police force in Greater Boston was then, and probably is still now, in Irish hands) what I wanted. I quickly gathered that my application had probably been misfiled somewhere, but this did not seem to matter.

Instead, the Chief looked me over, asked some perfunctory questions and without further ado began to type (with none too skilled fingers) the certificate which I needed, ending with a delightful phrase: ' ... he is neither tramp nor vagrant, and always bore a good reputation in our community.' Then he signed it, stamped it, and I was on my own—tacitly wondering how differently such matters are managed on both sides of the Atlantic. No lengthy correspondence and no search of files; simply a personal appearance was needed, a searching lookover, and you got what you deserved on the spot!

At any rate, all the necessary documents were soon in order, and early in May we repaired to Montreal in Canada, to change our visas from visitors' to immigrant ones—as the first step towards the American citizenship which we received seven years later.

## War Years at the Massachusetts Institute of Technology

In the meantime, however, war was inexorably approaching the American shores, and burst out suddenly, though not unexpectedly, on 8 December 1941 when the Japanese attacked Pearl Harbor; and until VJ Day on 15 August 1945, when the war in the Pacific ended with Japanese capitulation, the lives of all Americans were dominated by its twists and turns. I was fortunate enough that it did not interfere too much at first with my astronomical activities (except that, during the spring term of 1942, I was called upon to assist in teaching celestial navigation to large classes of future naval officers—a traditional wartime preoccupation of astronomers).

However, after the end of the academic year 1941/42 at Harvard, a more serious call for duty awaited me at the Massachusetts Institute of Technology, located about a mile from Harvard Yard on the shores of Charles River in Cambridge (see figure 4.12). The call was transmitted through Dr Shapley's office, which related that 'certain officers' would like to meet me at MIT that afternoon to discuss matters not fit for transmission by telephone. At the appointed time, I was directed to the 'Center of Analysis' of the Department of Electrical Engineering—so called because it possessed two differential analysers (one complete, the other (larger) in an advanced stage towards completion)—and there I was met by two naval officers from the US Naval Proving Ground in Dahlgren, Virginia and two colleagues of the MIT staff, and confronted with the following situation. A great expansion of the United States naval power required a parallel expansion in the construction of the firing tables for weapons employed by all fighting ships of the Navy, from the anti-aircraft guns to the 16 inch battleship rifles, many of which had not seen much previous use. They were being tested at Dahlgren; and the firing tables for their use were constructed in the past at the Proving Ground itself by men in uniform.

By now, however, this second task could no longer be carried out there because of personnel shortage; naval officers were needed at sea rather than behind computers on dry land. Therefore, a decision was taken to contract the work out, and MIT's Center of Analysis was chosen for this purpose, because differential analysers were available

192  *The American years*

there for the job, together with expert electrical engineers capable of making best use of them. However, what was lacking was mathematicians to use the data coming out of the computers to perform the actual work leading to the desired finished products. Would I be willing to undertake this job and build up a group of manpower requisite for this purpose?

**Figure 4.12** The Massachusetts Institute of Technology at Cambridge, Mass.

Needless to say, I was not sure; but my misgivings that I knew nothing about ballistics were brushed aside by a comment that no-one from Harvard Observatory was known to have failed on any job so far (obviously Dr Shapley, who was behind the scenes in all this, was more confident of my abilities than I was myself at that time); and if I knew nothing about ballistics, I could learn it as I went along. And so it came to pass: before I left MIT that afternoon, I was 'signed up for the duration' and asked to report in a week. By a special favour, I was given ten days' vacation (which I spent with my family at Nantucket) 'to learn ballistics' which I set out, indeed, to do.

It was our second vacation on that lovely island in two years; a year before, we had been there, in the hospitable house of Mr and Mrs Chapel on Union Street, when Germany attacked Russia; and from day to day we followed by radio the initial rapid advances of the German armies into the Ukraine and White Russia, which reached almost to Moscow. A year later the situation was even grimmer: the German armies were advancing towards the Volga and the Caucasus, and did not come to a halt till Stalingrad. The fate of the world still hung in the balance; and no-one needed to be shaken out of complacency!

On my return, the work commenced in earnest; and the first priority was to assemble the staff. My first collaborator became Dr Dorrit Hoffleit, likewise from Harvard Observatory, but half a year later she left for the US Army Proving Ground at Aberdeen on a similar assignment (after all, the Army and the Navy were war-time allies; although at times it was hard to believe that this was the case). However, in the meantime several other astronomers from Harvard Observatory came to join our staff—such as Dr Luigi Jacchia (my old colleague from AFOEV, who had left his native Italy and come to Harvard in 1939 when Mussolini turned anti-Semitic), Mrs Martha Shapley or Miss Virginia Brenton, who all remained with me at MIT until I left for Manchester in 1951; and so did Mrs Mary Howe Baker, a Vassar graduate with some postgraduate experience at Lick Observatory, who was married to Professor R H Baker of the University of Illinois (author of a well-known textbook on elementary astronomy). The fact that she was the daughter of the Louis McHenry Howe who until his death in 1936 was President Roosevelt's close confidant and political collaborator made his daughter a virtual member of the Roosevelt family and *habituée* of the White House throughout President Roosevelt's terms of office. Later in the war, our group at MIT was joined by Professor Richard Prager of Harvard Observatory, who, as a pupil once of Professor Julius Bauschinger, remained the last protagonist among us of logarithms versus machine computations. Other colleagues on our staff were Dr Alice Farnsworth, Professor of Astronomy at Mount Holyoke College (now part of the University of Massachusetts), and Dr Marjorie Williams of Smith College (and later of the Astronomy Division of the National Science Foundation in Washington).

These astronomers were, of course, joined in time by others, of whom two in particular remain in my mind since those days: Miss Jeannie R B Carmichael, a patriotic Scottish Highlander who retained the endearing characteristics of her native race even after many years in the United States; and Miss Katherine E Kavanagh (later to become Mrs Hanson), who for most of my years at MIT was my right-hand man (if this is the proper expression to apply to a girl who was as charming as she was talented) and teaching assistant on my courses on numerical analysis which I started giving there in post-war years.

There were, of course, many others of the younger generation whom I remember fondly to this day. At the peak of our activities between 1943 and 1945 our group (numbering over fifty members of different degrees of seniority) worked in two (and, at times, three) shifts around the clock. As I had to keep in touch with them all, my own hours were irregular; but this allowed me also to continue spending some time each week at Harvard Observatory (where I also continued to maintain an office). Whenever possible, however, I joined the staff for afternoon tea (introduced to our group as a social ritual by Miss Carmichael) or dinner—in the office, of course—at which the progress of work was discussed by all those present; and sometimes not only war-work, but also astronomy. Professor Prager, our senior, refreshed our connections with European astronomy north of the Alps before World War I when he was a student; while Luigi Jacchia with his Latin temperament presented the Italian point of view.

Mrs Shapley, of reserved nature, seldom said much; but whatever she did say was worth listening to. And so it often happened that, when I left the group later at night for home or the Observatory (before midnight by tram, as there was little petrol for driving cars, and after midnight often on foot!) a telephone would ring to bring me back to face some unexpected difficulty or snarl in schedules—and the distance to my home (at the end of Brattle Street in Cambridge) or the observatory was about three miles each way! Looking back at it after more than forty years it must have been a strenuous life; but I was then in the prime of my physical powers; and was not only able to do justice to the job, but (as the seat of the latter was within walking distance of Harvard Observatory) also to keep up, albeit at a reduced rate, my research activities unconnected with war-work.

To return to this work, I had to learn ballistics in parallel with the actual performance of the tasks immediately on hand. My immediate predecessor in charge of our job (then being done at Dahlgren) was (then) Commander William J Parsons, USN, whom I met only fleetingly at that time; as he was being recalled to sea. By 1945, as Captain Parsons, he made history as bombardier aboard *Enola Gay*, the plane that dropped the first nuclear bomb on Hiroshima. In post-war years he ended his career as Vice-Admiral in charge of the Naval Ordnance Laboratory in Silver Spring, Maryland; and I last met him at the symposium on the occasion of the opening of that Laboratory in June 1950, at which I was invited to give a paper (Kopal 1950).

Parsons was a career officer, whose knowledge of ballistics went back to his training at Annapolis in the 1920s. I had their textbooks and did not find in them very much of scientific interest; nor did I find much more in other books which came from a military pen.† Indeed, very

---

† Thus one compendium on ballistics which I located at the Widener Library of Harvard, written by a senior Italian officer, had a preface contributed by the editors which commenced with the words: 'General X. Y., having been placed on the retired list, became interested in ballistics . . .' This encouraged me to hope that what General X. Y. could do in his retirement, I was not too old to do myself, even though I was not a general, nor a descendant of Leonardo!

little work in this field seems to have been pursued between the two wars; and, by 1942, the state of the subject was still the same as in 1918. During World War I, the leading expert on exterior ballistics in the United States was also an astronomer, Professor Forrest Ray Moulton (1872–1952) of the University of Chicago, who, activated as a Lieutenant-Colonel of the US Army at Aberdeen in 1917 (the Navy had no separate ballistic establishment at that time), systematised methods to be used for the preparation of the firing tables for the artillery; and being a prolific writer of books, he summarised these in a slender post-war volume entitled *New Methods in Exterior Ballistics*, published by the University of Chicago Press in 1926. This was a book after my own heart; and so were numerous reports by Moulton and his associates, preserved at Aberdeen, which were made available to me in 1942.

In perusing all this literature I noted that the principal physical contribution Moulton made to the construction of firing tables during World War I was to take account, in computations of the trajectories traversed by projectiles of different calibre, of the fact that the density of air (and, therefore, the resistance which air offers to the motion of such shells) varies with altitude. In the generations preceding World War I this density was assumed to be constant (and the drag which our atmosphere exerts over the flight path of travelling projectiles depended only on their velocity). Moulton did away with this simplification; and in doing so improved the quality of his tables considerably.

In returning to these problems in 1943, I noted another phenomenon which neither Moulton nor anyone else had taken into account before: the effect of temperature on the drag exerted by our atmosphere. In more specific terms, it is well known that the resistance offered by the atmosphere (any gas) increases dramatically when the velocity of the projectile exceeds that of sound in the respective medium; and this latter speed is proportional to the (square root of) the absolute temperature. Accordingly, a flying shell will traverse the sonic barrier, and experience the corresponding change in drag, at different altitudes (depending on the temperature profile of the respective region), and this is bound to influence its subsequent path to the target.

That this should indeed be so was self-evident to a physicist but, unfortunately, not so evident to officers living by the book. When I mentioned this to our liaison officer at Dahlgren, Commander (later Captain) C C Bramble, I was encouraged to prepare a report on the situation, to be considered by appropriate committees of the naval Bureau of Ordnance. The report was duly submitted; but a response was slow to come—months, indeed years, went by without any action. To my inquiries Captain Bramble advised patience; for superior officers were no doubt preoccupied with more important issues (which was indeed true). In the meantime, the tables coming out of our hands were apparently doing good service (we got a special citation for their performance in the invasion of Sicily).

I recall a rather testy response by a very senior officer to my inquiry about the fate of our proposal, saying 'I wonder why you are so insistent, doctor, on bringing about such innovations? Aren't your present tables doing well enough?' Of course (as I heard one more junior officer say) an admiral is a man who is convinced that nothing should ever be done for the first time (one must attain five-star rank to overcome this handicap)—and it would probably never have occurred to him that we had already adopted the proposed innovations for many months off our own bat (my colleagues at MIT agreed with my strategy, even though we had to keep it a 'war secret'). At long last, shortly before the end of the war in the Pacific, the official wheels bestirred themselves to approve of our proposals; thus making me (albeit *in extremis*) an 'honest man'; and I am half convinced that they did so largely because the war was almost won, so that it would not matter what we did in the future! In this they were almost certainly right; for, after forty years of technological advances, the ordnance of the future will require a very different mathematical back-up.

And what is true of ordnance will be even more true of the computers servicing it. World War II came just too soon to take more advantage of the impending computer revolution. The differential analysers of MIT's Center of Analysis did not prove to be really adequate for our needs; the Bush mechanical analyser was too inaccurate for our purposes to begin with; and the Rockefeller electronic analyser (RDA 2) was only somewhat more accurate, and too slow. In fact, although we did not know it, the development of analogue devices for large-scale computations was running into a blank wall of inefficiency because of the type of analogues used for this purpose; and is likely to remain there until the introduction of other (probably optical) analogues (using photons rather than electrons) for this purpose in the future. Instead, the advent of relay-type computers (Mark I and its followers) at Harvard in 1944, and of electronic computers designed by von Neumann and his associates at Princeton (ENIAC) at about the same time, had cleared the way for further developments: the relay-type computers proved very quickly to represent another sterile offshoot of the mainstream of automatic computation, dominated firmly since the 1950s by electronic computers of increasing speed and memory store, but diminishing in size.

Yet in the days when we were engaged in preparing the firing tables, differential analysers were still the main showpiece of MIT's Center of Analysis shown to visiting VIPs. One of these (I cannot remember the exact date, but it was in 1944) was the Duke of Windsor—formerly King Edward VIII, and now Governor of the Bahamas—who visited

Boston on some business; and while his wife went shopping in Boston (or, perhaps, New York) he was (among others) shown around MIT. Meeting him at close range, I was surprised how really short he was (an impression not always disclosed by his official photographs; though lending verisimilitude to the way in which his lady-friends of pre-war years referred to him as 'the little man'). What he saw went, of course, wholly over his head (already his Oxford masters had decided that he was 'not bookish'). On the whole, he must have had a rather dull time with us—a fact which he did not try too much to conceal.

But, along more serious lines, with our main line of business well in hand, we received official encouragement to extend our interests also to theoretical work connected with the supersonic motion of artillery shells through air, and to investigate the resistance which the atmosphere offers to their flight. In order to strengthen our ability to pursue such work, our group was assigned as mathematical consultant no-one less than Norbert Wiener (1894–1964), Professor of Mathematics at MIT, master of Fourier transforms in the complex domain, and founder of modern cybernetics. By that time, Wiener was already engaged in giving expert advice to several defence projects at MIT, and he agreed to collaborate also with us to this end till the end of the war.

This is perhaps not the place to give a more detailed account of the personality of this outstanding mathematician, remarkable for his originality as well as the breadth of his scientific interests. Although he has been dead now for more than twenty years, his colourful personality remains very much alive for those who knew him. I was especially privileged in this respect; for his father (Leo Wiener) used to be Professor of Slavic Languages at Harvard University (it was he who equipped Harvard's Widener Library with a first-class collection of Czech books, which during our Harvard years remained a link with our home country). Professor Leo Wiener was a good acquaintance of President Masaryk of pre-war Czechoslovakia (Norbert told me that he and his father were more than once Masaryk's guests at Lány Castle); and although father Leo was no mathematician, Norbert inherited at least some of his father's linguistic talents; and several languages which he mastered included Chinese!

As a mathematician, Wiener was brilliantly original in subjects in which he happened to be interested; but to get anything useful out of his cooperation meant awakening his interest in our problems (otherwise he was apt to fall asleep in the middle of the day!†); and a part of my job was to keep that interest awake! As a teacher or expositor,

† I was told, however, by those who knew Wiener well that, under such conditions, the surest way to bring him out of his slumber was to whisper the (to him) magic words: 'Tauberian theorems'.

Wiener was often almost impossible: his usual pedagogical procedure was to cover the blackboard (not too legibly, and often without a word) with mathematical formulae which he then started to explain backwards from the end (he often found, in this process, that his notations were all mixed up!). I was also told by his students that, on occasion, he managed to walk into somebody else's classroom and begin to lecture on a subject then occupying his mind, but without anybody having any idea of what it was all about. This may have lasted sometimes for half an hour (with the regular lecturer sitting on the benches, and trying not to interrupt the performance), until Wiener realised that something was amiss; then without a word he would walk out on his short legs from the classroom without leaving anybody any wiser. However, in the course of a discussion which was of interest to him, Wiener could rapidly make a number of stimulating suggestions; but their elaboration he left to others.

When in good humour, however, Wiener was a delightful companion; but not safe to ride with in his car: for sometimes the ride would be stopped by a street-lamp or some other solid body which refused to give way. Fortunately for his fellow passengers, Wiener never drove too fast; and his inelastic collisions with solid obstacles released at times such interesting comments on his part that merely to listen to them was sufficient compensation for the risk.

The number of stories about Norbert Wiener in his lifetime (at least a half of which were probably self-made) are too many to be chronicled here; but one at least (of which I was the witness) should illustrate his way of life. It occurred on 19 March 1945 at a conference on supersonic hydrodynamics, held at the Institute for Advanced Study in Princeton, to which Wiener and I were sent by the Navy to represent our project at MIT. The introductory lecture to the subject was being given at a reasonable time of the morning by Professor von Neumann in his customary brilliant style; while Norbert Wiener sat in the audience in the front row (he would never settle for any other position).

But no sooner had the first slides appeared on the screen and the lights gone out than Wiener fell asleep—a fact which rapidly became evident to the audience through sonic effects which were bound to reach the speaker also. When the lights came on, Wiener awoke; and still blinking his myopic eyes he got up on his legs to ask the speaker: 'Pardon me, Sir; could you tell me what you have been talking about?' Von Neumann (a good friend of Wiener's, who knew what to expect), did not blink an eyelid; and with a straight face gave a brief succinct summary of what he had said so far. Thereupon Norbert Wiener said 'Thank you', sat back in his seat—and resumed his slumber! As to the significance of such extempores (whether the slumber was genuine, or only calculated to attract attention) opinions continue to differ; but there remained a good deal of the child in Norbert Wiener throughout his life. 'Except ye become as little children, ye shall in no wise enter the Kingdom of Heaven', stands in the New Testament; and so who would doubt it? But for those of us who still inhabit this Earth, cooperation on an intellectual level was not as easy with him as it could have been; and not many concrete results emerged from our collaboration which would have pleased the Navy. I understood, however, that in other fields (such as fire-control) Wiener's cooperation with others led to more satisfactory results. But all this time I could not help but wonder: were all great mathematicians of the past similarly eccentric? I doubt it.

But to return to the more mundane aspects of our work during those years, as a result of our growing activities it eventually became my lot to travel twice a month to Dahlgren to maintain a closer personal liaison with the Naval Proving Ground there. There I made the acquaintance of a number of outstanding officers (partly career, partly reserve), some of whom survived the war, but others did not. Of the former, two names in particular still stand out in my mind after a lapse of more than forty years: those of Dr (then Commander, USNR) Norris Bradbury, who in due course succeeded J Robert Oppenheimer in the Directorship of Los Alamos; and of Dr (then Captain, USNR) C C Bramble, our liaison officer, who in civilian life was Professor and Head of the Department of Mathematics at the postgraduate school of the US Naval Academy in Annapolis (subsequently removed to Monterey, California). In this capacity, when the war ended in the summer of 1945, Professor Bramble offered me a professorship of mathematics in his department at Annapolis. It was the first such offer I received in post-war years; and whether or not the Navy lost anything when I did not take it up is a moot question. During my last correspondence with Dr Bramble, he was no longer at the Academy in California himself, but chose an industrial career in the east.

But the commuting between Boston and the Naval Proving Ground (since re-named the Naval Weapons Research Laboratory) had, on occasion, its advantages also—such as at the time when, while at Dahlgren, I received a telephone invitation to tea with President Roosevelt and his family at the White House. I am sure it was not any outstanding reputation which earned me this undeserved attention, but rather a kind gesture towards a foreign national (since naturalisations had been stopped during war time I was still a Czechoslovak citizen at that time) participating in a joint war effort. Needless to say, I did so as a civilian; for the position I occupied was normally held by an officer of fairly high rank; but not yet being an American citizen,

I could not formally receive a commission at that time. This had, however, some advantages: for it made it easier to cut across red tape, in the interests of the work, without risking disciplinary action.

It was in November 1944 (the exact day I can no longer recall) that, towards the dusk of a cold autumn day and wearing a heavy overcoat, I approached the White House on foot (taxis were almost unobtainable in the streets of Washington in war time). The guards at the outer gate asked for my name; and when they located it in their records, they directed me to the entrance. I was not in any undue rush, for (according to my European instincts) I expected that the higher the rank of the dignitary one was going to see, the longer one had to wait to be admitted to the presence, and I began to compose a few appropriate remarks. But it was not to be! For no sooner had I approached the main entrance to the White House than its doors quickly opened in front of me, and a coloured butler led me without a word to the elevator which noiselessly took me to the second floor, and opened on the living room where the President's family and some friends (among whom I immediately recognised our Mrs Howe-Baker) were already having tea. The President was in his wheel-chair and playfully waved his hand at me: hello stranger!

But in so congenial a company I did not feel like one. The President asked about Czechoslovakia; and after he left (for a press conference) Mrs Roosevelt asked more. Who else from her family was present I can no longer remember; but I recall that their son Elliott (then a Brigadier-General in the US Air Force) was to return from Europe the next day. Nothing much was said about world events otherwise; and afterwards Mrs Baker took me around the second floor of the White House, and (in line with our current naval preoccupations) showed me many rare prints of historical vessels which were the President's personal property, and of which he was very fond; and (I recall) I was also shown the bed on which President Lincoln died on 14 April 1865.

After the visit was over, I remember walking to the Lincoln Memorial (which was again floodlit as in the years before the war), and down the Mall towards Union Station, from which I was to return by 'Federal' express home to Boston the same night. And while I wandered through the streets of Washington I wondered about many things. First, it was a cold day and I was wearing a heavy overcoat; but before I entered the White House to see the President, nobody took the trouble to ascertain whether I carried in its pockets any weapon with which I could do him harm; or, for that matter, if I was actually the person whom they expected. I was not asked by the guards to produce any badge of identification (I had one in my pocket, but nobody asked to see it). Could this have happened in any other capital of a nation at war? And to walk from the White House to the Union

Station at night today would be a very foolhardy act (with the chances of reaching the destination unharmed considerably less than one-half); while in those days all streets in the District of Columbia were still quite safe even after dark.

This was, incidentally, the first but not the last social contact between the Kopal and Roosevelt families. The second took place some fifteen years later, when our eldest daughter Georgiana (born on 30 August 1940 in Boston) was an undergraduate at Vassar College. It was Mrs Roosevelt's custom to invite foreign students from Vassar for tea at Hyde Park; and (since we by then lived in England) my daughter qualified to be among them. Encouraged by Mrs Roosevelt's kindness, Georgiana mentioned that her father was once the President's guest at the White House. Mrs Roosevelt said that of course she remembered him; although I am sure it was only diplomatic kindness; she was by then already well on in years (and, in writing these recollections, I know from personal experience how difficult it is accurately to remember events fifteen years in the past).

Chance willed it that I had an opportunity to see all of the 'Big Three' of World War II—Churchill and Stalin, as well as Roosevelt. Winston Churchill I met five years later, still in the United States, when he accepted an invitation from Dr Karl T Compton, President of MIT, to attend the celebrations of the Institute's 75th anniversary in 1949, and to deliver a speech on that occasion. Churchill (then the Leader of the Opposition) agreed to do so, and accepted also an appointment as an 'honorary lecturer' of MIT (by its Charter, the Institute is not permitted to award honorary degrees). What was, therefore, more natural than that the faculty should give a dinner for their newest colleague (actually oldest by age) before his evening lecture?

I was invited to attend this dinner as well (see figure 4.13); and although I did not speak with Churchill on that occasion, I had a chance to observe him at close range. Like the Duke of Windsor, Churchill in life was very much shorter than most of his publicity photographs made him out to be (especially as he did not hold himself particularly erect). In his 75 years he actually looked older than he was; but with a big cigar in his hand he smiled benignly, like a man without a care in the world. In contrast, Mrs Churchill at his side—a good deal taller (as well as younger) than her husband—seemed to watch each of his steps or motions with some apprehension; the readers of Lord Moran's book on Churchill (*The Struggle for Survival*, which appeared in 1966) will readily understand why. I do not remember the exact day in the spring of 1949 when the occasion I recall was held; but it was Mrs Churchill's birthday; she was presented with a cake (whose candles no-one counted), and I joined in singing 'Happy birthday to you' as did everyone else in the assembly faculty, whether they could carry a tune or not.

After dinner, the Churchill oration was delivered downtown in Boston in an arena that could accommodate several thousand listeners (he shared the platform with Harold Stassen, a Republican presidential hopeful of the day) with few seats remaining vacant. Neither of the speakers (Stassen or Churchill) said anything worth remembering at this time (it was a time when the hallmark of politicians addressing academics was to proclaim trivialities with distinction); though when Churchill got going, it was still the same voice we used to hear in 1940–41 when the fate of the world still hung in the balance; and the audience gave him a tremendous ovation.

**Figure 4.13** The author as associate professor of MIT, at the time of the 75th anniversary of the Institute's foundation in the spring of 1949.

I had as my guest on that occasion Dr (now Professor Emerita) Wilhelmina Ivanowska of the University of Torun (where Copernicus was born) in Poland, a former Vice-President of the IAU, who was visiting the United States at that time; and whom I had known since the 1938 IAU meeting in Stockholm. She was not at all impressed by Churchill's oratory; and when asked why, she responded: 'he sold us [Poland] down the river in 1945 as his predecessor Chamberlain did you [Czechoslovakia] in 1938'. There was some substance to this: but the real causes cannot be attached to individuals. In 1945, the British Government (with or without Churchill) was simply powerless to affect the course of events in Eastern Europe to any appreciable extent, and were forced by events to dishonour obligations assumed some years before—just as the French were forced by similar circumstances in 1938 to dishonour their obligations to Czechoslovakia at Munich. When it came to the crunch, they could not have done otherwise; because long before they had allowed themselves to be out-classed in quest of power by national units of different orders of magnitude; and history made the sequel inevitable.

However, the legitimate grudge which the Poles can hold against Churchill went back to the fact that, cornered as he was by Stalin (and Roosevelt!) at Yalta, Churchill chose to dress necessity in the garb of virtue. In order to uphold the fiction of a Great Power (which Britain no longer was at that time), he preferred to jump on Stalin's bandwagon ('if you can't stop them, join them')—a tactic which certainly did no one any good! The real reason was, of course, Britain's physical inability to do anything about it, and thus honour the pledges made to Poland in good faith at the beginning of the war, and accepted as such by the Poles.

That the latter did so, and accepted such guarantees at face value showed a lack of historical perspective for which the Poles have only themselves to blame—just like the Czechs, who allowed themselves to be lulled into a false sense of security by the French guarantees before the time of Munich. Nevertheless, it is a fact that the offer of such guarantees by the British Government (Chamberlain's, not Churchill's; though the latter could not have agreed more with Chamberlain on this score) at that critical time did encourage the pre-war Polish Government to take steps which proved near-fatal to their country—a poor consolation for those who had to bear the consequences, and may have to bear them who knows how long in the future!

As regards the third of the 'Big Three' of World War II, Josef Vissarionovich Stalin of the Soviet Union, I never could share with him my misgivings on the state of the world because I never met him in person (it would indeed have been highly risky to have done so). But they were very much in my mind as I saw him in Moscow at the time of the Tenth General Assembly of the IAU in August 1958, in the Lenin Mausoleum on Red Square—be-medalled and all, in his marshal's uniform at the side of Lenin, shortly before his mortal remains were removed to a less conspicuous place.

## Post-War Years in Cambridge

If anyone thought in 1945 that the termination of hostilities would bring things 'back to normal', he was of course mistaken; for a great many changes brought about by the six years of war proved to be irreversible; and a quest for some kind of equilibrium has been going on ever since.

Science was no exception. Certain spectacular successes achieved by men of science in the course of the war, such as the explosive release of nuclear energy on a macroscopic scale, which brought the war to its dramatic end, filled their fellow-citizens with awe (and fear); and made sure that a large fraction of scientists were smoked out of their pre-war ivory towers, not to return there to this day.

One (perhaps the main) reason was the fact that (partly as a result of post-war inflation) science had become an expensive preoccupation, which meant that few schools and universities could be adequately maintained from their own means; and in looking around for additional sources of support, they saw only one such source sufficient to meet their needs: the Federal Government—and, more specifically, its Armed Forces, soon to be united in one department that, in times of peace, quickly established itself as the principal dispenser of goverment largesse for science. It is true that this represented a temporary state of affairs; the Atomic Energy Commission soon joined the Department of Defense in providing federal support on a massive scale for one particular branch of physics; while the National Science Foundation did so somewhat later for many other branches of science (including astronomy).

However, all this happened beyond the period covered in this chapter. The National Science Foundation came into being in the United States after 1950; and it was not till towards the end of that decade that it became an important part of the picture, joined as it then was by the National Aeronautics and Space Administration which burgeoned after 1958 in the wake of the first successful Soviet *sputniks*, to land men on the Moon in little more than ten years, and to accomplish many other spectacular feats which should remain forever the pride of humanity on our planet. But by the time these became accomplished facts I had already been in Britain for several years; and although this did not prevent me from being connected with them and playing a certain role in these undertakings, an account of this belongs to a different part of this book.

As the war in the Pacific came to its end in the summer of 1945, some of the activities of my group at MIT (such as the construction of the naval firing tables) came naturally to an end; but not so other parts of our work. In particular, the Navy expressed continued interest in the hydrodynamical aspects of supersonic motions in gas; and as some of the greatest examples of such motions are observed in the sky, this was in line with our astronomical interests as well. As we accepted the naval support offered through MIT's Division for Industrial Cooperation, the staff of my group (eventually separated from the

Center of Analysis) far from shrinking actually kept expanding in post-war years (with all the old-timers, including Mrs Shapley, remaining in their places) until my departure in 1951 from MIT for Manchester; and the circumstances when that happened will be the subject of the next chapter.

However, one sad event should be recorded in this place: two months after the end of the war with Germany, our group lost its senior member when Professor Richard Prager passed away on 29 July 1945 after a brief illness. He remained on active duty till almost the end—I remember how proud he had been to become an American citizen a few months before—and in the last days of his life he had only one wish: to hear if his son Alfred (whom he had left behind in Germany in 1939) had survived the war.† Prager was already in the hospital from which he was not to return when the first news came from Alfred in Germany inquiring about his father. It was my pleasant duty to be able to deliver this news to Dr Prager in hospital only days before his passing; and I have never seen a happier man in my life.

The roots of our post-war work went back, however, to the war years and, to be more precise, to the summer of 1944, when the German V2's made their first appearance in the upper atmosphere (as harbingers of greater things to come); and with 'star wars' only forty years in the future, it suddenly became of operational importance to learn more about the structure of the upper atmosphere of the Earth. At that time, the principal class of bodies moving through these regions (at hypersonic speeds) was the meteorites—much studied photographically at Harvard Observatory for several years past. On the basis of these data, work on the subject began in earnest under the leadership of Dr Jacchia, who carried it on since the 1950s at the Smithsonian Astrophysical Observatory and became a real expert in it.

My own theoretical work was, however, increasingly directed towards supersonic hydrodynamics of the troposphere; and its first outcome was *Tables of Supersonic Flow around Cones*, published by the MIT Press in 1947 as a volume of xviii + 555 pages; followed by *Tables of Supersonic Flow around Yawing Cones* (xviii + 312; 1947) and *Tables of Supersonic Flow around Cones of Large Yaw* (xviii + 125; 1949)—the latter two in collaboration with Dr A H Stone of Cambridge, England and later of the University of Rochester in the US. These tables (reprinted in the 1960s by Edward Bros at Ann Arbor, Michigan) enjoyed for a time a fairly widespread use.

---

† Alfred's mother Käthe was an Aryan; and to save their son's life, the parents agreed to declare him illegitimate. The trick worked; and although Alfred had to spend several years as a German soldier on the eastern front as a result, he returned home alive from its holocaust.

Their users were, to be sure, mainly non-astronomers; and this led sometimes to amusing mis-identifications. Thus, when I was interviewed in 1950 for the Manchester Chair of Astronomy, the junior member of the committee, Professor M J (later Sir James) Lighthill was surprised that I was an astronomer; for from his knowledge of these tables he had taken me for an aerodynamicist. But I recall an even better story when, in the summer of 1960, I visited the NASA Ames Research Center at Moffet Field in California, to give a colloquium on planetary atmospheres. In taking leave of me on my departure, Dr Harvey Allen (then deputy director of the Center) added: 'and please give our regards also to your father.' I tacitly wondered how anyone at Ames could have known of the aging romance philologist in Prague, who had never visited America; but Dr Allen quickly supplied the clue: 'and tell him that his *Supersonic Tables* are still very much in use in these parts.' How could I respond otherwise than by saying that I should be glad to deliver the message, and that father would no doubt be very pleased (which, in due course, he was!)?

But this story again anticipates events by many years. By 1946 these tables were just off the press; but other work (still partly classified, so that it cannot be described in this book) made it necessary to commute between Cambridge and Princeton, Washington, the Naval Proving Ground at Dahlgren and the US Army Proving Ground at Aberdeen; and to meet many interesting people in the discharge of my duties: particularly in Princeton, where I had occasion to meet Professor John von Neumann, and of course Professor Henry Norris Russell at the Observatory. These were the last years of Russell's active service to his university, and during that time he used to visit Cambridge with increasing frequency. At Princeton, I also had an opportunity to meet Albert Einstein (1879–1955) in the company of his friend Philipp Frank (1884–1966) whom I knew well from Prague University (and later at Harvard); and a story of that visit is perhaps worthwhile recounting even after so many years.

Before doing so, let us mention that the first full professorship which Einstein held since March 1911 was in Prague; and although he did not stay there long (having gone back to Zürich in the autumn of 1912) the year and a half spent in Prague was the time when Einstein was close to the summit of his intellectual achievements: for it was then that he conceived the general theory of relativity; and his prediction of the deflection of light in the gravitational field of the Sun—one of the first phenomena by which the validity of the field equations of general relativity could be tested by observations—was published in the *Annalen der Physik* **35** 898 (1911) from his modest institute in Viničná ulice in Prague (which 25 years later became the academic home of Professor Freundlich).

When I met Einstein with Frank in Princeton, I ventured to ask him if there was any significance in the fact that his greatest intellectual achievement (i.e. general relativity) germinated in his mind during his stay in Prague. Einstein thought a little while about the answer; and then he said slowly: 'Ja' (the conversation was in German), 'in those years I was free from any administration.' Nor, we may add, was he yet the target of the 'publicity media'—no daily press took note of his doings, and television was still many years in the future. The only 'extracurricular' activity which Einstein pursued in his Prague days (as far as I heard) was music, for which that city has been renowned since the days of Mozart.

However, when we left Einstein's home on Mercer Street, Philipp Frank told me: 'You see, the old gentleman is obviously losing his memory. I shall tell you how it was. When Einstein resigned his chair in Prague, he recommended me as his successor; and his recommendation was, in due course, accepted. When I received my appointment from the hands of the Dean and confessed to misgivings on my ability to succeed so great a man, the Dean reassured me with (approximately) these words: "We do not require you to be exactly a genius; all I really want you to do is to answer at least the official correspondence from the Dean's office"'—which, by implication, Einstein had not! The identity of this good-natured Dean Frank could no longer remember; but whoever he was, he deserved well of posterity for letting Einstein get away with it! As is well known, this was not to be his fate too often in later years of his life. At Princeton, Einstein's principal value was one of publicity; but no dean in the United States worthy of his salt would have entrusted him with the administration of any money!

The only other unrecorded recollection of Einstein I can think of is second-hand, and dates back to his early years after World War I in Berlin, where he was brought from Zürich by Max Planck. At that time he was assigned by the Academy a young assistant, Dr Erwin Freundlich, of whom we have already heard in Chapter 3, and who became my professor at Prague between 1937 and 1938. Freundlich told me once that on one of his early appointments with Einstein he was delayed for some reason, and could not reach the agreed rendezvous till about half an hour after the appointed time. The place was in the open, and it rained. When, however, Freundlich started to present his apologies, Einstein waved them good-naturedly aside. 'After all', he was reported to have said 'I was not losing any time. I could do my work standing here—under the umbrella—just as well as if I were sitting in the office behind a desk.' And who would doubt it of one whose main mission in life was to think?

But the military (or, rather, naval) preoccupations connected with the Second World War and its aftermath, touched upon in the preceding section, did not claim all my time even while the war was still in progress; and certainly not in post-war years. The fact that my base of operations during this time remained in Cambridge made it possible for me to retain contact with the Harvard Observatory, and continue (albeit at a reduced rate) my pre-war work on eclipsing variables.

The first item of my programme as envisaged in pre-war years—namely, the completion of a technical monograph summarising the current state of research in this field—was first on the agenda; and its first draft was essentially completed during the war (in the autumn of 1944). It was intended for publication as a *Harvard Observatory Monograph* (of which five had appeared since 1930); but on completion of the manuscript the project began to run into difficulties (unexpected by me) which had their origin in Princeton with Professor Russell.

In those years, Russell was (among others) also a member of the Visiting Committee of Harvard Observatory, and Shapley considered it prudent to show him the text of my book in advance of publication. Russell's reaction was immediate and vehement. He had no objection to the inclusion of more advanced parts of the theory of photometric 'perturbations' arising from the distortion of the components of close eclipsing systems by rotation or tides. Of these Russell knew little, and cared even less. However, as far as an analysis of the light changes of eclipsing systems with spherical components was concerned, Russell firmly adhered to the view that nothing remained to be added to his work of 1912 (the tables for which were then being upgraded by John Merrill); and the mere inclusion of alternative methods, such as of the iterative approach which I had advanced in 1941 and developed in my text, he took almost as a personal affront.

Naturally I could not agree to such an authoritarian edict; and this created an impasse which it took Shapley more than a whole year to resolve. I still remember today how, one day, he came to see me in my office back from a meeting with Russell at Princeton and mentioned that he thought Russell could be induced to withdraw his objections if I were to dedicate my book to him! Naturally I did not object—I considered it, in fact, as a joke (Shapley probably felt the same)—and took one step further: I also included Shapley in the dedication!

So it came to pass. When the book was eventually published by the Harvard University Press in 1946 under the title of *An Introduction to the Study of Eclipsing Variables* (as *Harvard Observatory Monograph* number 6; x + 220 pp), it carried a dedication 'To Henry Norris Russell and Harlow Shapley—Pioneers in Research of Eclipsing Variables'. Russell was indeed more than satisfied by the gesture, and provided a brief but gracious Foreword to the little volume (dated October 1944). Moreover, while before that time he had styled me in our correspondence rather formally as 'Dear Dr Kopal', henceforth I became 'My dear Kopal'; and although Russell crossed swords with me (behind my back, rather than openly) more than once after that time, his way of addressing me always remained the same, up to the time of his death on 18 February 1957, at the age of almost eighty years.

Dr Shapley (who, with his sense of humour, must have regarded Russell's earlier hesitations with some amusement) perhaps got more out of this business than he really deserved: for having acted as an 'honest broker' he found it easy to induce the youthful author to accept a publication contract on a no-royalty basis (on the grounds that the publication of my book entailed a financial risk). Events proved that his professed concern was indeed groundless. The book had a very good press. A review by André Danjon (1946) called it *Un ouvrage fondamental sur l'un des problèmes les plus interessants de l'astronomie stellaire*; and commented favourably on the fact that, in my work, the shape of the constituent stars was treated as a resultant of the forces prevalent in the system, and not arbitrarily assumed as was the case in most previous investigations (in his younger years at Strasbourg, Danjon himself had observed Algol extensively; and he must have been keenly aware of the limitations inherent in previous treatment of such data).

The reviews of my book in many other journals were no different in tenor; and quickly relieved Shapley's mind of any concern about the financial aspects of the venture. Suffice it to say that within less than two years the book was out of print, and Harvard Observatory made a pretty penny on the venture, of which, alas, not one cent found its way into the pocket of the author!

The same, incidentally, proved also to be the case with *Harvard Observatory Monograph* number **8**, entitled *Computation of the Elements of Eclipsing Binary Systems*, published four years later (Kopal 1950), the sole difference being that its Preface was not contributed by Russell (whose dislike of my iterative methods was then beyond appeasement), but by Shapley himself.

The next step of our 1940 programme (special tables required to account for the photometric effects of distortion) caused no difficulty: with the support of a small grant to this end from the American Philosophical Society in Philadelphia the work on them started in my Computation Laboratory at MIT as soon as the pressure of war work subsided somewhat in 1945. The results were published in the *Harvard Observatory Circular* number **450** (Kopal 1947), and prefaced by an outline of the theory to which little was to be added since†.

---

† The original publication has long been out of print; but the essential parts of the tables were later reproduced in Appendices IV and VI of my book on *The Language of the Stars* (Kopal 1979).

But even before a chance came to put these tables to systematic use, another development occurred which influenced my work in this field for at least fifteen subsequent years: and that was intimately connected with the arrival at the Harvard Observatory of Dr Stefan L Piotrowski (1910–1985) of Kraków University, whose interests in the problems posed by eclipsing variables—acquired under the tutelage of the late Professor Th Banachiewicz (1882–1954)—went back to pre-war years (cf Piotrowski 1937).

Piotrowski's arrival in Harvard in September 1947 for one year's stay was indeed a very happy coincidence; for the next stage of our programme as planned in 1940 and only interrupted by the war was the construction of a new critical *Catalogue of Photometric Elements of Eclipsing Binary Systems*, to supersede the one contained in Shapley's 1914 doctoral thesis (Shapley 1915). It was, moreover, evident to both Russell and Shapley that methods of 1912 vintage for the analysis of light changes of eclipsing variables were no longer sufficient fully to exhaust the information provided by modern photoelectric observations of such systems; and even Russell agreed (albeit with some reluctance) that the iterative methods initiated in 1941 were probably the best for the purpose.

However, there was no denying that my earlier work on this subject (Kopal 1941c) still left much to be desired; and although its use of least-squares techniques should, in principle, furnish the most probable values of the desired elements, it could not specify the uncertainty arising from the finite rate of convergence of successive iterations. By 1948 this problem continued to baffle me; but it was not long after Piotrowski's arrival at Cambridge that our cooperation provided its complete solution (cf Piotrowski 1948, Kopal 1948), to which Professor Russell added the final seal of approval (cf Russell 1948); and thus a systematic work on the proposed *Catalogue of the Elements* could commence in earnest.

I should like to stress that this could scarcely have happened without Piotrowski's providential intervention at the right time. As a gifted pupil of Professor Banachiewicz, Piotrowski was much more familiar with least-squares techniques than either Russell or myself; scarcely anybody else but he could have accomplished the necessary work so well at the right time. It was a matter of sincere regret to us all that Piotrowski could not remain at Harvard longer than a year; but his native Poland (which had emerged only shortly before from the ravages of World War II) and old Professor Banachiewicz needed him even more at home, especially as Banachiewicz's second assistant and Piotrowski's colleague, Dr Karol Koziel, was in the United States at the same time, working at Yale Observatory with Professor Dirk Brouwer.

Fate willed it that all three of us, Piotrowski, Koziel and myself, had much to do with each other later in life and remained good friends ever after. As is well known, in the fullness of time, Koziel succeeded his teacher Banachiewicz in the latter's chair at Krakow University (which became vacant when Banachiewicz died in 1954); and our professional interests did not cross until after 1959 on the subject of the Moon (see figure 5.41); Piotrowski became (from 1953) Professor of Astronomy and Director of the Astronomical Observatory at the University of Warsaw and, in this capacity, he educated the whole post-war school of Polish astronomy, perhaps the best school in Europe at that time. The names of S Grzedzielski, A Kruszewski, W Krzeminski, O Pacholczyk, B Paczynski, S Rucinski, K Serkowski (died 1981), I Semeniuk, J Smak or J Stodolkiewicz (the first and last of whom also spent some time at Manchester)—well known to their professional colleagues all over the world—are all names of Piotrowski's former students! It is true to say (and how many others can truthfully say so about themselves!) that Professor Piotrowski veritably spent his life helping others, especially during the years when he served as Secretary of the Polish Academy of Sciences. I remember I once asked Piotrowski how he managed to educate such a brilliant pleiad of young astronomers in a country which suffered more than any other in Europe from the consequences of World War II. 'Oh, it is easy,' he replied; 'all that one should do is to select the students well to begin with—and then support them in what *they* want to do.' It is easily said; but how many among Piotrowski's contemporaries have succeeded in this effort as well as he did? I should only adjoin the words used to describe one of his famous post-war contemporaries (unfortunately, not an astronomer) Professor Albert Pražák (1880–1957) of the Charles University of Prague: 'His greatness was not to think of himself; his nobility—a sacrifice.' A society in which this is not the case is in decline.

But even before the road became clear theoretically for the construction of a new catalogue of the elements of eclipsing binary systems in the years immediately following World War II, Dr Shapley initiated in April 1946 steps to set up a 'Panel on the Orbits of Eclipsing Binaries', as a part of the American Section of the IAU, by the following circular letter:

Notification to H N Russell, Z Kopal, S Chandrasekhar, C M Huffer, Mrs C P Gaposchkin, A H Joy, G E Kron, D B McLaughlin, J E Merrill, J A Pearce, R M Petrie, N L Pierce, O Struve.

Acting on behalf of the members of the American Section of the International Astronomical Union, I am asking that you be an original member of the Panel on the Orbits of Eclipsing Binaries. The Panel was discussed and authorized by the Executive Committee and some other members of the Internationl Astronomical Union at the meeting in Copenhagen in March of this year. Professor H N Russell has agreed to act as chairman, and Dr Z Kopal as secretary. Further communication concerning the proposed work of the Panel will come direct from them.

It is intended that the membership include those who work on eclipsing stars spectroscopically, photometrically, photoelectrically, and on the theory. To keep the Panel workably small, it will probably be best that not more than two individuals be included from any one observatory. It is suggested, however, that additional members may be coopted by the original Panel membership.

In approving the plan for this small special organization, the Executive Committee of the IAU foresaw the probability that the Panel should be made a subcommittee of the Variable Star Commission (No. 27) of the IAU in 1948, and that its membership become international, providing the preliminary and exploratory work in America justified such a development.

There appears to be a clear opportunity in the near future to make significant additions to our knowledge of eclipsing binaries. I shall be very glad if you can cooperate actively with your colleagues in advancing the work in this fertile field.

Harlow Shapley, Chairman
American Section, IAU.


This Panel held its first meeting at Harvard Observatory on 28 December 1946; and arranged for a symposium in which H N Russell (then seventy years of age) delivered his first of the well-known 'Russell Lectures' on the subject of 'The Royal Road of Eclipses' (the text of which appeared in the *Harvard Observatory Monograph* number 7 1948; pp 181–209); with additional lectures presented by O Struve, S Chandrasekhar, R M Petrie, H Shapley, C H Payne-Gaposchkin and Z Kopal.

A continuous contact between the panellists was maintained in 1947–48 by *Bulletins* (six of which were issued at irregular intervals by the writer of this book in his function as the Panel's Executive Secretary) until the summer of 1948, when the Seventh General Assembly of the International Astronomical Union met in Zürich between 11 and 18 August 1948, ten years after the last General Assembly held in Stockholm in 1938. How much the world had changed in the meantime!

The 1948 IAU meeting in Zürich was the only one that I missed in the first third of a century of my membership of that organisation; but to me (albeit *in absentia*) this turned out to be a milestone. The IAU Executive of that time took due note of our activities in the US; but rather than making our Panel a sub-committee of Commission 27 on Variable Stars, they decided to create a new IAU Commission (number 42) on Photometric Binary Systems, and elected me its first President (a function which I held until the IAU General Assembly in Dublin in 1955).

When intelligence of this fact reached me in Cambridge, I could not but go back in my thoughts to 1931, when the Czech Astronomical Society in Prague had appointed me to chair its Variable Star Section seventeen years before; for, in both cases, I was called upon to preside over a body of which I was the youngest member! But, this time, there were implications; for (I learned later) it was H N Russell who expected that the function would come to him as a matter of course by right of seniority; and he did not take very kindly to the fact that this was not the case. Dr Shapley (who was in Zürich) noticed that immediately; and on his return to Cambridge urged me to do what I could to placate the old gentleman (in fact, he almost wrote for me the letter which I should send him); and I did so—alas, to little avail. The reply which came back was measured in tone; but it did not take long for me to hear by word of mouth that Russell was offended by the action of the IAU Executive, which he took as a slight to his dignity. Moreover, he apparently thought that I was co-responsible; and let himself be heard that '. . . the devil had to bring over to us this Czech b——d.' Although this expression (which I did not hear, but heard about) was hopefully meant as a joke, even so it would have been a pretty uncharitable term for the son of a clergyman to apply to a fellow-Christian!

But, in the meantime, that 'Czech b——d' was sitting in Cambridge and wondering if he should remain there, or return to his native country and his family (which, miraculously, survived the 1939–45 war unharmed). Soon after the end of the war postal connections with Czechoslovakia were re-established; and once more I was in contact with the Czech Astronomical Society and its Observatory in Prague, which after almost six years of isolation from the world needed some help.

When I left Prague in 1938, the membership of the Czech Astronomical Society (which we met in Chapter 2) was just under a thousand; but during the war years just past it had experienced a rather dramatic upswing. No wonder, perhaps, when almost any other form of cultural activity was *verboten* by the Nazis (then in occupation); during the blackouts enforced for many months the stars in the sky were the only lights which even Hitler could not extinguish, and which held hope for the future. At any rate, the citizens of Czechoslovakia responded accordingly; and by the time the war ended the membership of the Czech Astronomical Society had more than quintupled; and its

5000 members (out of a population of some fourteen million) were only too eager to catch up in astronomy what they had missed during the years just past.

The most significant help which I was in a position to render to my former colleagues when postal connections were re-established in the summer of 1945 concerned the astronomical literature. One aspect of this new activity which I remember with particular fondness concerned the Skalnaté Pleso Observatory of Slovakia, established during the war in the High Tatra Mountains (at an altitude of 1783 m) by Dr Antonín Bečvář (1901–1965) who directed it till 1950 (see figure 4.14). In prewar years I knew him but vaguely; but after the war he was one of the first to turn to me for advice as to what work he should embark on. Since his reputation for cartography was well known, I suggested that one part of the astronomical literature in grave need of updating was atlases of the stars for more general use: the well-known BD charts were almost a hundred years old, and most of their successors still went back to the nineteenth century.

**Figure 4.14**  Dr Antonín Bečvář (1901–1965) of Czechoslovakia—the sky-mapper of the twentieth century.

Bečvář agreed enthusiastically, but lacked sources on the basis of which new charts to meet the need of the day could be constructed. These I was fortunately in a position to supply (together with some plans for what the new atlas should contain); and I also got Charlie

Federer of the Sky Publishing Corporation (then located at Harvard) interested in bringing out an American edition of the proposed atlas (to which I was called upon to provide a Foreword). With all this behind him, Bečvář went to work, supported by a number of young and enthusiastic assistants, one of whom, Dr L Kresák, in the fullness of time was to become Vice-President of the International Astronomical Union. The result is well known: Bečvář's *Atlas Coeli Skalnaté Pleso* (1947) was an instant success, and has since gone through many editions; there is scarcely any active observatory in the world today where more than one copy of this work is not in constant use.

This work, followed by his *Atlas Eclipticalis* (1958), *Atlas Borealis* (1962) and *Atlas Australis* (1964), made Bečvář the sky-mapper of the century, and the author of contributions to astronomical cartography which are truly monumental; but their success was matched only by Bečvář's modesty, and lack of any trace of ego. When I congratulated him on the success of *Atlas Coeli*, he replied to me at Harvard (in a letter dated 17 December 1947) that he considered *me* to be the real author of the work, to which he had lent only his hands. In the face of so much modesty, it is perhaps not surprising that, in spite of world-wide renown for Bečvář's work, his personality has remained almost unknown to this day. Bečvář never travelled abroad, and even in his native Czechoslovakia his shy and retiring nature did not make him a familiar figure. The photograph of him reproduced in figure 4.14 is one of the very few which ever appeared in print; and although he has now been dead for more than twenty years, his work still very much lives on.

But in the years following World War II it appeared at times that Bečvář and I might be able to cooperate at a much smaller distance. Since 1946 I had been sounded by my *alma mater* on the possibility of returning to Prague to accept the chair of astronomy at Charles University; and in June 1947 Professor B Bydžovský, one of my old teachers of mathematics, and then Rector of Charles University, stopped at Cambridge (see figure 4.15) on his return from Princeton to discuss the particulars with me. Once more my future hung in the balance that summer; and who knows what would have happened if I had made a premature decision at that time?

Fortunately, events spared me the need; for while correspondence was still in progress, the democratic government of Czechoslovakia was overturned in February 1948 by a Soviet-inspired putsch; and this brought all negotiations about my return to Prague to an effective standstill. It was then that my wife and I applied for naturalisation in the United States; and, in due course, we became American citizens— an allegiance we have retained ever since.

Moreover, from the beginning of the same month which saw the

## 216  The American years

Communist takeover of Czechoslovakia, I was appointed Associate Professor of Numerical Analysis at the Massachusetts Institute of Technology (in addition to my honorary position at Harvard University). I actually started lecturing at MIT from the commencement of the academic year 1947/48; and, in the summer term of 1948, Dr Shapley invited me also to start a course at Harvard on the astronomy of double stars.

**Figure 4.15** Professor Bohumil Bydžovský (1880–1969), Rector of the Charles University of Prague, with Professor Philipp Frank (1884–1966) (left) and the present author (right) in front of the Physics Research Laboratories at Harvard University, during Bydžovský's visit to Cambridge in June 1947.

This I did with pleasure; but the course was not repeated, since the administrations of Harvard and MIT could not agree on the proper deployment of my time. However, in the spring of 1948 I lectured at Harvard to a distinguished group of listeners, several of whom were senior to me by age. One of the regular listeners was Mrs Shapley who may, in turn, have reported back to her husband (at least I recall that when I was once heading for the classroom in Building C of Harvard Observatory, Dr Shapley intercepted me with the admonition 'tell them only a half of what you have in mind, ... it will be enough'). The reason was that, in those days, I probably spoke too fast—which may have made it difficult for students to follow orally all the details. Another of the students who took that course was Dr Halton C Arp of subsequent quasar fame, and a young Indian student called Vainu Bappu, who had come to Harvard from Hyderabad shortly before.

Little did I know at that time that, in exposing this young and brilliant Indian scholar to some intricacies of the theory of close binary systems, I may have influenced the initial conditions of the professional career of a future President of the International Astronomical Union. This was the first (though not the last) time that I had an opportunity to do so; the second belongs to my Manchester years (see Chapter 5).

Returning to my years at MIT, these too I remember fondly for several reasons. First, because they held much more for me to learn on subjects which, at that time, used to form no part of the curriculum to which astronomers of my time were normally exposed. Under the terms at which the MIT's Center of Analysis was founded in the days of Vannevar Bush, I belonged to the Department of Electrical Engineering (which by that time formed almost an independent body within MIT) and I shall never forget the encouragement and inspiration received from its head, Professor Harold L Hazen (later to become Dean of MIT's Graduate School), who certainly knew much more of astronomy than I did of electrical engineering. It was indeed from Dr Hazen and his colleagues (among whom Dr Ernst A Guillemin should particularly be mentioned) that I probably received the first inspiration for an analysis of the light changes of eclipsing variables in the frequency-domain—a procedure familiar enough in electrical engineering; and a process which, after a long period of hibernation in the back of my mind, eventually led me to produce the best work of my life.

Another story of interest to astronomers should illustrate this fact. When Hazen himself was a graduate student in the 1920s, Vannevar Bush, then head of electrical engineering at MIT, was approached by Dr George Ellery Hale for professional advice about the mounting and control of the large telescopes to be erected at California's Palomar Mountain. Bush indeed agreed to do so, and as a preparatory measure commissioned young Hazen to visit the existing observatories in the United States, and report on the 'state of the art' of telescope control then in use.

Hazen's report (which, years later, I was privileged to see) was not very complimentary to astronomers, who were accused of nothing less than being at least a generation behind the times (and, in particular, completely unaware of advances made by naval engineers during World War I). Hazen did not perhaps fully realise at that time that the principal cause of this backwardness was a lack of money, and the traditional attitude of astronomers of spending the bulk of their funds on optics, leaving the control of the telescope largely up to the observer. However, Bush thought that such an attitude was ultimately self-defeating in terms of scientific returns, and Hale accepted his assessment: this was why the mounting of the 5 m Hale telescope was eventually entrusted to a US naval captain (Clyde S McDowell by name); and why the 200 inch reflector and the 48 inch Schmidt of Palomar Mountain became the first telescopes equipped with modern servo-control, whose launch in practical use truly marked the beginning of a new era in observational astronomy.

I saw this happen at the dedication of the 200 inch telescope at Mt Palomar (named, appropriately, after George Ellery Hale) in June 1948 (see figure 4.16) on my first visit to the American West Coast (where I was later to spend so much time). Up to 1948, all my travels in the US were restricted by circumstances to the country's East Coast, between Maine and Virginia; it was not till 1948 that this dedication of the 200 inch reflector provided the opportunity for a cross-country trip to the Pacific coast.

**Figure 4.16** The author at Palomar Mountain, on the day of the dedication of the 200 inch telescope in July 1948.

In those days, transcontinental travel was still largely confined to the rails, permitting one to sight-see *en route*. If one left Boston by train in the afternoon, one reached Chicago next morning; and the first interruption of the journey there enabled me to visit (for the first time) the Yerkes Observatory in Williams Bay, where I stayed overnight and was received very kindly by the Director, Otto Struve (1897-1963). A

## First visit to the West Coast

trip from Chicago to Los Angeles aboard a Santa Fe 'Superchief' train lasted two and a half days through the Mid-West, New Mexico and Arizona (where I interrupted the trip to visit the Grand Canyon and decend to the Colorado River on muleback).

At Los Angeles, I hurried from the Union Station to the Pacific Ocean at Santa Monica, twelve years after I had seen it from the opposite shores for the first time; and from the station to Pasadena one could still travel by tramway. On my return, I had an opportunity to visit the Lick Observatory at Mt Hamilton and meet for the first (and last) time in my life its Emeritus Director, Robert Grant Aitken (1864–1951), the legendary observer of binary stars (see figure 4.17), the like of whom we have never seen since. Shall I confess that, while in California and before I started from San Francisco back east, I also visited (likewise for the first time) the Yosemite Valley National Park—then very much better preserved in its pristine form than it has since become—and hiked there for a few days on mountain trails? Overnight I was lodged at the Ahwahnee Hotel at the bottom of the Valley; and recall that (as if they knew I was there) on the first day the local orchestra played for dinner Smetana's *Moldau*! Lastly, on the way back I stopped for a few days at Boulder, Colorado, to visit with Walter Roberts the Climax Station of Harvard Observatory in the high mountains (3450 m above sea-level), little knowing that, years later, Boulder would become the home of one part of our family.

**Figure 4.17** Dr Robert Grant Aitken (1864–1951), astronomer (since 1895) and Director (1930–1935) of the Lick Observatory at Mt Hamilton, California; and observer incarnate of visual binary systems.

After my return east, in the academic session 1948/49 I offered elective courses at MIT on numerical analysis for advanced undergraduates (of third or fourth year) and graduate students of science as well as engineering; they attracted a considerable audience. A part of this audience consisted also of officers of the US Armed Forces, from Captains to Colonels of the Army and the Air Force, or First Lieutenants to Commanders in the Navy, detached temporarily for academic studies preparatory to further promotion (a few of them did later attain the general and flag ranks). In addition, a certain number of registered students came from different industrial laboratories in different parts of the country; but while the officers were a highly selected group of outstanding ability, this was not always true of entrants from industry.

The classroom lectures on numerical analysis were accompanied by laboratory work, in which I received invaluable assistance from Miss Katherine E Kavanagh, who had joined my group at MIT in 1943; and who was affectionately referred to as K1 by colleagues and students alike (the symbol K2 being reserved for Miss Katherine Campbell, my secretary). Often, after hours, we (students and teachers alike) used to go to play tennis together on the courts of the Harvard Observatory or MIT, and generally had a jolly good time, especially when young officers were present. One of them, I heard, confessed (to K1) that he was not as intent on getting eagles on his shoulders (indicative of colonel's rank) as on having a chicken on his knees!

But several of my 'civilian' students did distinguish themselves in their subsequent professional careers. One who particularly comes to mind was Thomas F Jones, actually my senior by age, who ended his distinguished career as Executive Vice-President of the institution where he studied (partly under my supervision). Already in his student years Tom was a wonderful chap, always in good humour; and no party (especially at Christmas) would have been complete without him. One story he told me should not be allowed to fall into oblivion; it concerned his marriage. Tom himself was a typical Southerner (Georgian, I believe) who could have stepped out of the pages of *Gone With the Wind*; while his fiancée (an equally charming girl) was descended from an old New England family settled at Cape Cod.

When matters progressed far enough for Tom to be 'introduced to the family', the fiancée's grandmother (doubtless to impress Tom) showed him an old and rather battered tea kettle with an explanation: 'This pot came to these shores with my ancestors on the *Mayflower*.' 'Gee-whizz' responded that irreverent suitor from the South; 'don't you think it's high time to get yourself a new one?' It is more than thirty years now since we last saw each other (I recall it was on my 38th birthday, during one of my short returns from Manchester, when Tom entertained me for lunch); but I hope he and his wife have lived happily ever after and are still among us today. The same I hope is true of many other of my students from the MIT days, whose theses are still on the shelf of my office at Manchester. These are my ties with the institution at which I spent, during the last war and after, almost nine years of my life. I may add that, when my daughters had grown to maturity, two chose for their husbands graduates of MIT as well (though, I hasten to say, neither had been my student).

These were indeed the happy days of one's youth; but signs soon began to appear on the horizon that my association with MIT might represent a transient, rather than steady state. At the time of my appointment to the faculty in 1947, Dr Karl T Compton (1887-1954), then President of MIT, held out the possibility that our war-time computing centre might, in due course, become an inter-departmental Laboratory for Theoretical Astrophysics—a move supported by Dr Shapley (Compton's room-mate in their Princeton years, and his close personal friend); and I was invited to submit the requisite proposals (which are still in my files).

The beginnings were indeed auspicious; and for a year or two all went well. Our previous work on conical shock waves was extended to the systematic investigation of spherical shocks of self-gravitating configurations (with particular attention to similarity solutions applicable to stellar interiors), or to the vibrational stability of stellar configurations (cf Kopal 1950, Kopal and Lin 1951)—work generally supported by the Office of Naval Research of the US Navy Department, in which several of my students participated (cf Carrus *et al.* 1951).

I remember them all well to this day: Pierre Carrus, a dashing Free French war-time flier, then studying for an advanced degree at MIT under the GI Bill of Rights, later abandoned science for more lucrative terrestrial pursuits in his home country. Phyllis Ann Fox, an equally dashing lady, taller than Pierre (who, in size and agility, must have approximated Marshal Foch in his young years) from the State of Colorado, was last heard of (by me) on the staff of New York University; while the 'third musketeer' of our celestial exploits, Felix Haas, chose to follow a more sedentary career which led him (in the fullness of time) to become Dean of Science at Purdue University in Lafayette, Indiana.

I shall never forget my first close encounter with Phyllis Fox in 1949, when we were working together on spherical shock waves (Carrus *et al.* 1951). Her desk at MIT was next to my office. One day I needed (in her absence) some papers in joint use, and opened a drawer where Phyllis usually kept them. Great, however, was my surprise (not to say consternation!) when I saw on top of them a heavy revolver—a real western six-shooter, fully loaded! Needless to say, when Phyllis came I asked (with some circumspection) if she harboured friendly feelings towards the rest of us; and on receiving her assurance that she did, I explained the reason for my inquiry; and Phyllis readily explained the circumstances.

She was born high in the Colorado Rockies (where her father worked as a mining engineer); and when she reached the age to go to highschool (located about forty miles away along a very desolate road), her father gave her a car, taught her to drive (nobody cared much about driving age in those parts of the country), and a gun, with an admonition to keep it on her seat; and should anybody molest her *en route*, to use it! European readers may perhaps require an explanation to the effect that the Constitution of the United States guarantees every citizen of their country the right to carry arms to defend themselves (originally against the Indians—or the British!); but I am sure that Phyllis would have done so satisfactorily in any circumstances.† Indeed, Phyllis then added, the gun I had found in her drawer would be used that night for target practice on a shooting-range which MIT maintained for members of its Reserve Officer Training Corps on the campus; and I should not have expected her to score many misses!

These were my relations with the students; and with my colleagues they were no less friendly. Especially in the post-war years, MIT was being visited by a great many scholars visiting the US from all parts of the world for short stays or to give colloquia. It was in this way that I met some of my future Manchester colleagues (such as the mathematicians Michael Lighthill or Fritz Ursell) long before I crossed the Atlantic in the easterly direction myself; and at a joint Harvard–MIT colloquium on Leibniz I met for the first (and last) time Professor George Sarton (1884–1956), the great historian of science, whose works so moved and impressed me in my lifetime. Sarton (a Belgian by birth) came to the United States during World War I; and although he spent most of his active years at Harvard (in fact, in the Widener Library), he was too shy (and preoccupied) for any casual approach. Eventually I was introduced to him once by Professor Philipp Frank at one of the joint colloquia. I found Sarton's kindness to be equal to his modesty—in fact, he was so courteous and attentive to me that (I concluded) he must have mistaken me for somebody else, better deserving of his attention.

Unfortunately, as time went on, clouds began to gather around the

---

† Many years later, my daughter Zdenka and her family also made their home in Colorado; but I am afraid Zdenka's home education was not as comprehensive as that given to Phyllis by her father.

future of astronomy at MIT; and certain misgivings began to form in my mind within less than a year after my appointment to associate professorship. For, much to the sorrow of his friends, in 1949 Dr Compton suffered a serious setback in health, which soon compelled him (on the advice of his physicians) to resign from the Presidency of MIT; and this function passed on to Dr James R Killian (who had served for several years previously as Executive Vice-President under Compton).

Killian was certainly an outstanding administrator (and right-hand man to Compton while the latter was in office); but being neither scientist nor engineer (his highest earned degree was that of Bachelor of Business Administration), he was much more dependent than his predecessor on advice from others; and at the time of rapid growth of science in post-war years this was no advantage. † In addition, the time was approaching when Dr Shapley too would retire from the Directorship of Harvard Observatory, thus bringing to an end the 75 years of the Pickering–Shapley era; and the subterranean rumblings of a two-year 'war of succession' among the local *diadochi* (soon virtually to tear American astronomy asunder) were already audible to anyone who cared to listen.

It is at this time that an event of momentous significance—namely, the outbreak of the Korean war on 24 June 1950—took place which history may regard as the political watershed of this century. Following as it did only some months after the final breakdown of Kuo-min-tang China and the detonation of the first Soviet A-bomb on 3 September 1949 it rudely pricked the euphoria in which the people of the United States had lived since the summer of 1945 as undisputed masters of the world. The consequences of this unexpected challenge reverberated promptly among American educational institutions too, including MIT, which during the War and afterwards had particularly close relations with the Government. It meant, if anything, that the liaison between the Institute and different branches of the Defense Department would again be drawn closer; and that new civilian projects (such as the development of astronomy) would be downgraded in the Institute's further development plans in the foreseeable future.

Subsequent events confirmed this diagnosis (at least until October 1957, when the successful launch of the first Soviet *sputnik* again changed the whole situation completely); and caused me in 1950 to reconsider my plans for the future. Since the beginning of the academic

† The price to be paid for a policy of preferring mere administrators to more original thinkers transpired in the 1950s, when Dr Killian, in his capacity as Science Adviser to President Eisenhower, must bear his share of the responsibility for America's handicap of several years in the 'space race' which followed the launchings of the first Soviet *sputniks* in 1957.

year 1950/51, the influx of new work sponsored by the Defense Department had again begun to claim an increasing amount of attention in my department too; and the fact that this work was unlikely to diminish in the foreseeable future put me in a dilemma: should I bow to the wind and become once more a 'weekend astronomer' as had been the case for most of World War II, or should I follow my own inclinations at the risk of soon finding myself on the sidelines of MIT's main interests at that time? As I deliberated over such questions, a decision was already being made for me across the ocean; so that when the time came, all I had to do was merely to ratify it. But to outline how this happened will be the task of the next chapter.

# References

Betz Shapley M 1916 *Astrophys. J.* **44** 51–58
—— 1917 *Astrophys. J.* **46** 56–63
—— 1919 *Publ. Astron. Soc. Pacific* **30** 343–346
—— 1924 *Harvard Obs. Bull.* no. **797**
—— 1927 *Harvard Obs. Bull.* nos **843, 844, 845**
Carrus P A, Fox P A, Haas F and Kopal Z 1951 *Astrophys. J.* **113** 193–209 and 496–518
Chandrasekhar S 1980 *J. Astron. Astrophys.* **1** 33–45
Danjon A 1946 *Ann. d'Astrophys.* **9** 234
Eddington A S 1940 *The Observatory* **63** 187
Gingerich O 1975 in *Dictionary of Scientific Biography* (ed C C Gillespie) (New York: Scribner) **12** 345–352 (cf, in particular, p. 348)
Haramundanis K 1984 *Cecilia Payne-Gaposchkin* (Cambridge: Cambridge University Press)
Huffer C M and Collins G W 1962 *Astrophys. J. Suppl.* **7** 351–410
Jurkevich I 1970 in *Vistas in Astronomy* (ed A Beer) (London: Pergamon) **12** 63–116
Kennedy J F 1940 *Why England Slept* (London: Hutchinson)
Kopal Z 1941a *Ann. NY Acad. Sci.* **41** 13–48
—— 1941b *Astrophys. J.* **93** 92–103
—— 1941c *Astrophys. J.* **94** 145–158
—— 1941d *Astrophys. J.* **94** 159–170
—— 1942a *Astrophys. J.* **96** 20–27
—— 1942b *Proc. Am. Philos. Soc.* **85** 339–431
—— 1946 *An Introduction to the Study of Eclipsing Variables* (*Harvard Obs. Monogr.* no. **6**) (Cambridge, Mass.: Harvard University Press)
—— 1947 'Theory and Tables of Associated Alpha-Functions' (*Harvard Obs. Circ.* no. **450** 1–25)
—— 1948 *Astrophys. J.* **108** 46–52

―――1950 *Computation of the Elements of Eclipsing Binary Systems* (*Harvard Obs. Monogr.* no. 8) (Cambridge, Mass.: Harvard University Press)
―――1959 *Close Binary Systems* (London: Chapman and Hall and New York: Wiley)
―――1979 *The Language of the Stars* (Dordrecht: Reidel)
Kopal Z and Lin C C 1951 *Proc. US Nat. Acad. Sci.* **37** 495–506
Linnell A P and Proctor D D 1970 *Astrophys. J.* **161** 1043–1057, **162** 683–686
―――1971 *Astrophys. J.* **164** 131–136
Pickering E C 1880 *Proc. Am. Acad. Sci.* **16** 1
Piotrowski S L 1937 *Acta Astron.* (a) **4** 1–24
―――1948 *Astrophys. J.* **108** 36–45
Russell H N 1899 *Astrophys. J.* **10** 315–318
―――1900 *Astron. J.* **21** 25–28
―――1912 *Astrophys. J.* **35** 315–340; **36** 54–74
―――1913 *Proc. Am. Philos. Soc.* **51** 569–579
―――1942 *Astrophys. J.* **95** 345–355
―――1948 *Astrophys. J.* **108** 53–55
―――1956 in *Vistas in Astronomy* (ed A Beer) (London: Pergamon) **1** 1177–1186
Russell H N and Moore Ch E 1940 *The Masses of the Stars* (Chicago: Univ. Chicago Press)
Russell H N and Shapley H 1912 *Astrophys. J.* **36** 239–254, 385–408
Shapley H 1915 *Princeton Observ. Contr.* no. **3**
―――1968 *Through Rugged Ways to the Stars* (New York: Scribner) pp 24–52
Zwicky F 1971 *Catalogue of Selected Compact Galaxies and of Post-Eruptive Galaxies* (Zürich : Speich)

## Chapter 5

## The Manchester Years

*Arduus ad Solem*

It was in the latter part of May 1950 that, on returning home from MIT via Harvard Observatory, I found on my desk a brief letter from my old teacher Professor Freundlich (then at the University of St Andrews, and with whom I had been in occasional correspondence all these years), informing me that a new chair of astronomy was to be created at the University of Manchester in England; and asking if I would wish to be considered as a candidate for election. Although the letter was sent from St Andrews, I knew that Freundlich had connections with Manchester (he had given occasional courses on astronomy there in the recent past), and presumed that he had not written this letter merely off his own bat; and of this I soon received written confirmation.

For less than a week after I sent Freundlich an affirmative answer, a letter arrived from Manchester by Professor P M S Blackett, senior physicist of the Victoria University in that city, informing me officially of the position; and asking if I would allow him to put my name forward before the Committee of Electors due to meet in October 1950 at the commencement of the next academic session 1950/51. My answer was again in the affirmative; and from then on events took their speedy course: during the summer I received from Mr (now Dr) Vincent Knowles—then the University's Deputy Registrar (and soon to become Registrar for a term of service of many years, which almost overlapped with mine)—an invitation to visit the University for this purpose; and did so in the second week of October to attend the meeting of the Electors on 11 October (almost twelve years to the day since we had entered the United States in 1938).

## First visit to Manchester

I arrived in London by air from New York earlier that week (that was the flight in which, as mentioned in the Preface, our plane lost two engines *en route* from Gander to Shannon); and my first impressions in stepping on to English soil in London were rather depressing (no doubt partly because of fatigue after the long journey; and also because the autumn weather so contrasted with the Indian summer on the US East Coast). War-time damage caused by aerial bombardment was still much in evidence; and what impressed one even more sadly was the large number of invalids one saw in the streets, as mute but vivid examples of the horrors of the war years which elapsed since 1938.

But this was more than made up for by a happy reunion with my Prague schoolmates Karel Brušák and Gustav Jelínek (see pp 72–73) who survived the war in Britain. A few days later I went to revisit Cambridge, as a guest of the late Professor D R Hartree (1897–1958; whose acquaintance I had made at MIT during his post-war visits to the Institute), and of Dr M V Wilkes of the University's Mathematical Laboratory (whose book on the EDSAC computer I was then steering through the press in Boston). At last, on 10 October I repaired from London's Euston Station (then still in the full glory of its early Victorian Gothic) for Manchester—a city I had never visited before, and which was to become my professional home for the second half of my life. That the city greeted me on arrival with rain was perhaps only to be expected from its reputation; but, fortunately, the weather cleared up next day and remained pleasant for the rest of my stay in Britain—not only in Manchester, but also in Scotland, where I later repaired to visit Professor Freundlich before my return to London and back to the United States.

My visit to the University of Manchester and meeting with Professor Blackett and the Committee took place on 11 October to our mutual satisfaction. Nothing *official* could, of course, be disclosed about the outcome till the recommendation of the Committee was subsequently approved by the University's Senate and Council some weeks later; but the usual way to do so was by a 'calculated indiscretion'. In this case, this was done by the later Dr M Tyson, the University's Librarian, when in a conducted tour of the Science Library next day he asked casually 'which new books or journals I should like him to order when I come'.

Naturally, it behooved me still to keep a 'poker face' in waiting for more official word. After a day or so I left Manchester again in a northerly direction to visit Professor Freundlich (whom I had not seen since our wedding in 1938). This was my first visit to Scotland.†

---

† Although I did not know it then, some years later I was to go often to St Andrews as External Examiner in my subject.

I then flew back across the Atlantic (this time without any mishap) to Massachusetts, to await official word from Manchester. It was not slow to come: for less than two weeks after my return I received a telegram from Mr W Mansfield Cooper (the University's Registrar, and since 1956 Vice-Chancellor) conveying the news that 'The University unanimously invites you to accept the Chair of Astronomy'. Before the end of the month I accepted: and in November 1950 my appointment became official.

Appointments to new chairs at British universities were not as frequent then as they have since become; and the local press (as well as the national magazine *Nature*) gave it favourable publicity which earned me a certain number of messages of congratulation. One came from Professor Hartree, who in earlier years of his life held the chair of applied mathematics at Manchester (before he went to Cambridge to succeed R H Fowler); and his letter (of 27 January 1951) contained also a nice sentence: 'Congratulations also to Manchester! I think they were fortunate in the choice of their new Professor of Astronomy'—a prophecy which had yet to be borne out. A somewhat more cryptic message I received, however, from Professor Sydney Goldstein, who resigned the Beyer Chair in Applied Mathematics at Manchester in 1950 to return to his ancestral land of Israel. When I met him in the spring of 1951 at Cambridge (Mass.) at the International Congress of Mathematicians (at which S Chandrasekhar and I represented the American Astronomical Society), Goldstein told me: 'If you do go to Manchester, the only thing you can do there is to work!' This indeed did turn out to be the case; but I did not take it, then or later, as a discouragement.

I may add that I did not take up my new appointment in England immediately, but remained in office at MIT until the end of the 1950/51 academic session and continued by leave of absence during 1951/52 to wind up my remaining commitments (actually I spent the month of April 1952 there in doing so); but from that time my principal academic allegiance was to the University of Manchester until 1 October 1981, when I became Professor Emeritus. (Between 1964 and 1967 I served on the University's Court of Governors and for several years before my retirement was the senior member of the Senate.)

The universities in Great Britain can be divided into four fairly distinct geological strata, thus: the ancient universities (Oxford and Cambridge) going back to the Middle Ages; the Scottish universities (going back to the time of the Reformation); the 'modern' universities founded in the nineteenth century with the advent of the Industrial Revolution; while the last layer of universities was added after World War II (mostly since 1960); and their positions in the academic world remain still largely to be established.

In this hierarchy, the University of Manchester, founded in 1851 as Owens College, and recipient of a Royal Charter in 1903 raising it to full university rank, is the third oldest 'modern' university of the country (after University and King's Colleges of London, founded in 1826 and 1829 respectively).

There are many reasons which contributed to this end. Manchester can claim many titles to fame, and has a pedigree going back to Roman Britain. *Mancunium* ( = a camp of men), from which the present city derives its name, used to be the seat of a cohort constituting part of the famous Twentieth Legion (*Valeria Victrix*) of Chester, once commanded by Gnaeus Iulius Agricola, whose famous deeds were commemorated in the histories written by his son-in-law Publius Cornelius Tacitus; and remnants of the old Roman military road connecting Chester with Manchester are discernible in places to this day (as are the remnants of the Roman camp on the banks of the Irwell river).

After the breakdown of the Roman empire, the north of England vanished for many centuries from the world scene, until, in post-Napoleonic times, it recaptured its place in history as the cradle of the nascent Industrial Revolution. The reasons which brought this about are many, and wholly outside the scope of this chapter. Not so, however, its consequences; one of which was the realisation by its first-generation industrial elite that their undertakings could not properly prosper unless their families and staff secured access to a better education in science and technology than that offered to their alumni by the ancient universities—still concerned then with the past more than with the future. To be sure, the north of England had already given the physical sciences such priceless gifts as John Dalton (1766–1844) or James Prescott Joule (1818–1889); but these were solitary investigators (Joule, the discoverer of the mechanical equivalent of heat, was a brewer in Salford by trade) with no connection with any school, national or regional.

It was not until the advent of Owens College (later to become the Victoria University of Manchester) that the region acquired adequate scientific leadership. In more ways than one the institution became a forerunner of schools like Johns Hopkins University of Baltimore, by its emphasis on laboratory work and research rather than mere passive learning, as well as by the extremely favourable student-to-staff ratio prevalent in its formative years. The story goes that the young university had more professors than students, a fact which a well-known local newspaper, the *Manchester Guardian*, proclaimed as a hallmark of failure.

But under the intellectual leadership of such men as Sir Henry Roscoe (1833–1915), a distinguished chemist who held the chair of his subject at Manchester (between 1857 and 1887) for thirty years, or of the mathematician Sir Horace Lamb (1849–1934), born in Stockport and living in Manchester till the time of his death, who held his chair there even longer (between 1885 and 1919), the University rapidly acquired not only a national, but also an international reputation; and already by the end of the century had conferred its first degrees on such undergraduates as J J Thomson or A S Eddington—both sons of the north of England.

Astronomy was introduced to Manchester via the intermediary of physics in the early 1900s through the good offices of Sir Arthur Schuster (1851–1934), scion of a wealthy Hebrew family of West German industrialists who, like so many others at that time, transferred their extensive commercial interests from the Rhineland to the north of England as the next part of Europe awaiting development (especially since West Germany came under direct Prussian domination after 1870). The Schusters were only a part of the wave which brought to the Manchester area not only their wealth, but also culture. The well-known Hallé Orchestra came into being at about the same time; while another (and perhaps better known) family could boast an offspring by the name of Friedrich Engels (1820–1895), himself a student of social sciences, whose financial support (from money squeezed out of the Manchester working class in sweat-shops belonging to his father) helped Karl Marx (1818–1883) to acquire leisure in which to write *Das Kapital*. †

The Schuster family did not, to be sure, belong to quite the same league. Arthur Schuster was born in Frankfurt (on 12 September 1851) and educated in Switzerland. He received his doctorate in physics from the University of Heidelberg (under Kirchhoff) in 1873. It was shortly thereafter that he returned from the continent to Manchester for good, becoming in 1881 Professor of Applied Mathematics at Owens College, and in 1888 Professor of Physics in succession to Balfour Steward‡ (the mathematics chair then going to Horace Lamb). Schuster retained the Chair of Physics until 1907, when he retired (ahead of time) to enable the University to attract to Manchester a star

† Some descendants of the Marx family in the Manchester area now derive their livelihood from a large and prosperous Marks and Spencer department store in the city; but being very proper citizens, they do not like to be reminded unnecessarily of the nineteenth-century 'black sheep' in their pedigree.

‡ It may be of interest to note that when Schuster was being considered for the professorship at Manchester, his runner-up was nobody else than J J Thomson (1856–1940), who as a graduate of Owens College (and son of Manchester's Cheetham Hill) must have been a strong counter-candidate. Lord Rayleigh (one of the Electors) was apparently not sure that Schuster was the better man of the two (cf Strutt 1968, p. 413); but the local element (represented by Sir Henry Roscoe) in the end won for their candidate—not to the good of physics, perhaps, but certainly of astronomy (like his student Rutherford, Thomson too had no particular interest in our subject).

of the first magnitude—Ernest Rutherford, from Montreal—who held the Manchester physics chair with great distinction till 1919.

Arthur Schuster was not, to be sure, an original scientist of the same calibre as his distinguished successor; but he was a man of much wider interests—and these included astronomy! Soon after his return to Manchester, he organised and led several successful expeditions to observe the total eclipses of the Sun in Thailand (1872), Colorado (1878), Egypt (1882) and the West Indies (1886) (a tradition continued in the Manchester physics department to this day by Mr John James), the results of which are perhaps only of historical interest today. Among contemporary astronomers, Schuster's name is remembered in two connections: namely, for the discovery of a closed solution of the Lane–Emden function corresponding to the polytropic index $n = 5$ (cf Schuster 1883) jointly with Robert Emden and, more importantly, for the introduction of systematic methods of solution of the integro-differential equation of radiative transfer (Schuster 1905), discovered also (independently, but concurrently) by Karl Schwarzschild (1906)—a generalisation of which at the hands of Wick (1943), and especially Chandrasekhar (1944; this being the first of a series of papers published by Chandrasekhar betwen 1944 and 1948) brought this subject to its present state.† To this work and interests stemming from this source, Schuster remained faithful to the end of his life: from 1906 till 1934 (the year of his death) he served for sixty volumes on the Board of International Editors of the *Astrophysical Journal*; and in 1904 played also a prominent role in the foundation of the International Union for Cooperation in Solar Research, an organisation which lasted till the first World War.

Arthur Schuster was (at least in the days of his youth) a great traveller, and he later became a great committee man. At one time (between 1912 and 1919) he served as Secretary of the Royal Society of London; and after the war, when already in his seventies, he took an active part in the negotiations which led to the foundation of the International Astronomical Union. It was not, to be sure, the Union as we know it today, but one founded by Britain and France with their allies as an appendix to the Versailles Peace Treaty (that is why its 'official' languages were (and still are!) English and French) and from which (until 1935) Germany and its allies were duly excluded. Schuster, a native of Germany, but long naturalised in Great Britain, played a leading part in the preparation of such an organisation (see

---

† It may be of interest to note that like the Schusters the Schwarzschilds too hailed from an old Hebrew family settled long in Frankfurt (where Karl Schwarzschild was born in 1873); but whereas the Schusters went west, Karl Schwarzschild turned east (and lost his life in 1916 as a consequence of infection contracted on the Eastern front during World War I); his only son (Martin) still lives in the United States.

Hale 1935), though (as far as I know) he never attended any of its General Assemblies. He died in 1934 at the age of 83, in the fullness of years and honours, having outlived Karl Schwarzschild (22 years his junior) by 18 years, but not Schwarzschild's brother-in-law, Robert Emden, with whom he shared the laurels for the polytrope $n = 5$ solution, who survived (in Switzerland) till 1940.

Needless to say, I never met Schuster (who, as Emeritus Professor, continued to take part in the activities of the physics department till World War I; see figure 5.1); and I do not recall Eddington (Schuster's student at Manchester in his undergraduate years) mentioning him in any way. Years later I did, however, meet Schuster's widow (by many years junior to her husband), who was descended from an old Cumbrian family related to George Washington. Yet almost every day of my life, in passing through the vestibule of the Physical Laboratories at Manchester appropriately bearing Schuster's name, I pass by his bust (bearing a striking resemblance to his photograph reproduced in figure 5.1); so, to this extent, he continues to live among us.

**Figure 5.1** The third-year class of physics honours students of the University of Manchester in 1909, photographed in the Main Quadrangle of the University. Back row (left to right): W Eccles, S Kinoshita, R Rossi, W Kay, G N Antonoff, E Marsden, W C Lauisberry. Middle row: F W Whaley, H C Greenwood, W Wilson, W Brodowsky, Miss M White, E J Evans, H Geiger, T Tuomikoshi. Front: S Russ, H Stansfield, R L Slade, Emeritus Professor Arthur Schuster and his successor, Professor Ernest Rutherford, R Beattie, W A Harwood, J N Pring and W Makower.

Some would say, perhaps, that the greatest gift Schuster made to the University of Manchester was his premature retirement (at 66 years of age) for the express purpose of enabling the University to bring Ernest

Rutherford (1871–1937) back from Canada to England. But there was more: Schuster endowed (from private means) a chair for Rutherford, named the Langworthy Professorship (after a friend of Schuster's family; cf Eve 1939), with a salary attached to it which was larger than the combined salaries of the entire physics staff! In our day, no University could afford such a *tour-de-force* or 'brain-drain' in reverse, without getting into trouble with the Government and trade unions alike; but by the beginning of the century Britain was still rich, and its universities were private in deed and not only in name.

The 'Rutherford era' at Manchester between 1907 and 1919 (interrupted as it was by World War I) was comparable perhaps only with the 'Einstein decade' of 1905–1915 in its impact on contemporary physics. It was the time when the nucleus of the atom was both discovered (1911) and also split for the first time (1919) in the 'Old Physics' (later Schuster) building in Coupland Street of the University's campus. Memorial tablets in many of its rooms (installed there, I believe, under Rutherford's successor W L Bragg) commemorate the places of various important discoveries (in the room which was my office between 1962 and 1966, Rutherford and Royds identified in 1912 the $\alpha$ particles with the nuclei of helium atoms). Until 1966 work benches on the lower floors were still soaked with radioactivity, in testimony of the carelessness with which radioactive solutions used to be handled in the early part of this century.

Rutherford left Manchester for Cambridge in 1919 leaving the best part of his life behind; and died there on 19 October 1937 (see figure 5.2) a few months before I reached Cambridge myself (see Chapter 3). However, when I came to Manchester in 1951, senior colleagues were still on the staff who had served under Rutherford in the early part of the century and remembered him well. I recall, in particular, Dr J M Nuttall (of 'Geiger–Nuttall rules' fame). They brought those heroic days to life for us in their recollections—days which will never come back; for physics has travelled so far since that time; and, by now, they belong only to history.

But it is the history of physics to which they belong, and not of astronomy. For (unlike Schuster) Rutherford seems to have been devoid of any interest in astronomy—indeed, of any branches of physics other than nuclear physics (though willing to tolerate them as long as they did not require too much space). The same continued to be true under W L Bragg (1890–1971) who came to succeed Rutherford at Manchester in 1919, and at Cambridge in 1937. Indeed, astronomy as a part of physics, which had started budding hopefully at Manchester in the first decade of this century with Arthur Schuster, went back to sleep with his retirement, and remained in a dormant state until the end of World War II. It survived, to be sure, in mathematics, when

## 234  The Manchester years

Sir Horace Lamb was succeeded in 1919 by Professor Sydney Chapman (1888–1970), who held the Beyer Chair of Mathematics and Natural Philosophy till 1924 (when he went to the Imperial College of Science and Technology in London). Chapman was, in turn, succeeded in the same chair by Professor Edward Arthur Milne (1896–1950), who remained at Manchester till 1928, and resigned at the end of that year to become Rouse Ball Professor of Mathematics at Oxford.

**Figure 5.2**  Lord Rutherford of Nelson (1871–1937), Cavendish Professor of Physics at the University of Cambridge, in the last days of his life. This photograph was taken in the Cavendish Laboratory at the beginning of the academic session 1937/38, less than two weeks before Rutherford passed away (on 19 October) as a victim of sudden illness.

I met Professor Milne only once, very fleetingly, at the 1935 IAU meeting in Paris (though I saw more of his children during the war years at Cambridge, Mass.); but with Professor Chapman, in spite of the age difference between us, my acquaintance became much closer. When we started to hold regular astronomical colloquia, Professor Chapman was one of our first outside speakers, and he came more than once for this purpose from Oxford (where he was about to retire). But retirement had no real effect on Chapman; he merely transferred his activities across the Atlantic to the United States, and divided his time there between Boulder, Colorado and Fairbanks, Alaska.

I met Professor Chapman since then in many places: England, the United States, as well as Czechoslovakia, where at the time of the Thirteenth General Assembly of the IAU in 1967 I had the pleasure of conducting him to see some of the sights of the country. I recall, in particular, our excursion to the Karlštejn Castle, a jewel of Czech medieval Gothic architecture in the neighbourhood of Prague, in the company of another distinguished plasma physicist, Professor Hannes Alfvén. It was a fine excursion, but a long day (and, moreover, medieval castles were never intended to be easily accessible). It was also a warm day, of the type which (in August) are apt to end up with a thunderstorm; and one such did come (fortunately, as we were already on our way back). But Chapman (then 79 years of age) did not hear any of it; for, tired as he was, he fell asleep in the back of my car—a fact which gave Hannes an occasion to observe that 'Chapman missed on our excursion its most important point: namely the plasma effects.' When we returned to Prague, the plasma phenomena (both luminous and acoustic) fortunately subsided; and the end of the day was pure hydrodynamics (it rained cats and dogs).

**Figure 5.3** Professor Sydney Chapman (1888–1970), between 1919 and 1924 Beyer Professor of Applied Mathematics in the University of Manchester, here photographed in July 1969 when our university conferred upon him a doctorate of law (LLD) *honoris causa*.

I met Professor Chapman for the last time again at Manchester, when our University conferred upon him the degree of LLD *honoris causa* in July 1969 (see figure 5.3). He was my house-guest in Wilmslow at that time; and as my wife was in California (assisting our oldest daughter with the arrival of our grandson), Chapman and I spent some days together *en garçon*. I learned from him a good deal about life in

236  *The Manchester years*

the Manchester area after World War I, when the Chapmans lived in Prestbury, a village some sixteen miles from the University. But this distance did not prevent Chapman from bicycling to the city each day (perhaps there were not so many cars on the roads then to make cycling as hazardous as it has since become).

For all his life Professor Chapman was an energetic cyclist, and travelled on his favourite vehicle of locomotion not only between Prestbury and Manchester, but (a much more dangerous feat!) also from Pasadena to Los Angeles Civic Center down the Pasadena Freeway (an adventure, incidentally, strictly forbidden by the police) and, to crown it all, from Vancouver to Fairbanks via the Alaskan Highway, where sometimes one did not meet a soul for fifty miles! Professor Chapman's guardian angel must have had his job cut out to protect his charge in all such adventures, especially since (as his friends will remember) Chapman was a rather frail man of no prepossessing physique. But tough and wiry he remained till almost the end of his life. I recall that, after a strenuous day and our first dinner in a Wilmslow restaurant, Chapman proposed, 'And now let us have a walk.' 'How long a walk would you like to take?' I asked. 'Oh, let's make it five or six miles'—and we did, at a brisk pace along the Bollin river almost to Ringway Airport, to be back home before dark.

Incidentally, Chapman's fondness for the bicycle as the proper instrument of locomotion reminded me of Sir Arthur Eddington, for whom cycling was also his favourite sport. Everyone bicycled, of course, at Cambridge in those days; but when Easter came, Eddington used to take his bicycle for a trip from which he did not return for a week or so. Easter is usually the loveliest part of the year in England, with nature at its best. The trip would, of course, be carefully planned in advance; and a large map of English country lanes was on the wall of Eddington's study to segregate the roads already travelled from those still to be 'done'. These must indeed have been delightful times; it is hard to realise that they are only fifty years behind us!

But, to return to Manchester in our recollections, those were the times when astronomy was not represented on the University's senate. The situation improved but little when the (then) young Dr Thomas George Cowling (1906–   ) came to Manchester from Oxford as a lecturer in mathematics ten years after Milne left (to leave himself after the War for a chair in his subject at the University College of North Wales in Bangor and eventually Leeds (see figure 5.4)). In 1941 he was joined by Dr G L Camm from Oxford (who remained faithful to Manchester for the rest of his life). The work of all these distinguished scholars (in Manchester, and elsewhere) is too well known to need much recollection here; but suffice it at least to mention that the existence of a convective core of the Sun was born in the mind of Tom

Cowling at Manchester—in an office from which it was difficult to see the Sun even on clear days!

And then, in September 1939, World War II broke out, rapidly bringing most research which was non-essential for national defence to an effective standstill for the next six years. Professor Blackett (1897–1974), who succeeded W L Bragg in 1938 as Langworthy Professor of Physics and Director of the Physical Laboratories of the University of Manchester (see figure 5.5), departed almost immediately for the Admiralty where he headed the branch of operational research with great distinction; and most of his colleagues did the same in various other capacities. When they returned after the war to their academic home, astronomy at Manchester acquired its second (now hopefully permanent) lease of life.

**Figure 5.4** Emeritus Professor Thomas George Cowling of Leeds University (between 1938 and 1945 lecturer in mathematics at the University of Manchester).

It actually started with radio astronomy. The crucial role which radar played in the defence of Great Britain is so well known as scarcely to need emphasis and some of the country's best brains were concentrated in this field, in which both Australians and Canadians took their part, so that, after the war, many competent investigators with expertise in radio and radar methods were available to transfer their expertise to fields other than military. The best among them seem to have been captured by the lure of astronomy.

**Figure 5.5** Lord Blackett of Chelsea (1897–1974) during his Manchester years.

Radio astronomy as such had already been born, to be sure, before the War at Bell Telephone Laboratories in 1931, almost by accident (cf Jansky 1933a,b); and it underwent accelerated growth during World War II when it made history more than once, such as in 1942 when all aerial defences on the eastern approaches to the British Isles were mobilised by a powerful radio noise emanating from a source low in the sky, that was eventually localised in a large group of spots on the apparent disc of the Sun then quietly rising above the horizon; while at night, suspicious radar echoes were identified with meteor trails filling the sky (especially in certain times of the year).

A very active group of war veterans with considerable expertise in radio physics was concentrated at Manchester at the Jodrell Bank Experimental Station of the Physical Laboratories (since 1960 the Nuffield Radio Astronomy Laboratories), and the work which Drs A C B Lovell, J A Clegg and Tom Kaiser initiated in the field of meteor astronomy by radar methods soon attracted international attention, as did similar work by D W R McKinley and his collaborators in Canada. This was soon followed by pioneer work on galactic (and soon extragalactic) radio astronomy at frequencies of the order of 100 MHz by R Hanbury Brown, C Hazard and others.

It was indeed the best way to make use of war-time surplus equipment for astronomical purposes in a location proverbially unsuited for observations in the optical domain of the spectrum; and since the history of this development has already been written up more than once in recent years (cf Lovell 1968), it need not be repeated here. What should perhaps be stressed, however, is the fact that the cradle of post-war astronomy at Manchester was not only at Jodrell Bank, which initially started as an out-station specialising in radio and radar astronomy at long wavelengths, but also at the Physical Laboratories in Manchester directed by Professor Blackett; and it was he who brought Jodrell Bank into being.

The main line of research of these laboratories (introduced likewise by Blackett) was cosmic rays—a field in which Blackett gathered around him an excellent group of collaborators (of the calibre of G R Rochester, C C Butler, E H Elliott and others—Arnold Wolfendale, only recently President of the Royal Astronomical Society, was at the time of my arrival an assistant lecturer active in this field). Since, moreover, all cosmic rays are of extraterrestrial origin (whether solar, or galactic), their study belongs also to the domain of 'particle astronomy'—now an obvious fact (though perhaps not so obvious only thirty years ago). Between the two there was, of course, the vast domain of 'optical' astronomy and everything based upon it. It was this gap which Blackett would have liked to see closed at Manchester; and he encouraged me to do so, without, however, offering much tangible help.

It will be the aim of this chapter to give an account of my stewardship of the Chair of Astronomy which I was to hold for the next thirty years. Before I do so, however, I wish to take farewell of those senior colleagues of mine whom I met at Manchester at the time of my arrival; and the primacy among them goes, of course, to Patrick Blackett. What I can say here can, to be sure, contribute but little to what has already been written about this remarkable man (for his best biography, see Lovell 1975), largely because my personal acquaintance with him was to last but briefly—in fact, for only one year (the academic session 1951/52). For the next year Blackett was on leave; and before that year was over, he resigned the Langworthy Chair at Manchester for one at the Imperial College of Science and Technology in London.

When I had an opportunity to observe Blackett at Manchester at close range, he was in his mid-fifties, and (probably) past his prime as an original thinker; at least his view that the production of cosmic magnetic fields by rotation is indicative of the existence of a new law of nature (Blackett, 1947)—which was very much in his mind during his latter years at Manchester, and by which he set much store—is no longer tenable today. However, in spite of such occasional lapses, his abilities as an inspiring scientific administrator continued unimpaired at least until the end of his years at Manchester, where he went on running a first-class academic shop (or, as he was wont to describe it in naval terms, a 'happy ship').

Teaching as such was, however, not his *forte*; and he indulged in it only sparingly. I recall that, at the farewell dinner given for Professor and Mrs Blackett by the Manchester Student Physical Society in June 1953 (the year when I happened to be the Society's President), one of the officers, in counting his blessings, mentioned that during his three undergraduate years at Manchester he had heard Professor Blackett lecture three times!

To postgraduate students (whether supervised by him or by another of his colleagues) Blackett devoted a good deal more attention, and he often came to see them at work in the laboratory. To every question he always had (in the best tradition of the naval upbringing of his youth) a positive answer to hand (no-one I know ever heard him say, 'I must think it over'); and he had little patience with alternative views (their expression usually made him walk away). However, as a good scientist he did often think things over; and if, on second thought, his first reaction did not satisfy him after all, he would come back with another solution of the problem, equally positive (though without giving the reason which had made him change his mind). For, in accordance with his views of hierarchical society (and in line with his early upbringing) a captain on the bridge must always know, and be always right; anyone doubting this would, if not be disciplined, at least not remain too long on the right side of him!

This would probably not have been the case also with Rutherford (Blackett's teacher at Cambridge); but he too had his foibles. Thus Blackett's frequent assertion that 'all important formulae in physics could be written on the reverse side of a postcard' probably went back to Rutherford; though it would scarcely have been accepted by many theoretical physicists of his time.

Rutherford had other foibles too, which must have endeared him to students and colleagues alike, such as exemplified in the following story which came down to me through old-time Mancunians. A student came to Rutherford with the results of a laboratory exercise, with a report that the measured quantity was so and so much, within such and such limits of uncertainty, 'Go back, my boy,' replied Rutherford, patting the student on the shoulder, 'and repeat the measurements so that your result will be subject to no error'—an answer which a classically-trained astronomer would, of course, have received with incredulity! Rutherford was, to be sure, no astronomer (professionally concerned with the measurement of quantities which are at the limit of significance). However, from what I heard from his contemporaries who remembered him at Manchester, he was a very conscientious teacher, not only in lecture rooms, but also in the laboratories; while Blackett was already a forerunner of the times when administrative (and even political) involvements began to carry precedence over purely academic pursuits among senior members of the senate!

In later years, I met Blackett but rarely (for the last time at the inauguration of the new building for the Schuster Physical Laboratories at the University of Manchester in the spring of 1966); and I followed his subsequent career in London only at a distance. This career was, in turn, strongly modulated by the gyrations of national politics. On the political stage—and he was not one to avoid it—Blackett's stance was consistently to the left of the centre (though the radical left-wing element in the family was his wife who, being half-Italian, never betrayed the spirit of Garibaldi or Mazzini). If, however, anyone imputed the same views to Blackett himself, he was apt to find his fingers burned very quickly!

It did not take me too long to realise that Blackett's leftist political views were rather of the nature of an intellectual exercise, and that the veneer covering his solidly conservative core was rather thin. How could it have been otherwise? Born in the twilight of the Victorian era in Kensington as the son of a stockbroker, of an eminently respectable middle-class family associated (for two generations before) with the Church of England, it was on Blackett's mother's side that he had some military antecedents of empire-builders in his veins; and it was these that influenced his family's decision for Patrick to embrace a naval career, a course he pursued from the age of twelve through World War I, until the advent of peace (short-lived as it proved to be) turned his mind to more academic pursuits. His sharp mind and intellectual honesty no doubt saw through some of the self-serving ways of the Establishment into which he was born, and his leftist stance was probably a reaction against its hypocrisy; but it did not go very deep.

It would, however, not have been for Blackett to take a passive role in the political arena: for by every fibre of his personality (and in line with his early upbringing) he was a man of action—a doer rather than a thinker. Old-fashioned Labour leaders of Clement Attlee's vintage apparently only half-trusted socialist intellectuals of Blackett's class; and before the advent of the second Labour Government in 1964 the usual recognitions normally reserved for Blackett's accomplishments and social background were rather slow to materialise.

It was not until the years between 1965 and 1970 that Patrick Blackett at last came into his own. He was a member (albeit short-lasting) of Harold Wilson's first (1964) Cabinet, President of the Royal Society (1965–1970), he was made a Companion of Honour in 1965 and received the Order of Merit in 1967. He became a Life Peer as Baron Blackett of Chelsea in 1969, and enjoyed the influence which such recognitions bring in their wake. Alas, they came too late to enable him to accomplish what he could have done for British science had he remained longer in good health. Unfortunately, by that time his days were already numbered; and he passed away in July 1974, aged 77 years.

Even so, however, it was Blackett's influence in the last years of his active life that enabled others to carry out much of what was overdue, such as to detach the Royal Greenwich Observatory from the Admiralty (to which the Observatory had reported since 1675) and transfer it to the civilian Science Research Council; or to secure positions of authority for such gifted British astronomers as Fred Hoyle or Margaret Burbidge. But, alas, not all these reforms were of long duration: after Blackett was gone, the Royal Observatory did remain under the SRC (which suited many interests); but Margaret Burbidge, the Observatory's first Director under civilian rule, was soon eased out of Herstmonceux back to California (where her professional career continues to flourish); while Fred Hoyle was likewise eventually out-operated back into the wilderness of Cumbria (from which, according to some, he should never have emerged). A proper history of these upheavals, and of the fluctuations in the field of force which gave rise to them, remains yet to be written; but this is primarily a task for others.

Of the other colleagues I remember from my early days at Manchester, most have now gone to their reward; for at the time of my election, I was the second youngest member of the University's Senate; while in 1977 I became its senior. Professor Leon Rosenfeld (1904–1974) of theoretical physics received me very kindly: although by that time he taught mainly nuclear physics, he had also once been an astronomer, and had published several papers on astro-spectroscopy, some jointly with Mlle Yvonne Cambresier (who later became his wife). In 1937, together with his Belgian compatriot Dr Pol Swings (1906–1983), he participated in the spectroscopic discovery of the first diatomic molecule (of CH) in interstellar space.

At the time of my arrival, Professor Rosenfeld's personal interests had shifted, however, largely to the history of science; and as I was still a neophyte in that field, we did not have much opportunity to interact. The same was true of the mathematicians—Max Newman, M S Bartlett, Kurt Mahler or M J Lighthill—as these (with the exception of Lighthill) were too much on the 'pure' side for my liking (or comprehension). Alan Turing, of 'non-computable numbers' fame, flickers briefly in my memories till his ill-starred end in 1954. Of all the names mentioned above, none is left at Manchester, and only two are still alive.

In physics the turnover in staff since the middle of the century has been particularly pronounced. After Blackett left Manchester in 1953, his Chair was offered after some months of deliberation to Samuel Devons from Imperial College, whom I had known before in Cambridge (we are of the same age). He did not hold the Manchester Chair for long (in 1960, he resigned from it to continue his professional career at Columbia University in New York); but during one of his later infrequent visits to Manchester (where his oldest daughter and mine were schoolmates at the Withington Girls School in Fallowfield) he told his colleagues in the Manchester Physics Department a story too true to be forgotten by the historians of physics at the University. When informed of his election to the Langworthy Chair by the (then) Vice-Chancellor, Sir John (later Lord) Stopford, Sam professed his inadequacy to fill a Chair held previously by three Nobel Laureates in succession. 'Never mind,' replied Sir John in his good-natured manner, 'at least you will make it easier for your successors.' 'And I did, didn't I?' added Sam, looking with a twinkle in his eye at his successors sitting in front of him with none too happy expressions on their faces.

It is on this background that we can now proceed to outline briefly the fortunes and evolution of the astronomy department at the University of Manchester since 1951, when I came there to occupy my Chair. The first step, on which I started working while still on the other side of the Atlantic, was to build up at Manchester a professional library of astronomical literature for teaching as well as research purposes. What I found already available at the University's Christie (Science) Library was rather sketchy. A good deal of earlier but valuable literature (assembled no doubt on the initiative of Sir Arthur Schuster), such as, e.g., the almost complete series of more than a hundred volumes of *Harvard Annals* (received in the days of E C Pickering), or a complete run of the *Astrophysical Journal* from volume 1 (1895), was already on the shelves; and so were many others which would have been difficult to acquire, or replace in case of need.†

Moreover, I made some other fortunate discoveries on the library's open shelves, such as a priceless copy of the first edition of Copernicus's classic *De Revolutionibus Orbium Coelestium* of 1543, with extensive marginal notes by previous owners whose pedigree I could not trace;

---

† The University Library in Manchester suffered some damage by war-time bombardment at Christmas 1940; but, fortunately, no astronomical literature appears to have been the victim of it at that time.

but I saw to it that it was thereafter deposited with the rare books.†

An appeal for publications from observatories all over the world (which I launched in the spring of 1951 from Harvard) met with considerable success, both from institutions as well as individuals. I remember, for instance, that the late Professor F J M Stratton gave us his whole private astronomical library on his retirement; and so we were launched into space!

As far as staff were concerned, the first colleagues who joined me in efforts to launch astronomy at Manchester on an even keel were W H Ramsey (1922–1970) and F D Kahn (1926–    ). The first, a Bristol graduate and lecturer in Blackett's department at the time of my arrival, was an expert in solid-state physics, with prior interests in the internal structure of the planets, whose hypothesis of phase transitions in planetary cores continues to be under discussion up to this time. Unfortunately, a hereditary affliction cut short the life of this gifted scientist while still in his prime (in accordance with the biblical dictum that 'the sins of the fathers are visited on the children unto the seventh generation'); had he lived longer, I am certain that we would have learned more from him. I did learn from Bill Ramsey something myself too—he enriched my English vocabulary with such new terms as 'pub crawl', of which he not infrequently gave a demonstration (and of which he later became the victim).

The other of my first colleagues at Manchester, Franz Kahn, was a recent Oxford graduate (a pupil of Professor Chapman) and by 1951 assistant lecturer in applied mathematics at Manchester. In accordance with his interests (and with Professor Lighthill's blessing) Kahn was transferred (via an ICI Fellowship) to me, and remained with us till the time of my retirement thirty years later to become my successor. By that time, apart from the two professors (Franz received his personal chair in 1966), the department had acquired in due course two readers (Drs Dyson and Meaburn), in addition to more junior staff and research fellows whose number fluctuated somewhat from year to year. May it similarly grow in the future!

Last but not least, the third most senior member of the department joined my office in the spring of 1955 in the person of Ellen B Finlay, my secretary and personal assistant for almost thirty years (since September 1973 as Mrs Carling, having married a colleague of this name on our technical staff), well known not only to all past members and students of the department, but to the world astronomical com-

† There are, in fact, two copies of the first edition of Copernicus's masterpiece in Manchester: the other (a better preserved one) was at the John Rylands Library (since amalgamated with the University Library). That copy (to judge from its original *ex libris*) was once owned by the Earl of Crawford in Scotland.

munity at large as editorial assistant of our journals (see later)—a function which she has carried out with distinction since 1968.

As regards the teaching of astronomy over the past thirty years or more, the postgraduate school which has become our pride and joy took some time to build up. I used to give, to be sure, an elementary course on astronomy to our first-year undergraduates a few times in the 1950s to attempt to capture the interest of young students in their formative years (which I never had a chance to do at Harvard); but most of my teaching at postgraduate level in those years was done at Jodrell Bank.

As already mentioned, in the post-war years Professor Blackett managed to assemble there a very able group of radio-physicists, headed originally by A C B Lovell and J E Clegg, who were turning their eyes to the sky; and while there was little they did not already know about radio-physics, in astronomy they were still largely amateurs. One of my first commissions from Blackett was to make that group astronomy-minded—not only in the sense of bringing home to them what was known about the subject (printed literature can do that in less time than listening to lecture recitations), but to make the audience aware of the problems which could be solved with the aid of the newly opening channels of information.

I used to come to Jodrell in those years once a week for two hours (often more) in term-time (we lived in Nether Alderley in those years, in the country—much nearer to Jodrell Bank than to Manchester) to deliver lectures which every so often became discussion sessions; and although many years have gone by since that time, I remember those sessions with genuine pleasure, when I faced an audience in which Professor (afterwards Sir Bernard) Lovell was seldom missing, together with his brother-in-law R Hanbury Brown, J G and R D Davies, Roger Jennison of Strawberry Lane in Wilmslow, Tom Kaiser, Henry Palmer (who later all became professors in their own right) and many others.

Curiously enough, Sir Bernard seems to have forgotten all about this in his own recent autobiography (cf Saward 1984)—I hope at least that he did not forget the astronomy which I tried to teach at that time—but not so the others! At least Robert Hanbury Brown (a future President of the IAU) in his own 'Memories of Jodrell Bank' (see Sullivan 1984) brought me on to the stage of his narrative with a testimony that the '... Professor of Astronomy in the University of Manchester used to come once a week to teach us some astronomy. In those days, most radio astronomers were either physicists or engineers, and most of us knew very little conventional astronomy. Kopal was encyclopaedic and enthusiastic; he did a lot to bridge the local gap between radio and optical astronomy, and to make us feel that we were part of a wider community of astronomers' (*Op.cit.*, pp. 220–221).

Moreover, Hanbury went on, 'As far as I can remember, it was Zdeněk Kopal who first suggested that we should look for Tycho's supernova.' I too remember that, after one of our sessions on supernovae (in which I quoted the probable coordinates of Tycho's supernova of 1572 according to Baade 1945), Hanbury promptly located on his radio charts a source coinciding in position (within the limits of observational errors) with those of the remnant according to Baade (cf Hanbury Brown and Hazard 1952), the discovery of which was followed by radio detection of many other similar remnants in the years to come.

There were other sessions which ended with equally happy results (more about them later); but what pleases me most was that at that time I was able to teach some astronomy to a future President of the International Astronomical Union who (unlike some others) has not forgotten it up to this day. Moreover, Hanbury was not the only (future) President of the IAU whom I was privileged to teach. The first was, of course, Vainu Bappu at Harvard in 1948; he has given an expression of his feelings in a gracious Foreword to the *Festschrift*, published by my former students to commemorate my retirement from active service in September 1981, and edited by my colleague and successor Professor Kahn (cf Kahn 1981).

But now I return to another (albeit indirect) happy outcome of our sessions at Jodrell which had much to do with the Crab Nebula (the 1054 supernova) and that part (about 95%) of its light which exhibits a continuous spectrum. One week in 1954, just before one of our sessions at Jodrell, a publication was received from the Soviet Union (Vashakidze 1954), reporting that the continuous spectrum of the Crab Nebula is polarised up to 30%—a fact indicative of a synchrotron origin for the respective radiation. I promptly shared this result with our colleagues at Jodrell; and Hanbury took this news with him to Leiden, which he was scheduled to visit the next week. On his return, he reported that Professor Oort was sceptical and incredulous. As Hanbury had not taken the reference to Vashakidze's work with him to Leiden, he asked me to send it to Oort by mail. I did so, and received no answer; but some months later a paper appeared from Leiden (cf Oort and Walraven, 1956) confirming the Russian result and showing that my message had not fallen on deaf ears. In fact, as Dr Walraven told me at Manchester some time later (at the conference on Astronomical Optics in 1955), so excited was Oort by their discovery that he went night after night to the dome where observations of the Crab Nebula were in progress (which, Walraven told me, Oort did not do very often).

## Jodrell Bank

As time went on, however, my audience at Jodrell gradually began to change, as did the preoccupations which kept my colleagues busy. Ever since the money for the 250 foot steerable radio-telescope was raised (still largely by Blackett) in 1953, more and more of Bernard Lovell's time began to be claimed by administration or promotional schemes, and less and less by science. This was, however, not true of Hanbury Brown, the most original scientist at Jodrell Bank, whose departure for Australia in 1962 constituted an irreparable loss to the University of Manchester and to his country. His greatest contribution to science was, of course, the development in the late 1950s of the intensity interferometer for the measurement of stellar diameters; and while the original idea owed much to his friend R W Twiss, it was Hanbury who made it work; and this is the better part of the achievement.

I still recall vividly one of my visits to his flat at Alderley Edge (not far from Wilmslow, where we settled and have been living ever since that time) to hear (as one of the first) from Hanbury and Twiss of their design; and to answer their questions as to what a successful experiment of this kind could mean for astronomy—we have all learned the answer since!

As time went on, my ties with Jodrell Bank grew looser; and (apart from occasional colloquia) I gradually ceased to give regular lectures there, because our growing postgraduate school at Manchester University (see figure 5.12) began to claim all my time. Since the late 1950s I used to give lectures to postgraduates of 2–3 hours a week (4–5 hours in the 1970s, when students became more numerous, while my years were running out), and I kept at it, well past my official retirement age, till my 69th year.

Why this was so, and why my regular teaching at Manchester came to an end at that time will be explained later in this chapter. However, nothing deflected me, then or later, from returning to the work which had primarily led me back from America to Europe, and which an increasing weight of extracurricular involvements in the United States did not allow me to pursue there as vigorously as I would have liked. To do so in England was, however, easier said than done; for the new department at Manchester had first to be organised, and the loose ends of the past tied up, before one could turn over completely to a new leaf.

But before I take up this thread, my memories go back to the late summer of 1952, the first summer partly spent in Europe after so many years in the New World, and to a visit to Italy in September to attend the Eighth General Assembly of the IAU in Rome. Much water had flowed under the bridges of the Tiber since our brief visit to the Eternal City fourteen years before, on the way to the United States in 1938; and there was no longer any trace of Mussolini anywhere. Moreover, far from being a beginner in my field, as I was still at Stockholm in 1938, I was now an officer of the IAU, expected to take the Chair of its Commission 42 on Close Binary Systems, which had been created in Zürich in 1948, where (in my absence) I was appointed to become its first President (see pp 212–213).

**Figure 5.6** Father D J K O'Connell (1896–1982), left, and Abbé Georges Lemaître (1894–1966), father of the expanding universe, at the Eighth General Assembly of the International Astronomical Union in Rome, September 1952.

Some stories of this meeting in Rome (which I cherish in my memory second only to the 1935 IAU meeting in Paris) have already been mentioned in Chapter 3; but others should be added in this place. Of many happy reunions with professional colleagues which I recall, none was more pleasant than that with Father D J K O'Connell (1896–1982) (see figure 5.6), a life-long friend of eclipsing variables (and my successor, in 1955, in the Chair of Commission 42)—another 'Harvarder-in-Exile' (as Dr Shapley used to call us), whose acquaintance I had already made at Cambridge in 1948 on one of his visits from Sydney, Australia (where he directed, between 1938 and 1951, the Riverview College Observatory). A few months before the Rome meetings, Father O'Connell was called upon to succeed Father Johan Willem Stein (1871–1951) as director of the Specola Astronomica Vaticana at Castel Gandolfo (the premier astronomical institution of the Catholic Church), a post which he was to hold with great distinction for the next eighteen years.

In September 1952, Father O'Connell was already a 'power behind the throne' at Castel Gandolfo, and had much to do with the reception by Pope Pius XII at the Observatory (it was he who prepared the erudite address delivered by that Pontiff before the IAU). For us this was an especially memorable occasion; for when we were introduced to the Holy Father and he asked about our origin, we quickly established that we had one other acquaintance in common: the late Monsignor Otto Stanovský, canon of the Cathedral Chapter of Prague. Some readers may possibly remember his name from history: for as a young priest, Stanovský was chosen to educate the children of the ill-starred Archduke Franz Ferdinand (successor to the throne of Austro-Hungary), whose death in Sarajevo on 28 June 1914 was used by others as a pretext to unleash World War I more than a month later. It was the sad duty of Father Stanovský to have to inform the orphaned children (one of whom still lives to this day) of the death of their parents.

In post-war years, Mgr Stanovský continued his educational mission also in less exalted circles; and one of his former pupils at the *Gymnasium* now stood at Castel Gandolfo, together with the wife with whom Canon Stanovský had married him, in front of the Pope, who also remembered the good pastor. In fact, Pius XII probably saved Stanovský's life during World War II by intervening, on his behalf, with the Italian King Victor Emmanuel III (and the latter, with Hitler) to liberate him from the clutches of the Gestapo. Not, alas, for long; since after his release from German concentration camps (in which he was confined as a hostage) Mgr Stanovský departed for a better world before the end of the war.

This and much more went through my mind at the papal reception and later at Castel Gandolfo, where in the evening we were the guests of Father O'Connell for dinner at a villa not far from the observatory. It was a wonderful evening; and the villa was famous for its fine view from the terrace, with the dome of St Peter's visible on the horizon—a scenario worthy of the opening scene of Willa Cather's novel *Death comes for the Archbishop*. As we sat there after dinner watching the stars come out, lights started to appear also on land inhabited since this *campagna* was called Latium two and a half thousand years before. People still live there as they used to in the past; but in pointing out their villages Father O'Connell informed us with some indignation: 'and would you believe that every one of those villages today has a Communist mayor?' I tried to visualise so many Don Camillos and Peppones fencing with each other in contest for the welfare of their charges, but kept my peace; and I recall that Abbé Lemaître (figure 5.6), who was with us that evening, did likewise.

Fate willed it that, fifteen years later, when the International Astronomical Union met in Prague in 1967, at its close, our daughter Zdenka (figure 5.7) married a young American astronomer (she publishes now under the name of Z K Smith). The wedding took place in the same cathedral church of St Vitus where her parents had married 29 years before. But Canon Stanovský was no longer there to perform the ceremony; and neither was Josef Cardinal Beran, Archbishop of Prague, whose prerogative it would have been to perform marriages at the Cathedral's main altar. Cardinal Beran, himself a pupil of Mgr Stanovský from the Seminary, and one who during war-years had likewise carried out pastoral duties at Dachau, was in Rome at that time in temporary relegation. He sent, however, blessings to the newlyweds; and (in a letter to their father) added that '... negotiations for my return to Prague between the Vatican and the Czechoslovak Government are under way.' Before, however, these negotiations could come to a satisfactory conclusion, Alexander Dubček and the 'Prague Spring' of 1969 came along, to which the Soviet Government's only response was armed intervention on 21 August of that year, putting an end to many hopes.

**Figure 5.7** Comings and goings of the Kopal family at Ringway airport, Manchester, in June 1962. Georgiana Kopal (left) returns home after three years from the United States with a physics degree from Vassar College; her sister Zdenka (right) is soon to depart with her father (centre) for the Jet Propulsion Laboratory at CalTech, and eventually to receive the same degrees from Bryn Mawr College and Stanford University.

Cardinal Beran finished his life in Rome some years later, and was buried in the crypt under the main altar of the basilica of St Peter (next to the tomb of Pope John XXIII).

But to get back to science, on our return from Rome in 1952 we stopped for some weeks at the Observatoire de Haute Provence in southern France, for an attempt (jointly with the late Dr Arthur Beer) to get a glimpse of the K5 spectrum of the secondary component of the Algol system in the near-infrared, to specify its mass ratio. This attempt was only partly successful (Beer and Kopal 1954); for even with the aid of the observatory's 120 cm reflector and Kodak 1-N photographic emulsions, all we succeeded in doing was to establish the effect of the secondary's contribution to the hydrogen Paschen line profiles in the combined spectrum; but no radial velocities could be deduced from them. This feat had to wait for another quarter of a century at the hands of Tomkin and Lambert (1978), with more powerful optical means at their disposal.

But quite apart from such occasional preoccupations, during the summers of 1952 and 1953 I still used to return to the United States, to spend them at the Bendix Research Laboratories in Detroit, whose Associate Director, Dr Albert C Hall (later to become Assistant Secretary of Defense in the US Government) was a former colleague of mine from MIT. Al was very interested in seeing me finish my book on numerical analysis (summarising the contents of my postgraduate courses on this subject at MIT between 1947 and 1951), and generously placed at my disposal all necessary facilities to speed up its completion. Some of its material also passed through the mill of a summer course which I was invited to deliver at Wayne University in Detroit at that time.

By next summer (1954) the material was in reasonably complete form for publication, and went to the press under the title of *Numerical Analysis* (Kopal 1955b), to appear simultaneously in London (by Chapman and Hall) and New York (Wiley) the next year. The book attracted a good press: the first edition of 1955 was quickly sold out, and was followed by a second edition in 1960; while a Chinese edition appeared at about the same time in Shanghai. Incidentally, the completion of this task in the spring of 1954 interrupted my summer leaves in the United States for some time: the next three successive summers (1954–56) I spent with my family in Europe, embarking (at the relatively advanced age of forty) on my love-affair with the Swiss Alps—a romance which has never lost its lure, and to which I have remained faithful ever since (see figure 5.8).

The reader who remembers our family roots from Chapter 1 may wonder what produced this lasting attraction; and so do I. For as far as extant records disclose, our family stems from the lowlands, from which only on clear days can the distant chain of the Krkonoše mountains (mere hills in comparison with the Alpine giants) be sighted on the horizon. And yet it was Switzerland and Grenoble in France (not far from the Savoy Alps) that my father chose for his postgraduate studies of the French language, and it was in the High Tatra mountains in Slovakia that my brother Miloš lost his life on skis in March 1948, not yet 38 years of age (leaving behind a posthumous son of the same name). Still in my late forties, in the summer of 1962, I and my daughter Zdenka (figure 5.7) almost reached the summit of Mt Whitney in the California Rockies (at 4419 m, the highest mountain in the continental United States); and only the impossibility of making a bivouac overnight at that height in the open without adequate provisions deprived us of our final triumph. What was the urge that continued to propel us ever higher? The atavism inherited from our distant forefathers before they reached the Bohemian plains from ancestral lands, far to the east?

**Figure 5.8** In August 1981, the impending 67-year old emeritus professor on the Hörnli Ridge (3360 m above sea-level) on the north-eastern slopes of the Matterhorn. To the left, the glacier of the north face of the Matterhorn; to the right, the southern slopes of Dent Blanche. (Photograph by Zdenka Kopal Smith.)

My romance with the Alpine world had already commenced in 1953, in the course of an exploratory trip to the Continent, the main aim of which was to establish connections with our astronomical colleagues at institutions with which we were eager to collaborate. In the spring of that year, Dr W L Wilcock of the Physics Department at Manchester and I paid short visits to the Hochalpine Forschungstation at Jungfraujoch in the Bernese Oberland (where Swiss and French astronomers had set up a remarkable observatory at Sphinx, at an altitude of 3576 m, as shown in figure 5.9), the Astrophysical Observatory at Asiago (on the southern slopes of the Dolomites) of the University of Padua, and the Observatoire de Haute Provence in France.

*Galileo's chair*

**Figure 5.9** The high-altitude observatory at Sphinx, where Manchester astronomers worked on nebular photography in the 1950s, and where Peter Yorke Millns (figure 5.10) lost his life just before Christmas 1955.

Wherever we went, we were received with great kindness. I recall, in particular, our visit to Padua, where telescopic astronomy was born in the first decade of the seventeenth century, and has flourished ever since. In the course of a guided tour of that university (second oldest in Europe) Professor G Silva (in company with Professor Anti, then Rector of the University) took us also to see the chair from which (according to tradition) Galileo Galilei used to address his students. It was not so much a chair as a pulpit, from which professors used to lecture at elevated altitude to make themselves more easily heard at a distance (Galileo, in particular, had large classes), now protected against casual visitors by an appropriate fence. As a special token of friendship, this fence was opened for us and I was allowed to mount Galileo's chair to say a few words from it to my colleagues—without, to be sure, Galileo's rhetorical flair; yet mentioning a few things which would have been of interest to him as well.

But to revert to our scientific aims, at Manchester we were developing new methods for monochromatic photometry of faint celestial light sources, with the aid of interference as well as chemical light filters accepting wide-angle beams (cf Ring 1955, or Kopal and Millns 1956); partial results of this work were presented at a conference on astronomical optics, held at Manchester in April 1955 (cf Kopal 1956a). At Asiago we used a simplified version of Meinel's reducing camera mounted in the prime focus of the observatory's 120 cm reflector; and at Jungfraujoch, an 8 inch fast Schmidt camera (of focal ratio $f/0.7$). Dr Chalonge of the Institut d'Astrophysique in Paris kindly allowed us to mount our camera on his equatorial table. However, the work which commenced auspiciously in 1954 ended tragically the next year, when Peter Yorke Millns, one of my first students, who chose this line of work for his PhD research (see figure 5.10) lost his life on 19 December 1955 in a deep crevasse of the upper Guggi glacier a few hundred yards from the observatory†; and no one has come to continue his work since.

**Figure 5.10** Peter Yorke Millns (1933–1955), the young Manchester astronomer who lost his life in an unfortunate accident at the Jungfraujoch observatory on 19 December 1955, in the second year of his research on galactic hydrogen emission in H$\alpha$ light.

This did not mean, of course, that the topic itself was abandoned by Manchester astronomers; the opposite was the case. It is true that when Dr Ring left Manchester in 1962, he abandoned astronomy for industrial physics and was eventually lost to science. However, the work itself (extended more recently to Fabry–Perot techniques) was brilliantly advanced by Mr (since 1965 Dr) John Meaburn, one of the most original and creative research students I ever taught. His work on the detection and spectrometry of faint light rapidly earned John an international reputation in this field; and his book on this subject (Meaburn 1976) has appeared in more than one language.

As far as I was concerned, my own romance with the Alps also meant, incidentally, the end of the journey embarked upon in the

† For his obituary see Kopal (1956a).

summer of 1942, when the US Navy had summoned me to war duty, and kept me at sea-level for several years—not only geographically, but also mathematically. For apart from only isolated original contributions which I made to that field during 1957/58 when on sabbatical leave at the US Army Mathematics Research Center of the University of Wisconsin in Madison, and concerned with the application of operational methods in the calculus of finite differences—or, more specifically, the use of Padé approximants in numerical integration of differential equations (cf Kopal 1958; or Chapter IX together with Appendix II of Kopal 1960)—or the use of Tchebysheff polynomials for optimum interpolation (Kopal 1960, Appendix I), I never did any further work in numerical analysis.

The next item on the list of work brought over to Manchester in half-finished form was the *Catalogue of the Elements of Eclipsing Binary Systems*, a work on which Mrs Shapley and I had embarked at MIT in 1948 under the auspices of the American Philosophical Society (under the watchful eye of H N Russell); and which I now had to bring to an end myself. It was completed in 1956, and appeared on pages 141–221 of the first volume of series I of *Astronomical Contributions* published jointly by Jodrell Bank and the Astronomy Department at the University.

As could have been expected of work which took eight years to complete, much had been learned in the process of its construction that, because of its more general interest, transcended the immediate aims in view. The methods on which the work was based were previously summarised in a separate book, already referred to in Chapter 4 (cf Kopal 1950). However, no spin-off which resulted from this work exerted a more profound impact on research into close binary systems in the years to come than what transpired, at this stage, from a systematic application of the geometry of the so-called 'Roche model' to our problems; and this perhaps calls for a few words of explanation.

More than a century before, as a by-product of cosmological speculations on the origin of the solar system, the French mathematician Edouard Albert Roche (1820–1883; figure 5.11) discovered that the expressions for equipotential surfaces of self-gravitating massless envelopes which surround a revolving gravitational dipole in hydrostatic equilibrium can be described by closed algebraic forms (Roche 1849), which are, in fact, identical to the Jacobi energy integral of the restricted problem of three bodies in the absence of motion.

For more than eighty years, the Roche model led a shadowy existence on the periphery of cosmology as a mathematical museum piece of dubious application. It did not come to life until it transpired that (largely through the work of Eddington and his contemporaries in the 1920s on the internal structure of the stars) the Roche equipotential can simulate the actual form of the components of close binary systems to an astonishing degree of accuracy. It was the subsequent numerical work by Chandrasekhar (1933) on the equilibrium figures of distorted polytropic gas spheres that disclosed the full extent of the usefulness of the Roche model for such a representation; and when the present writer proved later that for stars of sufficiently high central condensation the same continues to be true, regardless of the actual run of the density distribution in their interiors (Kopal 1953), the value of the Roche model as a tool for describing the actual shapes of the stars constituting binary systems was established beyond any doubt.

**Figure 5.11** Edouard Albert Roche (1820–1883) of the University of Montpellier in southern France, father of the 'Roche limit'.

What it can do, in effect, is permit us to describe the hydrostatic shape of both components in such systems by three parameters only (the mass ratio of the system, and the values of the potential prevalent over the free surface of each star) and dispense completely with any semi-axes of the respective stars for this purpose: the latter are implied in the shape of the corresponding equipotential. If, moreover, the mass ratio of the system is known to us from spectroscopic observations, the values of the potentials specifying the surface in eclipsing binary systems can be inferred from the shape of their light curves.

In order to make this description of the system work we need, of course, to connect the geometry of the Roche surfaces with the physical parameters defining them. By 1948, it was not yet possible to do this because of a lack of sufficiently accurate numerical data on the Roche model. I set out to remedy this in my first years at Manchester: the computations necessary for a comparison of theory with observations were completed in 1953, and eventually appeared in print a year later (Kopal 1954a). Their application to observations went, of course, hand in hand with the preparations of the *Catalogue of the Elements*, from which the values of the potential were extracted; and as these progressed, very curious phenomena began to emerge from accumulating evidence which could be summarised as follows.

For each mass ratio, a limit (referred to as the 'Roche limit') exists beyond which no closed equipotential exists that is capable of containing the whole mass of the configuration. As long as both stars constituting such systems remain on the main sequence, their fractional dimensions remain well below their respective Roche limits. However, as soon as one component evolves away from the main sequence to become a subgiant, it quickly fills its Roche limit (the well-known system of Algol being a typical example); other binary systems (of W UMa type) exist where this appears to be true of both stars. These three groups I described as 'detached', 'semi-detached' and 'contact' (Kopal 1955a, 1956c)—terms which have since become familiar in the literature on the subject; and their existence has become the basis of all subsequent work on the evolution of close binary systems.

Up to a point, there seemed little room for doubt concerning the interpretation of such a situation. In particular, the observed clustering of fractional dimensions of the subgiant components of semi-detached systems around their Roche limits could only mean that these stars are secularly expanding towards this limit; and once this limit has been attained, a continuing tendency to expand should bring about a secular loss of mass. But from the nuclear theory of internal structure of the stars we know that the rate of evolution depends critically on mass: the larger the mass, the more rapidly the stars should evolve. According to this picture, subgiant components in close binary systems should be the more massive of the two, while observations disclose that the *opposite* is the case—an 'evolutionary paradox' which has been with us ever since, and which has not been satisfactorily resolved up to this time.

One line of approach followed from the beginning was the surmise that, once a component has attained its Roche limit, a continuing tendency to expand brings about a secular loss of mass by which the original roles of both components can be reversed. For instance, the present subgiant component of Algol, now 4.7 times less massive than its main sequence mate, would once have been the more massive of the two. But quantitative difficulties in the way of such an explanation of the 'evolutionary paradox' are rather formidable. All one can infer with confidence from the observed properties of binary stars is the fact that their components in the post-main sequence stage are bound to lose a large amount of their initial mass, but probably through high-velocity 'stellar winds' rather than by mere hydrostatic overflow.

A fuller discussion of these problems has subsequently been given by the present writer in Chapter VIII of his book on the *Dynamics of Close Binary Systems* (Kopal 1978) and more recently summarised in his Vainu Bappu Memorial Lecture at Bandung (Kopal 1984), to which the technical reader is referred for more information. The existence of 'semi-detached' or 'contact' systems among eclipsing variables was, however, already known to me before 1953 (the year when my detailed investigation of the geometrical properties of the Roche model went to press). But originally I was in no hurry to publish the findings, mainly, I suppose, because I continued to await the discovery of systems in which the evolved component would turn out to be still the more massive of the two—i.e., of the 'missing links' in the evolutionary picture which were then (as they are now) conspicuous by their absence in the sky.

That the subject was 'unveiled' in 1954 was an accident. In July of that year I attended the International Astrophysics Colloquium at Liège, at which Professor Bengt Strömgren delivered a summarising address on the theory of stellar evolution; and in the course of it he mentioned that subgiant stars are probably in the process of secular expansion. I thought that I knew of additional evidence showing that this is the case; and in the discussion which followed I presented the argument outlined in this section. Great was my surprise when Strömgren replied that his point was based on the same argument; and that he had heard of it already at Berkeley from Mr John A Crawford, a graduate student at the University of California working under the supervision of Professor Otto Struve. And, indeed, some weeks later Struve sent me a copy of Crawford's paper which was then being sent to the *Astrophysical Journal* (see Crawford 1955), containing the same idea as mine. In the meantime, Professor Swings (Chairman of the Liège Symposium at which I delivered my own remarks on the subject) invited me to write up these remarks for press; and these eventually appeared on pages 684–5 of the proceedings of the same session which contained Strömgren's introductory address (cf *Mem. Soc. R. Sci. Liège* **15**(1), 1954).

Crawford's investigation and mine were, of course, entirely independent; and we did not learn of each other's work until both were ready for release. That the components of Algol-type systems probably differ in shape had already, to be sure, been pointed out by the German astronomer K Walter in 1931; but prior to Crawford's work and mine no-one seems to have noticed that their fractional dimensions coincided (within the limits of observational errors) with their respective Roche limits. The evolutionary implications of this discovery, as developed in my second paper on this subject (Kopal 1956c) attracted considerable attention. Struve gave them a rather enthusiastic write-up in his

*Sky and Telescope* series of articles (cf Struve 1957); and from that time the literature on this subject has grown to many thousand of pages. There was much premature jumping to conclusions in those years, which later had to be abandoned; and some of this still continues to be true. But there is no doubt that, in the phenomena exhibited in the sky by close binary systems, Nature has provided many powerful clues to unlock the story of stellar evolution; and at present their use to this end is still far from being exhausted.

My work on close binary systems briefly referred to above filled almost a full decade (1950-1959) of my life, and culminated in my book on *Close Binary Systems*, written in the course of 1957, which appeared in London and New York in 1959 and for almost twenty years remained the standard source of information on its subject. But—I hasten to add—binary systems were not the only celestial objects of scientific interest to different members of the department. Already in the first six years of its existence, the postgraduate school of astronomy at the University of Manchester had grown by June 1957 to the group shown in figure 5.12.

**Figure 5.12** By 1957, the graduate school of astronomy at the University of Manchester had begun to gather momentum. Standing (left to right): F Holden, M J Elphick, A H Batten, J Hazlehurst, T W Olle, N J Woolf, J W Owen, G Fielder, W L W Sargent, J E Geake, J K W Davies. Seated: J Ring, R Kurth, M Hodgkinson, Z Kopal, E B Finlay, F D Kahn and W R Bradford. Several of the young scholars shown on this photograph now hold senior academic positions in Britain, Canada and the United States.

Some followed in my professional footsteps, such as Alan Batten, President (1982-85) of the IAU Commission on Close Binary Systems, Dr Miroslav Plavec (who came twice from Czechoslovakia to Manchester, as a postdoctoral fellow in 1957 and again in 1958 to complete his conversion from meteor to double-star astronomy); or

John Hazlehurst who likewise remained faithful to close binaries throughout the major part of his subsequent professional career. But others were attracted by different subjects: thus Franz Kahn, since 1966 Professor of Astronomy at Manchester (and, since 1981, my successor in the Chair of the department) made a name for himself in the field of hydrodynamics of the interstellar medium; and the same is true of Wal Sargent who, in the fullness of years, became Chairman of the astronomy department at the California Institute of Technology in Pasadena.

As time went on, our postgraduate school continued to grow: in later decades it more than doubled in size (attracting as it did an increasing number of students from different parts of the world (see figure 5.13)); and these still did not include postdoctoral scholars who came to work with us for months, and sometimes years (figure 5.14). But whatever their vintage three faces recur on them all without fail: these 'three musketeers' being Franz Kahn, Ellen Finlay (since 1973, Mrs Carling)—and myself!

**Figure 5.13** Postgraduate students and staff of the Department of Astronomy, University of Manchester, in November 1973. Third row: P Lamb (UK), R Carling (UK), S A E G Falle (Italy), D F Gregory (UK), M B El-Shaarawy (Egypt), C Knight (UK), G Tenorio-Tagle (Mexico), M Luheshi (Libya), F P Wheeler (UK), E Hamzaoglu (Turkey), C Mannion (UK), R Wilson (UK). Second row: M Dopita (UK), N Tebutt (UK), K H Elliott (UK), Dr J Meaburn (UK), Dr J E Dyson (UK), A A da Costa (Portugal), H Billing (UK), D Mason (UK), W Haymes (USA). Front row: M Akcyali (Turkey), H Livaniou (Greece), M Kamala Mahanta (Assam), Professor F D Kahn (UK), Professor Z Kopal (USA), Mrs E B Carling (UK), Mrs M Gorman (UK), H Panagiotopoulu (Greece).

**Figure 5.14** Dr (later Professor) Masatoshi Kitamura of Tokyo National University and Mrs Miwako Kitamura, our guests at the University of Manchester 1960–1962, on the wedding of Miss Georgiana Ludmila Kopal to Dr William Edwin Rudge IV in the parish church of Wilmslow on 24 June 1962.

The increasing size of the school called, of course, for a parallel expansion of our educational facilities; and these were indeed provided by the university. Up to 1966 our original home was in the Schuster Building of the Physical Laboratories on Coupland Street (a large part of which we gradually filled); but in the autumn of that year we were moved to a new building (see figure 5.15) the top floor of which we have been occupying ever since. This floor has provided adequate space (though no more than that) for staff and students, a reference library, computer outlets, room for measuring instruments and workshop for precision mechanics, as well as darkrooms for students and staff.

That no telescopic facilities for optical astronomy were planned for the Manchester area is perhaps understandable, in view of the Manchester reputation (exaggerated, to be sure, by oral tradition) for fine weather! However, collaborative arrangements were soon established with other European observatories (and, eventually, also those outside Europe), such as the Asiago Observatory of the University of Padua in Italy, Jungfraujoch Observatory in the Swiss Alps, and the Observatoire du Pic-du-Midi in the French High Pyrénées, under the terms of which we provided auxiliary instruments for special tasks to be used jointly at local telescopes. A particularly extensive collaboration of this type was maintained with the Observatoire du Pic-du-Midi for more than fifteen years (between 1956 and 1971), concerned mainly with work on the Moon (see the next section of this chapter), for which we also provided a special telescope of 107 cm free aperture with a Grubb Parsons mirror.

In later years, especially since my former student John Meaburn joined our staff to develop a flourishing group dedicated to nebular work with Fabry–Pérot spectrometers, we also gained access to even larger instruments in the southern hemisphere (such as the 3.9 m Anglo-Australian telescope at Siding Springs, the 1.94 m Grubb Parsons telescopes of the Helwan Observatory in Egypt, or the South African Astronomical Observatory). Since, however, I no longer took much of an active part in field work in those years, an account of it will have to await the reminiscences of my younger colleagues in the future.

**Figure 5.15**  The seventh floor of the building on the left has been the home of the Astronomy Department of Manchester University since 1966.

But with all this going on in the past thirty years or more, it is perhaps not surprising that our former students trained at Manchester have in the meantime gradually spread all over the world. In certain countries, such as Greece, Iraq or Turkey, astronomers with Manchester degrees may now outnumber those of any other foreign provenance; some of them have since become directors of their national observatories or presidents (rectors) of their universities. I am only waiting to see which of them will become the first Minister of Education in his or her country.

Moreover, more than once in those years we were called upon to host international conferences of experts or schools of advanced studies, such as a conference on astronomical optics (1955), on selenodesy and mapping of the Moon (1966) or on remote sensing (1979). In 1967, Manchester became the seat of the first of the IAU–UNESCO supported International Schools for Young Astronomers; and, in addition, my colleagues and I have taken an important part in subsequent such schools held in Hyderabad (India, 1969), Athens (Greece, 1975), and, most recently, Lembang (Indonesia, 1983). Twice in my years of service I was called upon to serve as Scientific Director of NATO Advanced Study Institutes—in 1971 at Patras, Greece (on the Moon) and 1980 at Maratea, Italy (on photometric and spectroscopic binary systems); and other conferences in which I was called upon to participate are too many to be recounted here in full. To attend these took me (since 1962) a dozen times around the world; and who knows how often I may still be called upon to do so in the future!

## Space Science and Astronomy of the Moon

But to return to the first decade of my activities at Manchester, its last years brought in their wake events which suddenly deflected my professional work to a very different direction: namely, to deal with problems which arose in connection with the sudden onset of the Space Age in the autumn of 1957.

The academic year of 1957/58 was the first in which I became entitled to sabbatical leave from Manchester; and from the offers I received at that time from the United States I accepted an invitation from the University of Wisconsin in Madison to spend ten months as the guest of its Mathematics Research Center then created within the University under the auspices of the US Army, and directed by Professor R E Langer (1894–1968; brother of Harvard's well-known historian, W Langer). The attractive feature of the offer was the fact that the University of Wisconsin had also a renowned astronomy department (to be housed in the same building as the Mathematics Research Center); and its former head, Professor Joel Stebbins (1878–1966), was a life-long student of Algol and other eclipsing variables, a tradition ably upheld by my dear old friend C M Huffer (1894–1981). Morse and Elizabeth Huffer spent their own sabbatical year 1948/49 with me at MIT in Cambridge; and I looked forward to an opportunity of returning the call.

It is not much more than a quarter of a century since the time I spent in Madison—but how few of the colleagues I knew there are still among us today? Of the mathematicians on the faculty of the University, Professors Langer, MacDuffie, Dean Ingraham are all gone; and so is Professor Joe Hischfelder, a renowned chemist, and his wife (a mathematician), with whom I spent many a pleasant evening in their home; or Julian Mack in physics! The same is also true of other foreign visitors to MRC—Arthur Erdélyi from Edinburgh, Louis M Milne-Thomson (Greenwich), Alexander M Ostrowski (Basel) or Eduard L Stiefel (Zürich), which whom I consorted in the office during daytime; and otherwise at the Claridge Apartment Hotel on West Washington Avenue (where most of us used to live during our visits). All gone now, and replaced by a new generation: *mortuos plango*!

But it is not only at the University of Madison that I often dwell in my recollections. Through Elizabeth and Morse Huffer I became acquainted also with other local personalities; and none was perhaps more important, and more lively of mind than Philipp LaFollette, former Governor of the state of Wisconsin, and son of Senator Robert M LaFollette, the legendary father of the Progressive Republican League in the first decades of this century (his son Robert, Philipp's brother, was also for many years a US Senator from Wisconsin).

By the time I got to know him, Philipp LaFollette was in his seventies and long retired from politics. But he was still very much interested in the world around us, and a living chronicle of US politics (and politicians) of this century—freely sharing his knowledge with those who cared to listen. The summer of 1960, when I last saw him (he died not long afterwards), was one of the Presidential campaign, in which John F Kennedy and Richard M Nixon were competing for first prize. Phil LaFollette knew them both; and his shrewd observations were sometimes not fit to print. But from what I recall, had he lived longer, neither Dallas nor Watergate would have come as very much of a surprise to him.

Another friend of the Huffers was Mrs Rauschenbush, daughter of the liberal Supreme Court Justice Louis Brandeis of the Woodrow Wilson era (who was, I believe, born in Czechoslovakia), whose sister Mrs Tuchman is a well-known writer. I sometimes took part in weekend excursions with a group to which Mrs Rauschenbush belonged, but seldom talked with her much; for she was (or if she lives, still is) a lady of very definite opinions which it would have been useless to question, let alone oppose.

In inviting me to join the MRC at Madison, Professor Langer was not very explicit as to what his Center expected of an astronomer of my professional qualifications; and I suspected first that it was connected with my previous work in numerical analysis (my book on this subject had appeared only two years before). This was (as I later found out) only partly the case. The main reason transpired on 4 October 1957—a memorable date for mankind—when the first artificial satellite (*Sputnik I*) was successfully launched into geocentric orbit by the Soviet Union as a forerunner of the more than 3000 spacecraft that have since followed in its wake.

That curtain-raiser of October 4, followed as it was by several others in quick succession, produced at that time an almost cataclysmic effect on humanity not sufficiently prepared for it; and gave rise to an unprecedented 'space rush' which drew many of us into its vortex. *Sputnik I* I never saw; for it became visible at Madison only in the small hours of the morning. *Sputnik II* (lauched on 3 November with the dog Laika on board) I did, however, spot in the evening sky while walking on the shores of Lake Mendota. *Sputnik III* (launched on 15 May 1958) I saw many times—the last occasion being, I believe, in the summer of that year at Tashkent airport while returning from Samarkand to Moscow (see p. 283). From that time, however, the population of *sputniks* in the sky began to grow by leaps and bounds; though most of these have been visible only from the south. While watching a *son et lumière* show at the Karnak temples of Luxor in upper Egypt in the autumn of 1977, I spotted at least a dozen of them crossing our skies in the course of a single evening performance!

Much of the early work on these man-made spacecraft is no longer significant enough to be recalled in any detail, but much of it was bound to benefit astronomy in the long run; and so it happened with certain work which I was encouraged to undertake concerning the equilibrium figures of self-gravitating celestial bodies in which their satellites revolve. I did so by a generalisation of Clairaut's theory correct to terms of second order in their departures from spherical form—Darwin (1900) or de Sitter (1924) had done so before only for effects arising from axial rotation—and the results appeared in a monograph published by the University of Wisconsin Press under the title *Figures of Equilibrium of Celestial Bodies* (Madison, 1960).

Subsequent generalisation of this theory to terms of third and fourth order in surface distortion was carried out in the 1970s by my students and myself at Manchester, and will be referred to later. However, by far the greatest impact exerted on my subsequent work during the time spent at Madison (where I again returned in the summers of 1959 and 1960) arose out of my increasingly close acquaintance with Professor Harold C Urey (1893–1981), then of the University of Chicago: probably the greatest man of science I got to know in my life at close range, and who exerted greater influence on much of my work of the next twenty years than anyone else.

It is scarcely necessary to introduce to our readers this truly outstanding man of science whose renown is revered throughout the whole world (see figure 5.16); but a few notes on his heroic career spanning almost a whole century may not go amiss. Harold C Urey was born on 29 April 1893 in Walkerton, Indiana, and his early interest in biology, which he developed as an undergraduate student at the University of Montana, soon drifted to the field of physical chemistry, a subject in which he received his doctorate from the University of California in 1923, and which remained the main field of his professional interest for more than his first fifty years. The discovery of deuterium and heavy water in 1934 at Columbia University earned Urey his Nobel Prize in the same year; and his work in subsequent years (including World War II) on mass spectrography and separation of isotopes (including uranium) made him one of the leaders who were instrumental in ushering nuclear physics and chemistry into the industrial age. In post-war years Harold Urey, then in his fifties, became an 'elder statesman' of atomic physics, adviser to the President, and a member of many international bodies planning the future. No fewer than 24 universities in many countries conferred honorary doctorates upon him; and eight academies claimed him as one of their own.

**Figure 5.16** Professor Harold C Urey (1893–1981) at a conference held in the Lunar Science Institute of NASA's Johnson Manned Space Flight Center in Houston, in the autumn of 1971.

Many scientists among his contemporaries with similar achievements to their credit preferred to rest on their laurels at this stage, and most of the early Nobel Prize winners have indeed been content to live for the rest of their lives off their past; but not Harold Urey! Instead, in the late 1940s his active mind was casting around for new problems worthy of his mettle. In retrospect, one could almost describe him as a 'steady developer'; for I have no doubt that, on purely scientific grounds, the future will value the contributions to science made (or inspired) by Harold Urey in the second half of his life even higher than those he made before the age of fifty. To be sure, he won his full Nobel Prize when he was only 41; but what for most scientists would have been the zenith of their professional career appears to have been almost incidental for Urey; and he would have qualified for the same award several times over later on, had his new fields of professional interest made him eligible for it.

Astronomers and students of the celestial bodies in general can only rejoice that, in mature age, Urey's interests began increasingly to run parallel with their own; and this is how it happened. After the end of the War in 1945, Urey did not return to his Chair at Columbia University, but left it to accept the Distinguished Service Professorship of Chemistry at the University of Chicago, where one of his new colleagues was Harrison Brown—a keen student of the meteorites. In the late 1940s, however, Harrison Brown left the University of Chicago for the California Institute of Technology in Pasadena, leaving behind a laboratory and students with interest in that field. Urey stepped into the breach as a courtesy to his departed colleague; but soon he found the field to be of surpassing interest to himself and, once attracted, he remained faithful to it for the rest of his life.

Urey's contributions to cosmochemistry, dating back to that time, were on a massive scale, and characterised by even greater originality than his earlier work in nuclear chemistry. Certainly, this time, his subject transgressed the doors of terrestrial laboratories and roamed freely through the explored parts of the cosmos!

In collaboration with his colleague Hans Suess, Urey made a fundamental contribution to our knowledge of the abundances of the elements in the solar system and was one of the first to underline the cosmological significance of carbonaceous chondritic meteorites. With his pupil Cesare Emiliani, Urey developed a new method for measuring the paleotemperatures of fossil seas (from the ratios of calcium isotopes in fossil sea-shells). With another group of students (Harmon Craig, Stanley Miller and Gerry Wasserburg) he pioneered the synthesis of amino acids by electrical discharge in a simulated primordial atmosphere of our planet. The possibility of life in the Universe outside the Earth exerted a lasting and powerful attraction on Harold Urey; and traces of organic compounds isolated by Nagy and others from carbonaceous chondrites greatly intrigued him. The origin of life continued to attract his mind till the end of his life; and my own last discussion with him (less than two years before he died) was concerned with this subject.

In 1953, Urey published a book on *The Planets, their Origin and Development* (based on his Silliman Lectures at Yale University the year before), which marked a milestone in the study of this subject. Prior to that time, solar system astronomy—that Cinderella of the astronomical family in the first half of this century—was still very largely based on outdated methods of observation, and drifted into the hands of amateurs. It was Urey who, by introducing modern physics and chemistry into the study of the solar system in the 1950s (as Eddington and others had done for the stars thirty years before), re-established academic respectability for this field. Moreover, he did so with such penetrating insight that the subject has never been the same since.

In these circumstances, it was only natural that when the Space Age was opened up so suddenly by the Russian *sputniks* in the late 1950s, with the possibilities it offered to extend lunar and planetary work further afield, Harold Urey was called upon to lend inspiration, wisdom, and professional counsels to the planners of the newly-created National Aeronautics and Space Administration. What he did, and how well he deserved of what has been accomplished in the early days of the Space Age was recorded by a witness eminently qualified to do so (Dr Homer E Newell, the Deputy Administrator of NASA) on the occasion of Urey's eightieth birthday in 1973 (see figure 5.17). Suffice it to stress that it was Urey's unflagging persistence in taking account of all evidence bearing on the subject (be this the Moon, the planets, or meteorites) and cautioning against explanations which rest on only a limited part of this evidence that made Urey uniquely qualified to perform the role of an inspirer as well as critic; and the way he discharged this role has already become part of history.

**Figure 5.17**  At a dinner on 1 May 1973, following a symposium to celebrate the eightieth birthday (on 29 April) of Professor Urey at the Lunar Science Institute in Houston. The jubilant can be seen on the left, and the present author (who served as the symposium chairman) is at the microphone. The face in the right-hand corner is that of the late Dr Homer E Newell (then Deputy Administrator of NASA).

I met him for the first time in the late 1940s, when he once came to give a colloquium on meteorites at Harvard Observatory; but our personal acquaintance did not properly begin until 1956 in England—the year when Urey came to Oxford as Eastman Visiting Professor. It did not take long for us to get together again on this side of the Atlantic. Urey's first colloquium outside Oxford on 'Diamonds, Meteorites and the Origin of the Solar System' was given at Manchester in the autumn of 1956; and from that time we remained in close touch till the end of his life.

The next academic year (1957/58) it was my turn to cross the Atlantic and spend a sabbatical year at the University of Wisconsin, not far from Chicago, which was then Harold's academic home. We visited each other several times that year (once, I recall, I walked for our rendezvous all the way from Chicago's Midway Airport to the University on foot!); and almost every time we ended up in front of a map (or some photograph) of the Moon in Urey's office, with myself listening to what Harold thought the stony sculptures on the surface of our satellite signified for its past, and for the past of the solar system in general.

It was, to be sure, not the first time that I beheld the face of the Moon in some detail, and was spellbound by its enigmatic relief. My acquaintance with it went back, in fact, to the enchanting months of the autumn of 1928 with the aid of my primitive home-made telescope (see Chapter 2); and, in more recent years, I had an occasional opportunity to snatch for this purpose a few moments with the 60 cm refractor of the Observatoire du Pic-du-Midi (see figure 5.19). Going back to my teens, I vividly recall having copied by hand (with futile excess of zeal) Nasmyth's map of the Moon (of 1874 vintage) from a book entitled *The Realm of the Stars* (1895) by Professor Gustav Gruss (1854–1922), which my father borrowed for me from the Prague University Library (and of which I later acquired a copy, still in my hands today). In later years I got hold of better maps of the Moon (in particular, the *Mappa Selenographica* by Karel Anděl, published in Prague in 1926), and followed (up to a point) the more modern literature on this subject, so that I was not completely ignorant of the facts which occupied Urey's mind at that time; or unaware of the problems arising in this connection.

However, when Harold and I were together at that time, in Madison or Chicago, and were looking at the maps and photographs of the Moon which filled the walls of Harold's study, I was at times unable to follow him in all his views—not because they were unreasonable (I never met anyone in my life whose views deserved greater attention), but because the observational data on the three-dimensional relief of the Moon then at our disposal were generally less accurate than he thought.

## 270  The Manchester years

In my opinion, the most promising method of improving these would be cinematography of sunrise or sunset on the lunar surface, as pioneered by McMath *et al.* (1937) twenty years before, using a sufficiently large telescope located at a site of superior seeing. We already had some experience with such work at Pic-du-Midi Observatory (see figure 5.18), where Bernard Lyot moved in 1942 a 60 cm coudé refractor of 18.22 m focal length (figure 5.19) and founded its reputation for the excellent quality of its images, permitting a resolution of half a kilometre on the lunar surface for about 5 per cent of the time. At the time of sunrise, horizontal measurements of plates taken with that telescope accurate to half a kilometre could thus enable us to distinguish altitude differences on the Moon of the order of only a few metres; and a continuous tracking of the length of the shadow would permit us to record the unevenness of ground on which the shadows were cast to the same order of accuracy.

**Figure 5.18**  Observatoire du Pic-du-Midi in the French Pyrénées (2860 m above sea-level) where a large part of the Manchester Lunar Programme between 1960 and 1966 was carried out. The dome containing the 60 cm refractor is on the left of the picture; that housing the 107 cm reflector is on the right.

I kept telling Harold this during our sessions in his office in the early spring of 1958; but once he countered my expositions with the words: 'Why don't you do something about it?' This was, in effect, the beginning of the Manchester Lunar Programme. I was tempted to accept the challenge, but hesitated on the gounds of the cost which such work would be bound to entail. Harold, however, waved such qualms aside: 'If you are willing to undertake this work, I shall help you to raise money which may be necessary for this purpose.' And he was as good as his word: some weeks later I was visited at Madison by Mr Charles Campen of the Air Force Cambridge Research Laboratories who offered contractual support for such an undertaking; and when all the details of the proposed arrangements were worked out by January 1959, I was greatly pleased that our Project Officer, Lt Col Trakowski of the US Air Force Research and Development Command in Brussels, happened to be a former student of mine from MIT, ten years before (cf p. 220)!

**Figure 5.19** The 'equatorial coudé', of 60 cm free aperture and 18.22 m focal length, in the western dome of the Observatoire du Pic-du-Midi (dismantled after 1970), with which much of the Manchester Lunar Programme (consisting of some 60 000 ciné frames taken on 9 inch films) was carried out. (Pin-hole camera photo by Charles Lowe.)

And so we embarked on our work, originally intended only to explore the full capabilities of the ciné-method as a tool for three-dimensional mapping of the lunar surface; but its scope did not remain so limited for long. As is well known, in January 1961 President Kennedy proclaimed (in his 'State of the Union' message) a manned exploration of the Moon as one of the national goals for the 1960s; and although the President himself did not live long enough to see its final triumph, its attainment (on schedule) between 1969 and 1972 will belong among the glories of human history in our turbulent century;

and we at Manchester were soon called upon to play a role (albeit modest) in advancing towards this goal.

In response to the increased urgency of the task which we had initially agreed to undertake as a piece of research, and in recognition of the promise borne out by it, early in 1960 we were 'upgraded' to a task force to keep pace with other phases of the grand design which culminated in the Apollo landings of 1969 onwards.

First, the sponsorship of our project was extended from the Cambridge Air Force Research Laboratories (AFCRL) to include the Air Force's Aeronautical Chart and Information Center (ACIC), which was commissioned by the National Aeronautics and Space Administration (NASA) to provide the maps for all lunar landings; and to this day we think with genuine pleasure of our liaison officers, Dr John W Salisbury and Mahlon Hunt of AFCRL, as well as of Mr R W Carder of ACIC, who made our collaboration with their organisations so

**Figure 5.20** Participants at the first conference on 'Problems of Lunar Topography', held at Bagnères de Bigorre in France, 19–23 April 1960. Identification key (from left to right): Dr G von Schrutka Rechtenstamm (Vienna), Dr J Ring (Manchester), Major A G Kearns (USAF), Miss Zdenka Kopal (Manchester), Dr A Dollfuss (Paris–Meudon), Dr H Camichel (Pic-du-Midi), Miss Ellen B Finlay (Manchester), Professor K Koziel (Krakow), Messrs C A Campen and Mahlon S Hunt (both AFCRL), Professor Jean Rösch (Director of the Observatoire du Pic-du-Midi and host to the conference), Mr R W Carder (ACIC), Mr G de Gentili (Pic-du-Midi), Dr Th Weimer (Paris), Mr T W Rackham, the author, Mrs Kopal and daughter Eva (all from Manchester).

smooth and fruitful. And we all have been very grateful to our French colleagues at Bagnères and Pic-du-Midi (see figure 5.20), headed by the Observatory's Director, Dr (later Professor) Jean Rösch (1914–    ) for allowing us access to their unique site for our work which extended for several years. Jean Rösch and I are close contemporaries (with myself only three weeks the older); and an additional bond arose between us from the fact that we are both interested in mountains—on the Earth as well as on the Moon!

The work entailed in our joint programme for the study of the Moon was organised as follows. To obtain the actual photographs of the Moon on the Pic on 9 inch films with aerial cameras (since 1964, with a Data Camera specially built for this purpose) was the sole responsibility of the Manchester astronomers: the 60 cm refractor (figure 5.19) (and, since 1964, also the 107 cm reflector (Figure 5.21) built for this purpose as a joint Franco-British venture) was manned on all clear nights by our staff, among whom the stalwart Drs Meirion Jones, Thomas Rackham, Patrick Sudbury and Mr Brian Temple deserve an especially honourable mention. The work with the 60 cm refractor was the particular responsibility of Dr Rackham, and that with the 107 cm reflector of Dr Sudbury, who were assisted by a number of students and junior employees who changed from time to time. At the peak of our activities between 1964 and 1966 our total staff exceeded fifty colleagues of different degrees of seniority; and an example of their work can be seen in figure 5.22.

Figure 5.21 The photographic camera for lunar work, attached to the Cassegrain reflector of 107 cm free aperture at the Observatoire du Pic-du-Midi. (Manchester Lunar Programme.)

274  *The Manchester years*

**Figure 5.22** Sunrise over the lunar crater Tycho (87 km across) and its surroundings (north to top), taken by P V Sudbury on 30 March 1966 with the 107 cm reflector of the Observatoire du Pic-du-Midi, as part of the Manchester Lunar Programme. The definition of the image (corresponding to about half a kilometre at the distance of the Moon) is about as good as one can attain with any terrestrial telescope anywhere in the world.

These were indeed exciting times—because of the exciting tasks ahead of us, and the historical significance of the effort of which we were a part. The work itself was also exciting, both at Manchester and on the Pic. The travel from our beclouded north to the Pyrénées did not take long: from Ringway by air to Paris, and by overnight train to Tarbes, where we arrived in the morning to reach the Pic before noon, in time to meet the observers of the previous night getting up from their morning (not overnight!) rest, and partake with them of an excellent déjeuner for which the cuisine at the Pic was renowned. When dinner-time came, the domes as often as not were already open, and observers had to take turns, not only at the telescope, but also at the table.

Sometimes to get to the Pic from Manchester even faster, we flew from Ringway to Dublin, and thence to Lourdes in southern France (in close proximity to the Pic-du-Midi) by Aer Lingus flights. These flights always used to be full of ecclesiastics, with at least one or two bishops among them on their way to the tombs of the Apostles or some other ecclesiastic business (in company of lesser priests and monks or nuns). Their company gave us confidence that should, perchance, something go wrong with our plane, it would not crash, but only go

Astronomy of the Moon 275

up, and we could try to gate-crash (together with our fellow-passengers) into the Kingdom of Heaven. We never had a chance to put this theory to the test; but we did verify that (on the right days), by choosing this route, one could breakfast at home in Manchester, and lunch on the Pic.

But to return to our work, up to several hundred photographs taken on 9 inch films (just about wide enough to record the entire image of the Moon formed in the focal plane of the 60 cm refractor) were obtained each clear night when the Moon was above the horizon; but to develop all these on the Pic would have been impossible. Instead, the harvest of the preceding night was transported each morning by cable-car to La Mongie (of Tour de France fame), where a military vehicle was waiting to transport them to the US Air Force Base at Dreux (near Paris), and from there by air to the Headquarters of the Aeronautical Chart and Information Center at St Louis in the United States where they were developed, processed and deposited for subsequent distribution. Under ordinary circumstances, the difference

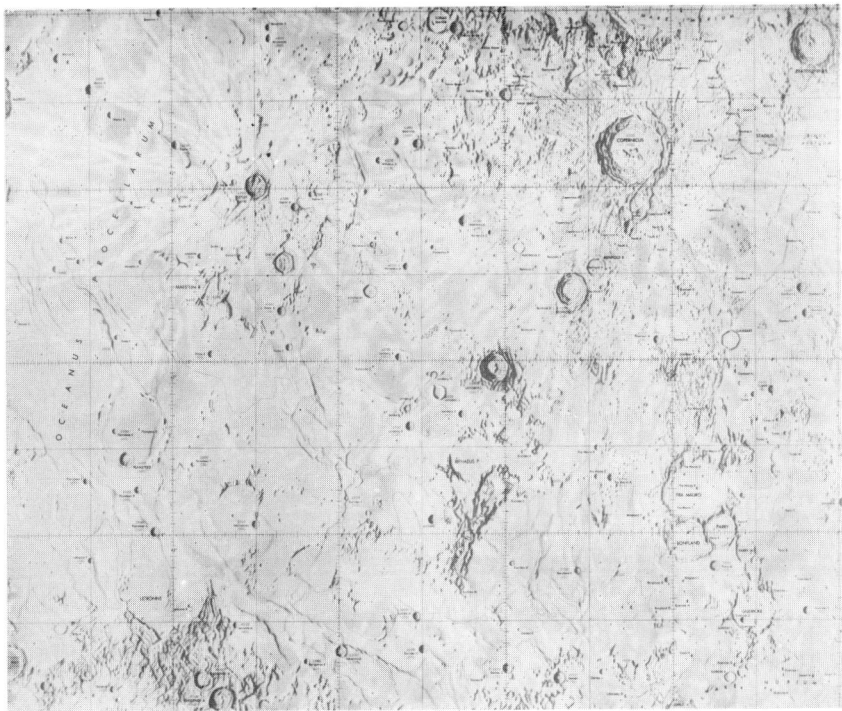

Figure 5.23  An example of the Lunar Air Force Chart of the region of Oceanus Procellarum on the Moon (south-west of the craters Copernicus and Kepler), constructed on the scale of 1:2 000 000, in collaboration with the University of Manchester.

between the time of exposure of a film on the Pic and its development at St Louis did not exceed 1–2 days.

Between 1960 and 1967 over sixty thousand photographs of the Moon were obtained at Pic-du-Midi, to serve as a basis for the well-known LAC charts of the Moon (an example of which can be seen on figure 5.23), constructed by ACIC for NASA to serve for a preliminary selection of Apollo landing sites. While the European end of the operations underlying this venture was primarily the responsibility of the present writer (ably assisted by Dr Rackham), with headquarters at Manchester, the American operations at ACIC were under the jurisdiction of Mr Robert W Carder (figure 5.24), a true friend of the Moon if ever there was one. A summary of this joint work was subsequently written up in a book (Kopal and Carder 1974). While a large majority of the photographs secured for the entire programme at Pic-du-Midi remained on deposit at ACIC in St Louis, several hundred of them found their way back to Manchester, where they served as a basis for the development of the methods for their reduction and actual construction of the maps themselves.

**Figure 5.24** Mr Robert W Carder, Chief of the Aerospace Charting Branch of ACIC in St Louis, Missouri (photograph taken in May 1974).

To do this entire job 'in-house', our Manchester facilities would, of course, have been utterly inadequate; but even so the total expenses of the Manchester end of the joint operation between 1959 and 1969 (provided by the US Air Force as well as NASA) exceeded one million dollars—not too large a sum, perhaps, in comparison with the costs of building and operating spacecraft; but for astronomers previously accustomed to support from more conventional sources this was a significant windfall. These years (and, in particular, 1960–1966) were one time of my life when administration occupied most of my time—a function in whose discharge I had the ready cooperation of the University's financial authorities (I recall with particular gratitude the help of Mr James Hanson, the University's Deputy Bursar and Finance Officer); and, within the Department, I enjoyed the able assistance of Miss Ellen B Finlay (figure 5.25), whom the University promoted officially to the rank of my Personal Assistant. But even so the number of my original publications in those years shows a conspicuous dip, which it took a second wind and some years to fill.

**Figure 5.25** Miss Ellen B Finlay of the Department of Astronomy, University of Manchester, sorting out lunar films for microdensitometric measurements in April 1962.

Needless to say our participation in lunar mapping at that time attracted also a certain amount of attention in the Manchester area; and in the spring of 1960 we were encouraged to mount a small exhibition of our photographs of the Moon from Pic-du-Midi in the University Library. The exhibit was opened on 28 March by Lord Hailsham, then Minister of Science (who happened to be visiting the University on other business), together with Professor W Mansfield Cooper (the University's Vice-Chancellor).

The Minister evinced considerable interest in our work, and in particular, in the site of Pic-du-Midi. In an obvious effort to be helpful, he made a suggestion which rendered us fairly speechless. 'I am also somewhat of a mountaineer; and know how clear and quiet the air can be in high mountains. But why go with your telescopes to their summits? Can't you pump down some of this clear air to the valley to photograph the Moon there in greater comfort?'

I confess that this had not occurred to us before, but I did not dare to explore such a possibility with the Minister at that time. At the lunch which followed I sat next to the Minister's secretary, who felt it incumbent upon him to cover up for his chief's fertile imagination; but added that 'otherwise the Minister is a very clever man.' What else could I have replied than 'I surely hope so'!

All this was more than 25 years ago. I have never met Lord Hailsham since, to discuss with him lunar mapping or any other subject. However, I still watch him sometimes (on television) when, in his present capacity of Lord Chancellor, he descends backwards from the dais after having handed the Queen her speech from the Throne; and think to myself 'you will not see the Moon through the clear air at the top of Mt Blanc any more'.

But to return to my own work, one reason why I published relatively little in those years was the fact that, administration apart, much of my time was spent in the development of new methods to carry out our tasks, the results of which are scattered in various reports of limited circulation. One of these developments may perhaps be of interest for the more general reader. As is well known, the most effective way of determining relative altitudes (or level differences) on the Moon is based on a triangulation of the shadows cast by any unevenness of the surface on the surrounding landscape in the oblique rays of the rising or setting Sun. At the time when the entire disc of the Sun (of half a degree apparent diameter) just clears the lunar horizon, any unevenness of the surface is magnified $\cot 0°\!.5 = 115$ times in the length of the shadow cast by it near the centre of the apparent lunar disc; and exceeds this degree of magnification in the penumbral zone illuminated by the only partly risen Sun. If, therefore, direct photography from the distance of the Earth can resolve shadows (say) a half to one kilometre in length (a feat practicable with the 60 cm refractor at Pic-du-Midi), these should permit us to identify vertical altitude differences of the order of only 5–10 metres, and down to one metre from photographs taken when the obstacle intercepts the first (or last) rays of the rising (or setting) Sun.

Prior to 1960, most photographs taken from the Earth were ill suited for this purpose; for they were taken with exposures optimised to record the maximum amount of information on the lunar disc visible

Astronomy of the Moon 279

at any particular phase; and this automatically meant that the penumbral zone, where shadows are longest, would be hopelessly underexposed; for the exposures needed to bring out information in such a zone imply, of course, gross over-exposure of the rest of the image, as demonstrated by the examples in figures 5.26 and 5.27. This was not fully realised prior to our own work in the 1960s; and the methods developed then (cf Kopal 1965, 1967) should permit Earth-bound photography to compete successfully with spacecraft in the three-dimensional mapping of the lunar surface up to the present time.

**Figure 5.26** Sunset over Mare Fecunditatis, taken at the $f/15$ focus of the 107 cm reflector of Pic-du-Midi on 2 March 1964. Exposures: left, 0.025 second, centre, 0.5 second, right, 2 seconds; all within the same minute. Note increasing definition of the terminator with increasing exposure time.

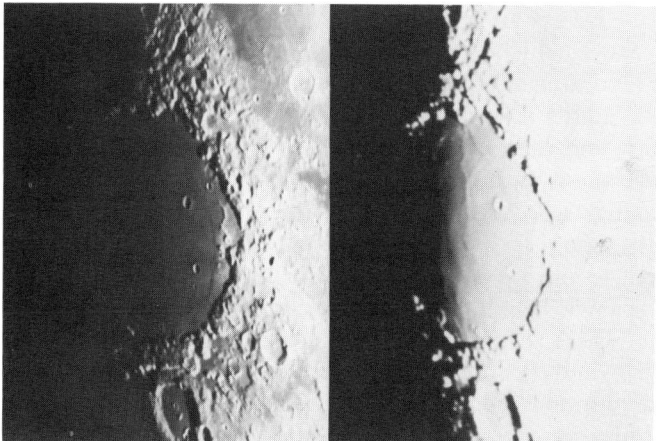

**Figure 5.27** Sunset over Mare Crisium, photographed on low-contrast plates with the 60 cm refractor at Pic-du-Midi on 2 January 1965. Exposures: left, 1 second, right, 50 seconds; both within the same minute.

This, of course, remains true only of very special tasks in lunar studies: for most others the advent of the spacecraft shifted the centre of gravity of lunar research in the 1960s very much closer to the Moon. As far as the mapping of the Moon was concerned, the usefulness of ground-based lunar photography effectively ended in 1966, with the advent of the NASA-sponsored Lunar Orbiting Satellites, built and operated by the Boeing Company in Seattle. Altogether five such photographic satellites were launched, all successfully, between 14 August 1966 and 5 August 1967; and the data supplied by them marked a new epoch in the degree of intimacy of our acquaintance with our satellite.

I too had something to do with these spacecraft and their use for lunar exploration; though not from Manchester. The extramural part of these activities commenced, in fact, in 1959/60, when after my first sabbatical leave at the University of Wisconsin I returned to its Mathematics Research Center (the building of which was damaged in the summer of 1964 by a bomb detonated by student 'activists') in the summers of 1959 and 1960. However, between 1961 and 1964 (when I became a member of the Lunar–Planetary Committee of the US National Space Board, then chaired by Harold Urey) my activities shifted westward. Instead of heading for Madison during my academic vacations I travelled (three times a year) to the Jet Propulsion Laboratory (JPL) of the California Institute of Technology (the rapidly growing national centre of lunar and planetary research in the United States) on consulting missions, the last of which (concerned with lunar *Surveyors*) was not completed until spring 1964.

The years immediately preceding (1961–1963) turned out to be some of the busiest of my life. For it was not only JPL or the US Army and Air Force which were in need of my occasional services, but also research laboratories of several American aerospace companies (Boeing, Chance-Vought, Douglas, Lockheed or North-American Aviation) which now and then needed advice (in their work for NASA) which I could give them within the limits of the time available.

In looking back years later on all these consulting activities, I recall them with mixed feelings. By feeding aerospace companies new ideas on what it would be desirable to do from the scientific point of view, men of science helped them to seek (and often obtain) Government contracts which earned their industries many millions of dollars. However, when it came to the reduction and interpretation of the data obtained with new instrumentation in space, aerospace companies gave the scientific community but very little help. There were large profits in building hardware, but not enough in the interpretation of the results. And so it happened that, in the 1970s, most aerospace companies in the US closed down their scientific research laboratories set up in the Apollo era (the Government did little better—only more slowly); with the result that a sizeable fraction of the scientific data gathered in the 1960s was not only not properly utilised, but actually misplaced or otherwise rendered unusable.

In this connection I recall (in particular) my visits in 1962 to the Chance-Vought Laboratories in Dallas, Texas, in the interest of the (later abandoned) Project Dyna-Soar, a distant predecessor of the space shuttle. I was once allowed to enter its mock-up training model and operated its controls for a simulated landing at Edwards Air Force Base in California. I did my best; and it was impressed on me that I should follow the landscape below shown on realistic film in real time, until the metallic voice of the loudspeaker behind pronounced: 'and now you are dead', meaning that I had already committed errors of navigation beyond repair and was bound to crash. I was thinking of these dread moments when, years later, I watched (on television) John Young piloting down to the same base the *Columbia* shuttle with obviously a much surer hand than I possessed at Dallas.

It must have been after one of these occasions (though not as a tribute to my navigational skill) that on 27 August 1962 I received at Amarillo, Texas, a certificate (under the seal of the State's Governor Price Daniels) making me an 'honorary Texas citizen'. I shared this honour with Dr Edward Teller, introduced on the platform as the 'father of the hydrogen bomb'; though Teller disclaimed this modestly, and acknowledged only being the father of three children! As for myself, although less than ten years later I was to spend almost a year in my newly-adopted state at the NASA Johnson Space Flight Center in connection with the Apollo 16 mission, I still have not to this day acquired a proper ten-gallon hat to make me look like a real Texan!

But more about this later. In 1963, while in California, I was allowed to enter the lunar landing module for one of the Apollo missions to the Moon, then being built by the North-American Aviation Corporation at Downey, Los Angeles. I was confined inside it for some minutes, and found the experience oppressive; but was told that this would be so only in the terrestrial gravity field; once in orbit, the claustrophobia would vanish. Later I was told by the astronauts who travelled in these capsules for many hours that this was indeed the case.

Those were the days when I used to be spending more time in jet planes flying from place to place than behind my desk, and no wonder. My yearly mileage used to be around 100 000; and friends figured out that during academic vacations (when most of this travel took place) my average speed used to be about 40 miles per hour (day and night!). During the first flight I ever took in my life, from Prague to Trieste in September 1938, the crossing of the Alps in an unpressurised DC2 was so painful on the ears that it took another twelve years before I ventured to take to the air again (for my first trans-Atlantic flight in October 1950); and in 1952 an Atlantic crossing aboard a Boeing Stratocruiser still took more than twelve hours to accomplish non-stop.

My first jet flight aboard a Russian TU 104 from Prague to Moscow for the Tenth General Assembly of the IAU in 1958 (and a flight from Moscow to Samarkand after these meetings) marked the beginning of an era of fast travel which made it possible for me to attain the 100 000 miles per year record (mainly aboard Boeing 707 jetliners); but it was not until 1979, on my first crossing of the Atlantic aboard a Concorde flight from London to Washington, that I experienced a passage through the sonic barrier and overtook the Sun in its diurnal motion.

That first jet flight from Prague to Moscow in the summer of 1958 brings back also memories of the Tenth General Assembly of the IAU in that city which I had not revisited for 22 years. Much had changed in Moscow since the time of the 1936 eclipse in Japan; and much had changed also in the organisation of the General Assembly. Unlike in the past (when the meetings commenced in the morning), the Moscow meetings, in accordance with local customs, commenced after 9 PM; and official speeches (in different languages, but overlapping in content) lasted till the small hours of the morning, while assembled astronomers were dozing off in their seats—dreaming, perhaps, of what the same magnificent Hall of Columns (the Club of Nobility in the Tsarist days) must have witnessed in the time of Tolstoy and before.

But we saw Moscow and its surrounding countryside also in full daylight on our excursion to the Monastery of Zagorsk, as well as at night during a reception to which the Soviet Government invited the Union in the Kremlin; and at which we were addressed (not in person, but through loudspeakers) by Chairman Krushchev.

By 1958, the membership of our Union had already become so large that no single hall inside the Kremlin could have taken us all; and we had to be distributed in several. My station was in the Georgievsky Hall, the walls of which were covered by thousands of gold-lettered memorial tablets of dazzling white marble, each commemorating an officer of the Order of St George killed in battle in defence of his country; in the part of the room where I was stationed for dinner, most of these tablets commemorated the victims of the battle of Borodino in 1812, which broke the power of Napoleon. In each of these halls the host was a senior officer of the Soviet Government; and in Georgievsky it was, appropriately enough, Marshal Rodion Malinovsky, hero of two world wars—a short but wide-shouldered officer (whose tunic was covered by more decorations that I have seen anyone else wear), radiating tremendous power. He stood there, aloof and motionless, as if he were alone in the room (wondering, perhaps, what he was doing there among so many astrologers); and nobody, Russian or foreign, dared to approach him with any small talk.

After the meetings in Moscow we were given an opportunity, still rare in those days, to visit Samarkand in central Asia and see the excavated remnants of the famous observatory of Ulugh Bek from the first half of the fifteenth century (see Chapter 7). Not much of it has survived to this day; the principal cause of destruction was not so much the varying fortunes of war since that time, but mainly earthquakes which are quite frequent in that part of the world. Professors André Danjon (then President of the IAU), Minnaert and I were invited by Professor V P Shcheglov (Director of the Tashkent Observatory, and our host at Samarkand) to pay a flying visit to the International Latitude Station at Kitab, located some 200 km (and behind a chain of mountains) south of Samarkand.

The trip was as exciting as it was instructive. Our local hosts were about to install Danjon's new *astrolabe à prisme*; and after lunch (when our senior colleagues were resting in the shade) Professor Shcheglov and I took a bath, *à la Russe*, in the waves of the Aksu-darya river, within sight of the northern outposts of the Pamir mountains. However, our return journey in the late afternoon turned out to be more trying: on the way down our little plane had been rather badly buffeted by air currents across the mountains; but on the way back the turbulence became severe, so that some of our fellow-passengers (including Mme Danjon) expected the worst any minute. But not so Professor Danjon himself, formally dressed as always, who never lost his habitual composure; though even he must have been secretly relieved when our plane touched ground at Samarkand Airport (this was still before the days of concrete runways) none the worse for wear.

The return journey from Tashkent to Moscow on a TU 104 jet did not lack its element of excitement. The weather on departure from Tashkent was clear (I recall that *Sputnik III* was just crossing the sky as we were walking across the tarmac to board our plane); but it was a different story in Moscow. Thunderstorms are quite frequent there in the summer; and one of them descended on the airport at the scheduled time of our landing. The pilot at first hesitated to try his luck, but later he had to (not to run out of fuel). In coming down he handled his plane as a dashing cavalryman would handle his steed: he descended through the thunderclouds with lightning flashing right and left everywhere around us; but fortune favours the bold, and we made it!

We were glad to have done so, and reached the terminal building safely (albeit wet); but our anabasis that night was not yet over. No bus was on hand to transport us to Moscow (apparently they did not expect us to land); and so we had to wait till one could come for us from the city. More than one hour later, as we were dozing in easy chairs at the airport and worried about the arrival of our bus, I recall that Professor Urey (who took part in this expedition) woke up from his slumber and, in somnolent state, murmured a statement which most people around would have found incomprehensible: 'You know, Zdeněk, the only thing which worries me is Wargentin'—by which he did not mean Per Wilhelm Wargentin (1717–1783), once director of the Stockholm Observatory, but a crater named after him on the Moon, remarkable for its uplifted floor, a feature which was as much of a puzzle for Harold Urey in 1958 as it has remained for the rest of us ever since.

This was my first post-war visit to the Soviet Union on astronomical business. The last one took place in the summer of 1974, to attend an international conference on the Moon in the post-Apollo era, convoked by the late Academician Vinogradov at the University of Moscow. During this conference most of us were housed in downtown Moscow in the new large 'Rossia' Hotel, within walking distance of Red Square. The days were long at that time of the year; and one day, Dr John Adams (a geologist friend from JPL) and I took an after-dinner stroll to Red Square. John is a quiet-faced tall man, and, like every good American, carried a large camera in the hope of an occasional snapshot. The evening was still bright, and Red Square full of people. As we strolled along, a young mother with a boy some five to seven years old was walking towards us on the same side of the street. The boy, a charming youngster, was, however, an observant little fellow: for on spotting John he tugged on his mother's sleeve, and pointing with his little finger at John, exclaimed excitedly: '*shpion*'! It must have been John's camera, or his poker face, which earned him this epithet. The boy's mother, in some embarrassment, tried to calm down the child (who may have been disappointed by a lack of credit for this discovery). However, the little incident brought home to us the fact that a systematic indoctrination of Soviet children against a stereotyped image of foreign spies must commence at a very early age.

Most of this concerned my travels outside Europe during the 1960s; but at least one of them could have ended fatally, and not by a failure of transport. In February 1961, when my connections with the Jet Propulsion Laboratory at CalTech were particularly close, Dr Albert R Hibbs (then Chief of JPL's Space Sciences Division) summoned me to attend an urgent meeting in Pasadena at short notice; and I booked a non-stop flight from London to Los Angeles to do so. Virtually at the last minute this meeting was, however, called off; and the news reached me shortly before I was due to depart from home for Manchester Airport.

And just as well that it did; for only hours later I began to develop symptoms which our doctor, Dr Ross Martyn (and God bless him for that!), identified as due to acute appendicitis. To make a long story short, before many more hours I found myself in hospital for an emergency operation, and none too soon; for if the Pasadena meeting had not been providentially called off, I would have been in the air at that time some seven miles above the North Canadian shield, aboard a Boeing 707 jet; and it was unlikely (as I was told subsequently by the surgeon) that I would have reached Los Angeles alive. If, therefore, you doubt that we all have a guardian angel, think again! I would not have been the first astronomer to meet an untimely end due to this cause; another one, well-remembered in Czechoslovakia, was Dr Jan Frič (1863–1897), brother of the founder of the Ondřejov Observatory. In my case, however, all ended happily; for only a week later I was back home at my desk, happy at the thought that inasmuch as the human body has only one appendix, no further interruption of my schedule caused by it could happen. When the delayed meeting took place at Pasadena some weeks later, I was present, none the worse for wear.

But to return to Manchester, in March 1961 (no sooner than I got out of hospital in an appendix-less state), the late Sir Harrie Massey, in his capacity as Chairman of the British National Committee for Space Research, invited me to organise (and chair) its sub-committee for Lunar and Planetary Exploration—a function which I held for the next three years. In this capacity, I served on more than one committee of the nascent European Space Research Organisation (ESRO), as well as on the scientific committee of COSPAR, chaired by the late Professor Pol Swings; with Professors L Biermann and M Minnaert among its faithful members.

The early 1960s were also the years when students of lunar research felt a greater need than before to meet at frequent intervals, to discuss their common tasks in person, and make their progress and views known also to absent colleagues through the published proceedings of such conferences. I had more than my fair share of such work as chairman (or member of the organising committee) of several such occasions, such as at Bagnères de Bigorre in France in April 1960 (see figure 5.20), or at the IAU Symposium in Leningrad in December of the same year (figure 5.28), the proceedings of which appeared in book form in due course (cf Kopal and Finlay, 1961; or Kopal and Mikhailov 1962). The last opportunity I had occasion to do so was in 1966 when the second International Conference on Selenodesy and Lunar Topography was held at the University of Manchester between 30 May and 4 June of that year. For a group photograph of its 42 participants from 13 countries see figure 5.29, and for the publication of its proceedings, see Kopal and Goudas (1967).

It should be added, in this connection, that our assignment at

Manchester within the framework of the general effort sponsored by AFCRL and ACIC was not limited to providing the observational data and developing methods for studies of three-dimensional topography of the lunar surface, but also to studying the global form of the Moon and its relation to the lunar gravitational field in which men were soon to orbit around our satellite. In order to gain new observational data needed to this end, we helped to plan, jointly with the Astronomical Sciences Laboratory of the Lockheed California Company, to send out two expeditions to observe the annular eclipses of the Sun by the Moon observable on 31 July 1962 from Kheur Soce, Senegal in Central Africa, and on 25 January 1963 near Oudtshoorn in the Republic of South Africa.

**Figure 5.28** At IAU Symposium number 14 on 'The Moon', sponsored by the USSR Academy of Sciences at the Pulkovo Observatory near Leningrad, 5–11 December 1960. Front row (left to right): G P Kuiper, Mrs Zdenka K Mikhailov, E M Shoemaker, H C Urey, Z Kopal, T W Rackham, A C Hibbs, A Dollfuss and A A Mikhailov (Director of the Pulkovo Observatory and host of the symposium). Second row: J Rösch, K Koziel, T Gold, J G Davies, J Ring, V V Sharonov, A C Mason, N Bonev and M S Zverev (?).

Both expeditions, led in the field by Dr Larry G Stoddard (1912–1966) of Lockheed California, were equipped to film both events with long-focus optics (see figure 5.30) and both were highly successful. The differential measurements of lunar cross-sections (exposed to view from the Earth at the time of eclipse) with respect to the solar limb (theoretically a circle at all times) were equally successful (cf Davidson *et al.* 1967). The Manchester field participant of the expedition to Senegal was our technician Jim Hallows (who had some

**Figure 5.29** Second International Conference on Selenodesy and Lunar Topography, held at the University of Manchester, 30 May–4 June 1966. In front, from left to right, are Mrs B Sudbury, Miss E B Finlay, Mrs J Finlay-Reid and Mrs A Rifaat. Sitting: Professor S Miyamoto (Japan), Professor J Rösch (Pic-du-Midi), Professor G Colombo (Padua), Professor K Koziel (Krakow), Mlle O Calame (Paris), Professor A H Samaha (Cairo), Mr A Nowicki (AMS, Washington), Professor J Hopmann (Vienna), Professor M J G Minnaert (Utrecht), Dr R H Stoy (South Africa), Professor R V Karandikar (Hyderabad). Standing: Dr M E Davidson (Manchester), Dr P V Sudbury (Manchester), Dr (now Professor) M D Moutsoulas (Manchester), Dr J Maslowski (Krakow), Mr A Orszag (Paris), Dr Th Weimer (Paris), Dr (now Professor) C L Goudas (Patras), Dr D H Eckhardt (AFCRL), Dr R Julian (Hughes Aircraft), Professor Z Kopal (Manchester), Col C S Downie (USAF), Mr J Sasser (NASA–Houston), Mr M S Hunt (AFCRL), Col W H Chappas (ACIC), Dr J Mietelski (Krakow), Mr R W Carder (ACIC), Dr J Lorell (JPL Pasadena), Dr G A Mills (Manchester), Mr D W G Arthur (Tucson), Mr D Meyer (ACIC), Mr L Klages (ACIC), Dr T W Rackham (Manchester), Mr M Hamdy (Cairo).

previous experience with lunar photography from Pic-du-Midi), since also deceased—a sad fact, not (yet) true of the Manchester expeditionary to South Africa (who was myself!). This was the first opportunity for me to see, not only the annular eclipse of the Sun in daytime, but also the southern sky at night: a chance to behold (in January) the constellation of Orion 'upside down', and the Magellanic Clouds high above the horizon.

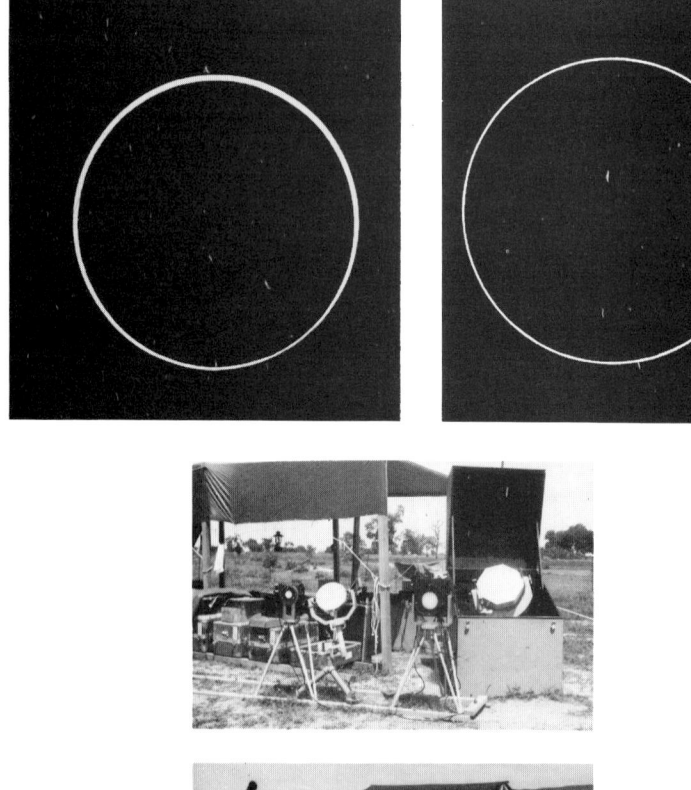

**Figure 5.30** (top) Photographs of the annular eclipses of the Sun on 31 July 1962 at Kheur Soce in Senegal (left) and on 25 January 1963 near Oudtshoorn, South Africa (right), secured by Lockheed–California expeditions with Manchester participation. The eclipse camps at Kheur Soce and Oudtshoorn are shown centre and bottom, respectively.

As is usually the case, the expedition itself was not without some adventures. Its first leg, a Comet flight from London to Johannesburg, ended inauspiciously, when something went wrong with our engines when crossing the Alps that required a day in Rome for repairs. The airline arranged, to be sure, some sightseeing for its passengers during the stop-over; but this proved to me only that the January weather in Rome can be as cold and wet as in Manchester.

It was not till the next night, 24 hours behind schedule, that we took off safely from Leonardo da Vinci airport for our long journey, crossing the north African shores by daybreak, and heading south across Egypt and the Sudan for central Africa. The skies over Africa were clear; and from hour to hour we could observe a gradual transition of the land below from sandy desert to savannah ground and tropical forests. Around noon (before landing at Nairobi) we got a glimpse of the Ruwenzori (Ptolemy's 'Mountains of the Moon') summits on the starboard side of our plane; and after take-off we flew within sight of the snow-capped summit of Kilimanjaro. By mid-afternoon we were in Salisbury; and in late afternoon we reached Johannesburg for the first night in South Africa.

The continuation flight next morning to Port Elizabeth on the shores of the Indian Ocean, the southernmost point of my journey (as far south of the equator as Tunis is to the north) took less than two hours; but thereafter the pilgrim's progress by surface slowed down considerably: the train on the track from Oudtshoorn got derailed, and the time needed to repair the damage was almost as long as to fix the Comet's engines at Rome. As a result, that train ride from Port Elizabeth to Oudtshoorn lasted many hours more than the Comet flight from London to Johannesburg.

When I finally reached our destination, my Lockheed colleagues were already there; and nothing but friendliness awaited us in this Afrikaans town of Dutch farmers whose main produce was ostrich feathers. One of them, Mr Paargeeter by name, very kindly offered a part of his land for us to camp on; and the only creatures about whose feelings we could not make up our minds were our immediate neighbours, the ostriches. These kept observing us with curiosity at close range; and should they attack (which they did not) we were advised to lie down on the ground quickly (in which case they would lose interest in the interlopers); though any more defiant posture could (we were told) entail serious consequences.

At any rate, we managed to maintain friendly relations with all living creatures encountered in South Africa, the weather on the eclipse day was perfect, and all systems worked; so that the aim of the expedition was fulfilled according to plan. Larry Stoddard and Don Carson (the Lockheed part of our expedition) remained at Oudtshoorn for some days to pack up the equipment and despatch it back to California, while I departed with two stop-overs: the first at Port Elizabeth, where I visited (among other sights) the local museum with its true-to-life model of *Brontosaurus* (a giant reptile which used to live in South Africa some 200 million years ago), and the second in Athens. This was my first visit to the country which was to become my second spiritual home.

I left South Africa at the height of its summer; but when we arrived at Athens less than twelve hours later, the weather was not much better than it had been in Rome on the way out. In spite (or, rather, because) of this, some of the first experiences of my short stay on Greek soil at that time turned out to be unique. To begin with, it was my first opportunity to visit the Akropolis of Athens, a place which I had looked forward to seeing for so many years, and to which all heirs of Hellenic culture should pay a pilgrimage, as Mohammedans do to Mecca. I did so with a group of French tourists led by a professional guide; and as the group was not large (January is not exactly the peak of the tourist season even in Athens), I had ample opportunity to ask questions. Our guide (a lady) did not, however, seem to enjoy being asked too many; and presently confronted me with a statement which did not sound like a compliment: 'Monsieur, you must be German!' I asked what she based her surmise on; and she said that only the Germans came to Greece, with their Baedekers in hand, intent on not missing any sights to which they think they are entitled. As I had no Baedeker even in my pocket, I thought the evidence was inconclusive; and confessed to being an American. 'An American?' The guide just burst out laughing. 'I do not believe you!'; and imitating as well as she could the broad Mid-western accent, she asked: 'Where are your keemeras?' Now I was fairly caught in un-American behaviour, and tried to explain that not all Americans carry their cameras all the time; but to no avail. Can you imagine an American in Athens without a camera over his shoulder? Impossible.

The second day, after a good night's rest, I took an all-day excursion bus to Mycenae. The weather did not improve: it actually started to snow on that famous site, as I can prove on film (since, mindful of my experience on the Akropolis, I did not leave my camera in the hotel this time). It was not difficult to secure this evidence; for a veritable blizzard soon gave the entire region an appearance not inappropriate to the landscape beyond the arctic circle. On the way back (via Epidauros), beyond Nemea the accumulated snowdrifts brought our tourist bus to a complete standstill; and it remained stationary for several hours, before a rescue expedition arrived with snow-chains to render the bus mobile once more. We eventually reached Athens, but only in the small hours of the next morning.

Next day (after a long night's rest), I had an occasion to watch from the windows of the Amalia Hotel the arrival of the Danish Princess Anne-Marie, betrothed to the young King Constantine. In spite of the weather, which continued inclement, enough people turned out in the street to bid the princess a friendly welcome, which the young royal couple returned with as much grace as they could muster, little knowing that the days of the monarchy in Greece were just about over; and that the young bride would be destined to spend most of her life with her husband in exile.

In the meantime, the pressure of work connected with manned landings on the Moon was on the increase; and my own part in it in the United States had shifted northwards on the country's West Coast from Pasadena to Seattle, the seat of the Boeing Company, whose Aerospace Division was awarded the NASA contract for the construction and operation of five *Lunar Orbiter* satellites (and, eventually, the integration of the entire Apollo mission). The Division was headed by the Group Vice-President George Stoner, as brilliant an engineer as ever reached as far north-west as Seattle (by a great good chance, his Swiss wife's aunt, Mme Babier, taught our daughters French at Withington Girls School in Manchester).

In addition, the same division of Boeing won at that time a NASA contract for the feasibility study of a Manned Orbiting Telescope; the contract monitor happened to be our former Manchester student (not mine, but Lighthill's), Dr Leonard Roberts; and I was invited to become its astronomical adviser. This study (the first to be commissioned by NASA for this purpose) had no immediate follow-up (the bulk of NASA's budget at that time was still committed to the Moon). It was, however, revived later at NASA's Marshall Space Flight Center (with Dr O'Dell of the Yerkes Observatory as astronomical adviser); and eventually transferred to Johns Hopkins University where, under the leadership of Dr Riccardo Giacconi, the project is now heading towards its triumphal conclusion in the near future. I have long since stood only on the sidelines of this effort (my last contribution to the subject was my semi-popular book on *Telescopes in Space*, which appeared in London and New York in 1968). However, in the academic year of 1964/65 I was due for another sabbatical leave from my University; and the Boeing Company invited me to spend it at their Scientific Research Laboratories in Seattle. With so much going on there at that time how could I resist?

And so it happened that the major part of the academic year 1964/65 I spent (partly with my family) at Seattle, an attractive city with an intellectual atmosphere, surrounded by landscape of striking beauty, with several National Parks (Rainier, Olympus, Glacier) accessible within a few hours' car ride; and in which, weather permitting, I usually spent one day each week exploring the primeval forests. At that time I was, however, already in my fifties and no longer in shape to climb Mt Rainier (4370 m); the best I could manage was to walk up from Paradise (at 1500 m above sea-level) to Muir Hut (at just under 3000 m)—about the same walk as from Mürren to the summit of Schilthorn in the Bernese Alps. I did, however, once make it, with my faithful pupil and friend Constantine Goudas, almost to the top of Mt St Helen, south of Seattle, whose summit cone was (some ten years later) to be disfigured by a spectacular volcanic eruption.

Although I never set foot on the summit of Mt Rainier, at least I flew over it! Mr Guilford Hollingsworth, Director of the Boeing Scientific Research Laboratories at that time, who combined his qualifications as an aerodynamicist with the dashing qualities of a sportsman (and whose accomplishments included flying), maintained at Boeing Field a small private plane which he used with great skill. One summer (I believe, in 1966) he invited me and my daughter Zdenka to take a flight with him to Mt Rainier early in the morning (before the air gets rough) and be back before the Laboratories opened for the day; and we accepted with alacrity. I recall that when word got around in the Labs of our intentions, not all our colleagues shared our enthusiasm: in particular, Dr Burton H Colvin (Head of Mathematics Research, and my host) approached us with a sad face and parting words: 'it was nice to have known you'.

But, come what may, next morning at 6 AM we met Guil and mounted his four-seat plane for our expedition, in spite of the fact that (as I recall) its fuel indicator stood at zero—a fact that Guil waved aside with the confident statement that *some* fuel must be left in the tanks because the engines did start. And so did we. The flight to Rainier (a trip which, by road, would have taken at least two hours) lasted no more than twenty minutes; and from a cruising altitude of some fifteen thousand feet we saw inside the crater and down its fourteen glaciers; but the sheer beauty of the entire horizontal panorama with its half dozen volcanic cones from Mt Shasta in northern California to Mt Baker at the Canadian frontier just defied imagination.

Twice we circled around the majestic mountain, but no more; for Guil may have had second thoughts in the back of his mind about the amount of fuel remaining in the tanks, and turned his plane back for Seattle again. We did make it under our own power, and to the surprise of our colleagues who were arriving at the Laboratories at about the same time; though I would have been a bit more concerned had I known that our pilot (apart from his many undoubted talents) was also a diabetic. Although younger than myself, Guil Hollingsworth has now been dead for several years (though his end was not due to any flying accident).

These were our occasional relaxations in and around Seattle; though my main preoccupation there was to work with an excellent group of Boeing scientists and engineers on tasks connected with the Moon. With some of them (like Dick Shorthill, co-discoverer of the 'hot spots' on the Moon (which are really cold) during the well-known lunar eclipse of 19 December 1964; or Tano Ronca, an old acquaintance from AFCRL) I have remained in professional touch ever since—which is, unfortunately, not true of Dick's friend and fellow-expeditionary to the Land of the Pharaohs in 1964, Jack Saari (1924–71) who died under tragic circumstances before the end of the Apollo era (for his obituary see *The Moon* 2 405, 1971).

However, all was well at Boeing in the spring of 1965 when I returned to Manchester, only to start commuting between Manchester and Seattle via the Polar Route twice or three times a year until 1969. On 20 July 1969, the first men (N A Armstrong and E E Aldrin) stepped on the surface of our satellite; and before the end of 1972 their astronaut colleagues repeated the same feat with full success no less than five times.

That 20 July, a truly memorable date for mankind, was also the greatest day I ever experienced in my life, and I do not expect to see its like in the future. I had awaited the launch of Apollo 11 on 16 July in Seattle; but as soon as it became clear that the lunar excursion module would reach the Moon two days later, I took off by air for Los Angeles where another duty would soon await me: to report (with others) to the American public on television what the astronauts found out about the Moon on the spot, and to interpret the scientific significance of the event with a group of colleagues with whom we had met before the TV cameras on previous similar occasions.

While waiting for the cue to 'go on the air', we had an opportunity to watch many other parts of the programme prepared by CBS for that historic day; and none awakened distant recollections from my youth more vividly than excerpts shown then from an old German film of the early 1930s (made in the UFA studios in Berlin), called *Frau am Mond* (*Woman on the Moon*). Few readers of this book will remember this movie today, but in its time it was of historical significance; for its scenario was based on the technical advice of Hermann Oberth (1894–    ), pioneer (with Robert H Goddard (1882–1945) and Konstantin Eduardovich Tsiolkovsky (1857–1935)) of spaceflight, whose 1926 book on *Rockets for Interplanetary Space* attracted a good deal of attention. The movie in question gave Oberth a chance to translate some of his ideas (such as, for instance, the design of multi-stage rockets!) into at least studio hardware—a task which he carried out with the aid of his (then) very young assistant Wernher von Braun (1912–1977); and its box-office success helped Oberth to raise the money needed to support his more serious experiments, before these were eventually funded by the German *Wehrmacht* under Colonel (later General) Walter Dornberger.

The human plot underlying this movie was rather naive (as most films used to be at that time), but its more technical aspects were not; and it made a great impression on my young mind in Prague in 1930, yet dreaming but little that the essentials of the story shown on the silver screen would become a reality before I reached my father's age (though among the twelve astronauts who actually descended on to the Moon between 1969 and 1972, there was no woman!). Even less did I foresee at that time that not only would I have a chance to meet some of these astronauts in person, but also the German pioneers of spaceflight—Oberth, von Braun, as well as Dornberger—happily reunited in post-war years in the United States with the astronauts who translated the pioneers' dreams into reality. Two of them are dead now; but their spirits fly through space aboard other spacecraft to ever-increasing distances from our terrestrial home.

But, to return to my own performance on that memorable 20 July 1969, it was not the first time I was called upon to assist in such a function. Three years before, in June 1966, shortly after the first *Surveyor* spacecraft had soft-landed on the Moon, Tommy Gold from Cornell University, Gerard P Kuiper from the University of Arizona, Eugene M Shoemaker from CalTech and I had been invited by CBS to report on a coast-to-coast television programme from Los Angeles on the significance of lunar exploration with their aid; and on this 'day of days' we were convoked once more (this time with Harold Urey) to do the same in the wake of the first steps made by men on the Moon. We were interviewed to this end by Walter Cronkite (the nation's leading commentator and equivalent of the British Richard Dimbleby) from New York, on the other coast of the United States—a performance watched (reportedly) by some 80 million viewers. Never before (nor since) did I have a chance to speak to so large an audience. What we actually said now escapes my memory completely, so much was the attention of us all fixed on the exploits of the astronauts on the Moon!

I recall only that I interrupted my return journey to Seattle in the afternoon by a stop-over at Palo Alto, to spend the evening there with the family of my daughter Zdenka, while Georgiana's family from San José came to meet me as well with her children. There I faced another and more difficult interrogation: my oldest granddaughter July (aged five years) wanted to know how she could travel to the Moon herself; or at least how to get inside a television studio. Only my youngest grandson Robin (whom I beheld in my arms then for the first time) did not ask any questions; for he was only 45 days old.

In the meantime, preparations for future Apollo missions continued at full speed (Apollo 12 was launched on 14 November of the same year, Apollo 13, which ended in near-disaster, on 11 April 1970, and Apollo 14 on 31 January 1971). The year 1969 was, however, the last in which I took part in this project from Seattle; and Apollo 13 was the first such mission I watched on television during our vacations in Switzerland.

Of particular connection with me, however, became the Apollo 16 mission in April 1972, which I had the good fortune of being able to observe from a ring-side seat. It so happened that, for my next sabbatical year of 1971/72, I accepted a visiting appointment as senior staff scientist of the Lunar Science Institute of NASA's Johnson Manned Space Flight Center in Houston, Texas. With dozens of astronauts (several of whom had already been to the Moon, and seen its far side with their own eyes) living around with their families at Clear Lake; with the receiving laboratory for all rocks (eventually, 382 kg of them) brought back from the Moon; as well as a depository of all photographic and other data secured by successive Apollo missions, the Johnson Space Flight Center had at that time become the veritable hub of lunar research, and an ideal place from which to take stock of all aspects of our knowledge of our only natural satellite accumulated up to that time. More important still, I was invited to participate in the formulation of the scientific goals to be fulfilled by the Apollo 16 mission scheduled for April 1972; and this brought me, in due course, in contact with the astronauts selected for this mission.

This process could, of course, be advanced not only in the laboratories during office hours, but also during many social occasions which flourished off-duty. The then director of the Lunar Science Institute and our host, Dr J W Chamberlain, lived in the house at Clear Lake previously occupied by Colonel Aldrin, the second man to step on the surface of the Moon. 'Buzz' Aldrin no longer lived at Clear Lake, though we often heard his voice on the radio, extolling the virtues of Volkswagen cars to their prospective buyers for some dealer in the Houston area. It was also under Joe Chamberlain's roof that I made the personal acquaintance of Captain J A Lovell, USN, the hero of the Apollo 13 mission in April 1970, who had twice seen (aboard Apollo 8 in December 1968, and once more as Commander of the Apollo 13 mission) the Moon's far side while in orbit around it, but who was deprived in 1970 of a descent to its surface by a technical malfunction of the spacecraft hardware. A naval flyer of rare skill (it was his leadership that converted the mission from a near-disaster to a triumph of space navigation and guided it safely home), off duty he turned out to be a genial companion over a glass of gin-and-tonic. On direct interrogation I verified that he was half-Czech (having been born in Chicago, the largest Czech city in the world after Prague, of a Czech immigrant mother).

My wife and I came over to Texas to take up residence in Bay House at Clear Lake, within walking distance of the Institute of Lunar Studies, and found ourselves at once in congenial surroundings. Our next-door neighbour was the senior Canadian astronomer C S Beals (1899–1979), whose professional interests had run close to mine in the past in the field of Wolf–Rayet eclipsing systems; and I met with him now on the Moon as well (his wife, a very charming neighbour, was, I believe, sister-in-law of my English colleague R O Redman).

Furthermore, across the corridor from us on the same floor lived an obviously competent and business-like young man who did not have much time for social intercourse: he was nobody else than Dr Harrison H Schmitt, the professional geologist who was to become the first scientist–astronaut (of the Apollo 17 mission) to descend to the lunar surface on 10 December 1972 (see figure 5.31). Dr Schmitt had also a political career before him. He later became Senator for the State of New Mexico, but before he embarked on this career we met once more, two years later, in August 1974 in Moscow, at a Lunar Science Conference organised by the late Academician Vinogradov.

**Figure 5.31** Dr Harrison H Schmitt, scientist–astronaut of Apollo 17, examines lunar rocks in the neighbourhood of the Taurus–Littrow landing site on 11 December 1972.

While in Houston, my contacts with Dr Schmitt were limited to an occasional nod when we ran into each other on the stairs (we talked about the Moon for the first time in Moscow; though about its origin—hot or cold—we were as far apart as the Democrats and Republicans are on the floor of the US Senate). I had, however, a better chance to get acquainted with the Apollo 16 astronauts, John W Young (commander of the mission, and future pilot of the *Columbia* space shuttle), Charles M Duke (the landing module pilot), and especially Tom K Mattingly (the command module pilot) who, in April 1972, was to keep a lonely vigil for more than 73 hours in lunar orbit before his fellow-astronauts (who successfully descended on to the surface of the Moon) could rejoin him on the homeward journey to the Earth.

When it came to the start of this mission, I was allowed to watch the take-off of my friends from Cape Kennedy on 16 April 1972. The afternoon before I flew from Houston to Orlando, Florida; and not being able to find a roof over my head any nearer, I stayed overnight at a motel near the airport to start by daybreak for the Cape by car. It was about one hour's drive through Florida's swamps on both sides of the road as the eastern sky was brightening with the advent of the great day; and it took some hours more before I could park my car at the appointed place and be transported to the VIP enclosure some 3 km from the place of the launch (offering haven at that time to such celebrities as Vice-President Spiro Agnew and King Hussein of Jordan). *En route* we met the ambulance-like bus transporting the three astronauts to the place of their launch, and waved good luck to them!

The take-off took place under near-perfect conditions a few minutes after noon. What a thrill it was to watch the giant Saturn V rocket ignite its engines and, with a temendous din (reaching us with a time-lag of several seconds) rise gently from the surrounding vapour clouds upwards with ever-increasing speed (see figure 5.32). How tiny the spacecraft carrying my friends and perched on top of the 109 m configuration seemed to be at our distance! Burn-out and separation of the first stage took place some 60 km above the Earth (an event which we could still see in the deep-blue sky with unaided eyes). Before we could get away to our parking place through the huge crowds, to return from the Cape back to Orlando and eventually Houston the same night, the loudspeakers announced that the second stage of the spacecraft was again overhead after the completion of its first orbit around the Earth. One certainly could travel much faster through space than on the ground!

On my return to Houston, I was privileged to watch the astronauts on the Moon from the mission's control room at the Johnson Space Flight Center (see figure 5.33), and to welcome them back home from their long journey at Ellington Air Force Base, where they landed after a successful splash-down in the Pacific after an absence of almost one week. I shall never forget that evening, when a B-52 Air Force plane with the astronauts aboard came to a standstill, and John Young stepped out of the cabin with the simple words, 'Well, here we are; and there is *she*!', pointing out to a nearly full Moon just rising in the east to join in the homecoming celebration of its latest group of cosmic visitors.

**Figure 5.32** Anchors aweigh! A Saturn V rocket of total height 109 m and carrying the Apollo 16 spacecraft takes off from Cape Kennedy for a round-trip to the Moon.

**Figure 5.33** Astronaut John W Young, commander of the Apollo 16 mission, at the Descartes landing site on 16 April 1972, next to the lunar excursion module.

Shortly thereafter, after an absence of more than eight months I returned to Manchester again; but some months later I was presented with a priceless souvenir of that great adventure: namely, a diploma with an American flag which the Apollo 16 astronauts had brought back from the Moon, and which they signed for me, adding some gracious remarks.

My real farewell to Houston did not occur, to be sure, in May 1972 on our return to Manchester, but a year later, when Harold Urey, a great friend of the Moon as well as of (some of) her terrestrial students reached his eightieth birthday. Joe Chamberlain decided to celebrate that occasion by a special symposium convoked for this purpose, and invited me to serve as its chairman. This symposium took place on 1–2 May, 1973, with the participation of 45 scientists from 12 countries; and on the day preceding it (Harold's birthday fell on 29 April) a memorial volume of 57 papers by 135 authors, edited by the present writer, was presented to the jubilant at a special dinner in his honour (cf figure 5.16). As the circle of those who wished to pay homage to Urey at that time was world-wide, my preface to the volume (Kopal 1973) was written in Latin, the traditional language of the scholars. This caught Harold unawares; but not to be outdone, he responded to it (in a letter dated 4 May 1973) in the following words:

Carissime Sidoni:
    Haroldus, mitis et humilis, fratri meo caro, salutationes in rebus lunaribus.
    Dedicatio ex calamo tuo me vertit mutum. Quid dicere possum? Gratias tibi ago, amice, propter verbis tuis floreis et honorificis de operibus meis lunaribus. Tanta blandimenta recipere ex viro astronomico tam egregio quam te maxime me placet. Multas gratias.
    Datum Tucsonensi, die $4°$ maio, in octagesimo anno meo.

Haroldus

Harold's eightieth birthday in 1973 was, unfortunately, the last occasion at which many of his friends saw him still at his best, as he was known to us for so many years past. A year later, aged 81, he had to undergo an operation from which he never fully recovered; but it was not till almost seven years later that he took leave of us all for a better world on 6 January 1981. I too slowed down after 1973 in my travels to places where lunar science was cultivated; and several years elapsed before our trajectories again crossed, when I came to the US West Coast to attend the Second International Conference on Mars in Pasadena, California between 15 and 18 January 1979; and on my return was invited by my old friend Hannes Alfvén to give a colloquium at the University of California at La Jolla (where Harold Urey and his wife Frieda were spending their years of retirement).

The subject of my colloquium was to deal with the origin of the solar system, a subject in which Harold had been interested for many years; but (contrary to his earlier intentions) he was no longer able to attend. Instead, he invited us after the colloquium (together with Hannes and Kirsten Alfvén, and Michael Moutsoulas from Greece) to his home.

It was on that occasion that he and I met for the last time. Harold (by that time 86 years of age) could no longer walk freely about the house, and his voice was also no longer what it used to be in the past. But one part of our conversation touched me deeply, when Harold suddenly asked: 'Zdeněk, do you think there will be life for us again after death?' The question surprised me greatly; for I knew that, in the past, Harold had regarded the problem of life purely as one of physical chemistry, and would have come down like a ton of bricks on anyone who thought otherwise.

Not knowing what Harold wanted to hear, I limited myself to saying: 'I do not know.' 'I don't know either,' he responded; 'but if there is'—and he feebly gripped my hand—'I shall wait for you there.' These were almost the last words we exchanged in this world; and, almost ten years later, I can only add that ... if there is, Harold, you will not have to wait for me much longer. Once (in 1982) I was already on my way to you (in the Alps); and if I did not then make the final crossing, it must have been because there was still work for me to do in this world; some time I shall tell you about it.

The parting with Harold Urey at La Jolla in 1979 was one of my acts of farewell to lunar studies; another had been the hundredth anniversary of the foundation of the Observatoire du Pic-du-Midi the year before in July 1978. This offered me the latest opportunity to visit the observatory with which we had collaborated since 1956, and possibly also the last; for my old friend Jean Rösch was to retire from its directorship soon thereafter; and the Observatory's 60 cm coudé refractor (with which we had done most of our work on the Moon between 1959 and 1970) was already dismantled to make room for solar research. It was also the first (and last) opportunity for me to take in person the Chair of the Fondation Internationale du Pic-du-Midi, to the Vice-Presidency of which I succeeded in 1964 (after the death of Sir Harold Spencer Jones, the late Astronomer Royal), and whose Presidency I assumed in 1977 on the death of Lord Blackett.

Finally, an account of my services to the cause of the Moon in the course of the fifteen years between 1959 and 1974 would not be complete without at least a brief reference to the number of occasions on which I was called up to serve our satellite, not only at the telescope, but also with a pen in my hand. Since the first particles of terrestrial matter (in the form of the Soviet spacecraft *Luna 2*) crash-landed on 13 September 1959 on the surface of our satellite in the plains of its Mare

Imbrium, the interest in the Moon on the part of the more general public escalated dramatically; and it was not long before I was called upon to make good of some of the gaps in the existing literature on the subject.

As far as popular writings are concerned, I attempted at an early date to satisfy the curiosity of the public in recent lunar affairs by a small book on *The Moon—Our Nearest Celestial Neighbour* (London and New York 1960), the second edition of which appeared in 1963, and which went through Russian and Swedish translations.

In the meantime, the professional literature on lunar studies began likewise to grow by leaps and bounds; though only that in which I was directly involved will be mentioned. Such were the proceedings of the IAU Symposium 14 on *The Moon*, held at the Pulkovo Observatory in December 1960 (cf Kopal and Mikhailov 1962; a Russian version appeared a year later). A compendium on *Physics and Astronomy of the Moon*, edited by the writer (Kopal 1962; second edition 1971), appeared likewise in a Russian translation in Moscow in 1972. The proceedings of the conference on *Problems of Lunar Topography* (Kopal and Finlay 1961) held in Bagnères de Bigorre the preceding year, and on *Selenodesy and Lunar Topography* (Kopal and Goudas 1967) held at Manchester in 1966, have already been referred to.

My principal contributions to lunar studies in book form, addressed mainly to professionals, appeared between 1966 and 1974, under the following titles: *An Introduction to the Study of the Moon* (1966), *The Moon* (1969), and *The Moon in the Post-Apollo Era* (1974); which together with the *Mapping of the Moon* (Kopal and Carder 1974) represent my main legacy to this field. By now all these books are out of print; and their sequels will have to be written by others. The same is also true of my smaller booklet on *Exploration of the Moon by Spacecraft* (Kopal 1968), which two years later appeared in a Swedish translation in Lund; and of the two *Atlases of the Moon*. The first of these, written jointly with J Klepešta and T W Rackham, was published in 1965 by Academic Press in London; and the second (under the title of *New Photographic Atlas of the Moon*) appeared in New York in the spring of 1971 with an introduction by H C Urey. The first of these works was based solely on photographs taken as part of the Manchester Lunar Programme with the 60 cm refractor at Pic-du-Midi; while the second (much better) atlas also contained photographs taken with the 107 cm Pic-du-Midi–Manchester reflector, in addition to some taken by spacecraft.

I may add that, since the 1970s, when the emphasis shifted to planetary missions (which throughout the 1960s played second fiddle to the accomplishments of lunar craft), I was called upon to extend my literary coverage on a semi-popular level to the realm of the planets as well: and out of this arose my little book on *The Solar System* (Oxford

1972, New York 1973; translated into Japanese in 1974); and, in the late 1970s, I wrote *The Realm of the Terrestrial Planets*, which appeared in London and New York in 1979 (and in Czech and Bulgarian translations in 1985).

Throughout my life I have written only one popular book covering the subject of astronomy as a whole, which appeared under the title of *Man and his Universe* in London and New York in 1972 (and in a Czech translation in 1976 in Prague). Although now (at least partly) out of date, it had a reasonably good reception (in Czech alone it sold thirty thousand copies in less than two months), as attested (perhaps better than by printed reviews, which are accessible to anyone) by private opinions which contemporary experts shared with my publishers (and which the latter shared with me as well). Thus Wernher von Braun (whose name we have already encountered, and who was NASA's Deputy Associate Administrator in Washington when my book appeared) wrote in a letter of 17 March 1972:

> Professor Kopal's new book, *Man and his Universe*, just arrived and I'm looking forward to a quiet weekend when I can curl up with it.
>
> I've always been an avid reader of Professor Kopal's excellent books on astronomy. I think they are so well and clearly written that even an engineer like myself can understand them, and being on the transportation end of the ancient science of astronomy, I've always been interested in where we are going.

I should like to quote another opinion, by Professor Hannes Alfvén, Nobel Laureate 1970 (see figure 5.34), which reached me through the same channel:

> *Man and his Universe* by Zdeněk Kopal is a fascinating book on the universe and man's place in it. It is a true expression of the author's personality: he knows humanities and art almost as thoroughly as he knows astronomy. Moreover, he can transfer his knowledge and his enthusiasm to both astronomers and laymen.
>
> Most surveys of astronomy concentrate so much on distant objects like pulsars and quasars that our closest neighborhood, the solar system, is neglected. Kopal is one of the rather few astronomers who fully realize that space research is rejuvenating solar system astronomy to such an extent that it may be the leading field in present-day astronomy. He devotes about half of his book to this most modern of all fields. This is also appropriate with regard to his general purpose: to find what place man occupies in his universe.
>
> The last chapter deals with life in the universe and our possibility of coming in contact with distant civilizations. It demonstrates that Zdeněk Kopal also is an historian and a very wise man: He tells us what to do if the 'space phone rings'.

*Literary talents* 303

**Figure 5.34** Professor Hannes Alfvén at the Lunar Science Institute of NASA's Johnson Manned Space Flight Center in Houston, in 1973.

Perhaps I may be permitted to reproduce the above excerpts in this place, not to bolster the sales of the book (which has, indeed, been out of print for some time), but because of Professor Alfvén's reference to my occasional excursions into humanities—and, more specifically, to the history of science. This interest was present in my mind since my student years; but it was not till my Manchester years that it began to take a more concrete form. In the latter 1950s my senior colleague Professor Michael Polanyi informed me of the current work of his friend and well-known writer, Arthur Koestler (1905–1983), which eventually resulted in his book *The Sleepwalkers*, which dealt with characters whom Koestler chose from the ranks of astronomers; and asked whether I would be willing to examine its text from the technical point of view. Naturally, I agreed; but in scrutinising its original text I came across a number of slips (not all of which were removed before final printing). Yet, by Koestler's originality of approach to many age-old problems his *Sleepwalkers* made truly fascinating reading; and it is greatly to be regretted that its author did not find time to prepare its second edition before his premature passing.

To me Koestler's book provided an incentive to return to some parts of the subject in a work of my own. The result was a book entitled *Widening Horizons*, a story of man's quest to understand his place in the

304  *The Manchester years*

Universe, which appeared in London in 1970; while its New York edition came out in 1971. It has now been out of print for more than ten years; and it has been my earnest hope to replace it, in due course, by a new and more substantial text. But, alas, with advancing years it seems that this hope (like so many others) may remain unfulfilled.

### International Travels and Adventures

Lest the heading of this section sound too much like stories unconnected with science, let me set out to show that this is far from the case. Indeed, its sole aim will be to describe my contacts with the international community of astronomers which started to ramify after my return to Europe in the 1950s.

The first major geographic hiatus in my entire career was, of course, our transfer across the Atlantic in 1938 to the New World. The circumstances by which it happened, and what I learned during the American years have already been dealt with in the preceding chapter; while the aim of the present one will be to outline subsequent ramifications of these as they have developed from Manchester, eventually to encompass many parts of the world.

To begin with, let me first take the reader back to the country I originally came from; and to the resumption of my personal contacts with Czechoslovakia after an absence which lasted almost nineteen years. On my return to Great Britain as an American citizen, I needed, of course, a visa to visit my old country; and in the days when the whole of central as well as eastern Europe lived under the dark shadow of Stalin (and also for some time afterwards) it was hopeless even to apply for such a visa, let alone obtain one. It was not till well after the post-Stalin thaw set in that the Czechoslovak Academy of Sciences found it feasible to invite me to visit Prague as a guest of the Academy in the spring of 1957. And so on 12 April of that year I stepped once more on the soil of my native country at Ruzyně airport, the place from which in September 1938 we had left for America (which became the scene in November 1939 of the massacre of Czech students by the Nazis; and again of the Soviet invasion of Czechoslovakia in August 1968) to meet my father and many other old friends from the Czech astronomical community; though no longer my brother Miloš (who had perished in March 1948 in the High Tatra Mountains) or mother (then recuperating from pneumonia in a sanatorium).

This first visit to post-war Czechoslovakia was not too long, but had some more lasting consequences. If there was a branch of human activities which had blossomed during the two decades since I left my native country, it was indeed science; and astronomy was no exception. In particular, the Ondřejov Observatory was about to undergo a major reorganisation. This observatory was founded in 1898 as a private establishment by Dr Josef J Frič (1861–1945), an enlightened amateur, as a memorial for his prematurely departed brother Jan (1863–1897), and given to Charles University in 1928. In my own student days we used to call it the 'sleeping beauty' (because it was located in a lovely landscape, some 50 kms from Prague, but lacked astronomers willing to work there at night). In 1949, it was attached to the Czechoslovak Academy of Sciences as one of its research institutions; but its largest instrument was still an 8 inch refractor (figure 5.35), unchanged since the days of my youth. However, under a new conscientious and dedicated director, Dr Bohumil Šternberk (1897–1983), supported by a whole new generation of able young astronomers (among whom Drs Bumba, Kleczek, Plavec and Švestka later attained international renown), the Ondřejov Observatory began to spread its wings to become the premier astronomical institution of the country.

**Figure 5.35** The old dome of the Ondřejov Observatory in Czechoslovakia, going back to the beginning of this century, and housing an 8 inch refractor, the objective of which was made in 1854 by Alvan Clark (1804–1887) for the Reverend William Rutter Dawes (1799–1868), a distinguished British observer of double stars. After the latter's death, the objective was acquired by Professor Vojtěch Šafařík (1829–1902) in Prague, an equally distinguished variable star observer. After his death, his widow, Mrs Pavlína Šafaříková (herself an astronomer of no mean distinction) donated it to Mr J J Frič for the Ondřejov Observatory, where it spent another half century in the service of astronomy. Now a museum-piece, the Clark 8 inch objective was still in use there during the author's student years.

When I visited Prague in the spring of 1957, plans were being considered to re-equip the old observatory with several new instruments of moderate size for different types of work—somewhat the same as the Polish astronomers did later in Krakow and elsewhere. To me this did not, however, seem to be very reasonable: instead, I urged my colleagues to consider the acquisition of one large reflector (as large as the means on hand would permit) to serve as the focus of the Observatory's activities, which could place Ondřejov in the front rank of European astronomical institutions.

Dr Šternberk and his colleagues did not need much persuasion to adopt the plans which I proposed to them at that time (and which we kept discussing together till the time came for my departure; see figures 5.36 and 5.37).

**Figure 5.36** At Ruzyně airport in Prague (25 April 1957) on the author's departure from his first post-war visit to the country of his birth. From left to right: Professor V Nechvíle, the author, Dr M Plavec, Professor Kopal senior and Professor Emil Buchar.

However, to realise these plans it was necessary to clear them through the Presidium of the Academy which alone could supply the necessary funds. And in this matter too I was, by a long arm of coincidence, in a position to offer some help: for the President of the Academy (and politically a very influential person at that time) was Professor Zdeněk Nejedlý (1878–1962), a native of Litomyšl, with whom my father had been on terms of personal friendship since I was born. It was, to be sure, most unlikely that Nejedlý would have remembered me as a tot. However, he had heard about me from the Russian astronomers during his war years in Moscow, and even more from Dr Shapley in the Kremlin during that convivial occasion already described on p. 160.

Figure 5.37  On 25 April 1957 at Ruzyně airport, Prague, with Mr František Kadavý (1896–1972) of the Štefánik Observatory (right). Kadavý was the co-author of my first book on variable stars.

It was Nejedlý who (in his capacity as Minister of Education of Czechoslovakia between 1945 and 1948) was behind the move then afoot to bring me back to Prague. Apparently he did not hold it against me that I did not come (probably forgot all about it in his advanced age). Moreover—and this was much more important—his professional field (history of music) made no particular claims on the Academy's financial resources. At the request of my astronomical colleagues in Prague (Drs Šternberk and Perek) I wrote, therefore, to Nejedlý in the summer of 1958 in support of our plans; and whatever the effect of my letter may have been, I heard shortly thereafter that our proposals had indeed been accepted. Although Nejedlý himself died three years later, the future of observational astronomy in Czechoslovakia was thereby assured; for the Presidium of the Academy approved of the acquisition (from Carl Zeiss of Jena) of a 2 m reflector, which (together with its sister-telescope of the same origin in Tautenburg, East Germany) has remained the largest telescope in Europe north of the Alps ever since.

As is well known, this new reflector (figure 5.38) was installed at Ondřejov in 1967, and dedicated to service on the eve of the Thirteenth General Assembly of the International Astronomical Union in Prague.

(It was on the day after the General Assembly closed that our daughter Zdenka was married in St Vitus Cathedral to a young American astronomer, Dr Dean F Smith, almost 29 years to the day after her parents had gone through the same ceremony, in the same place, as mentioned on p. 153.) On the day of dedication (30 August 1967) President Nejedlý was no longer with us (he had died five years before); but the new President, Academician František Šorm (see figure 5.39), told me, on that occasion, that had he been President in 1958, the astronomers would have never acquired the telescope which he came to inaugurate; as an organic chemist, he would no doubt have seen other uses for the money.

Figure 5.38  A wide-angle snapshot of the Zeiss 2 m reflector of the Ondřejov Observatory, in service since August 1967.

These were the best years of post-war Czechoslovakia, a year before the invasion by Soviet tanks in the summer of 1968. My wife and our daughter Eva were in Prague at that time (I was in Seattle), and shared the forced retreat of the American visitors to the frontier, including Professor A G W Cameron of Harvard, the Co-editor of *Astrophysics and Space Science* (about which more later), and the former child film star Shirley Temple. That invasion was an event which set the clock back for peaceful developments in Czechoslovakia for many years to come.

I was, however, to meet with Academician Šorm once more in his life: namely, in April 1969, when at the COSPAR meetings in Prague I received from his hands a Gold Medal conferred upon me by the

Czechoslovak Academy of Sciences (see figure 5.40); the other recipient at that time was the American astronaut Frank Borman, commander of the Apollo 8 mission to the Moon during Christmas 1968. Soon thereafter Academician Šorm (who at the time of the Soviet invasion of his country the year before sent a protest against that wanton act of aggression to the USSR Academy of Sciences) was suspended from most of his public functions; he died in 1983. As a consequence of the political events which followed in the wake of the Soviet invasion of Czechoslovakia in August 1968, several of the best Czech astronomers left Ondřejov for various parts of the world (as did Hungarian astronomers in 1957, or Polish astronomers after 1980); and, as the 1970s went on, my own contacts with the old country became increasingly restricted. In the spring term of 1974 (a year after my mother passed away at the age of 89 years) I was still invited to lecture for one term as visiting professor at Charles University in Prague; but thereafter my contacts with professional colleagues in Czechoslovakia remained largely limited to correspondence.

**Figure 5.39** The day of dedication (30 August 1967) of the Zeiss 2 m reflector of Ondřejov Observatory. In the foreground are Miss Eva M Kopal (1949– ) with her father and President František Šorm (1913–1983) of the Czechoslovak Academy of Sciences. Behind Dr Šorm can be seen Dr Bohumil Šternberk (1897–1983), the Observatory's Director (at that time IAU Vice-President), and Professor Dr Emil Buchar (1901–1979), Chairman of the Czechoslovak National Committee for Astronomy.

In the meantime, Dr Šternberk, senior of Czechoslovak astronomers, and the first to graduate in our subject after World War I (on the basis of research carried out at the Berlin–Babelsberg Observatory under Professor Paul Guthnick) and who, in the fullness of years (between 1958 and 1964), served as Vice-President of the IAU and its host in Czechoslovakia in 1967, and who had directed the Ondřejov Observatory for almost twenty years following World War II, retired after the inauguration of its large new reflector (see figure 5.38) that year. He lived, however, for another sixteen years till death claimed him in 1983 at the age of almost 86 years. His contemporaries of pre-war vintage—Professor Emil Buchar, Mikuláš Mohr and Vladimír Guth, all born in the first decade of this century—left this world between 1979 and 1980 in their late seventies; and after Dr František Link died in Paris in September 1984 at the age of 79, I found myself to be the last survivor of those who graduated in astronomy at Charles University before World War II—for how long?

**Figure 5.40** Receiving the Gold Medal of the Czechoslovak Academy of Sciences in May 1969 (at the time of the COSPAR meeting in Prague) from the President of the Academy, Academician F Sorm (right), in the presence of (from left to right) mother Kopal (then aged 85), the Secretary of the Academy, and Dr Zdeněk Svestka (acting chairman of the Academy's section on astronomy).

Speaking of those who have already departed, this account would be incomplete without a testimony of my respect and affection for two of them in particular: Professor Vincent Nechvíle (1890–1964) of Charles University, and Professor Emil Buchar (1901–1979) of the Technical University in Prague. We have already heard about Nechvíle on pp 93–95; while Professor Buchar did not have a chance to blossom out till after World War II, when I no longer lived in Czechoslovakia.

Of the generation of astronomers who graduated after World War I, Buchar was without doubt the most talented of those mentioned so far in this chapter, and a kind man of the character reminiscent of

Vincent Nechvíle; but while Nechvíle was largely of passive nature, Buchar knew how to stand up in the interest of a good and just cause. A master (like Nechvíle) of celestial mechanics and other branches of classical astronomy (and, like Nechvíle, with postdoctoral work in France to his credit), Buchar had also to lecture at the Technical University on spherical astronomy to large classes with many foreign students who came to Prague sometimes but incompletely prepared from their home countries. This was especially true of those coming from Eastern Europe or the Balkans; and it was sometimes a tiresome business for the teaching staff to segregate chaff from wheat.

In carrying out this process, Buchar earned for himself the reputation of a just and kind man. When a foreign student presented himself for an oral examination, in order to set his mind at ease Buchar was known to ask, first, which country he had come from, and then how he liked it in Prague; with technical questions following only afterwards. And so it happened that a student once presented himself who confessed to be an Albanian, liking it in Prague all right: but when it came to technical questions he failed completely, and Buchar had to send him down for at least one term. This was, however, not to the liking of the Albanian would-be scholar; and in going out he complained to Buchar's assistant—nobody else than Dr Miroslav Plavec, now of the University of California at Los Angeles (who told me the story)—in the following words: 'Isn't the professor unjust? He asked me three questions: the first two I answered well, and only the other I did not know; and yet he failed me!'

But before I bring my recollections of Prague in bygone days to a close, let me return once more to Dr Šternberk; for there is one private joke between us which perhaps deserves to be told here. When, in 1977, Šternberk attained the age of eighty, in my congratulations on this anniversary I stressed the merits of his wise guidance which had brought the Ondřejov Observatory to its present level. But, with his characteristic modesty, the jubilant disclaimed much credit for the deed. He admitted that it was not always easy to nurse the project in difficult post-war years to a successful conclusion; but, he added, he had done so only within the limits of the old proverb of Roman jurisprudence: '*Mater certa, pater incertus*; and on this latter role you should know more than I.'

Next, let us turn to Poland, the country adjacent to Czechoslovakia, which I first passed through in 1936 on our journey to (and from) the Far East. Chance willed it that I should not re-enter that country again for almost a quarter of a century, until I stopped at Warsaw again *en route*, in December 1960, from the IAU Symposium on the Moon at Pulkovo, to visit my old friend (now, alas, also deceased) Stefan Piotrowski from our Harvard days (see figure 5.41) who was then

312   *The Manchester years*

Professor of Astronomy and Director of Warsaw University Observatory. Pan Stefan had visited us at Manchester some years before (under the auspices of the British Council); and it was a pleasure to renew also my acquaintance with younger Polish astronomers of his school (Drs Grzedzielski and Stodolkiewicz) who had also previously spent some time with us at Manchester.

A guided tour through Warsaw at that time was still a shattering experience: the tremendous amount of destruction wrought there by

**Figure 5.41**   At the Eleventh General Assembly of the IAU in Berkeley, California, August 1961. (Above; from left to right): Professor Stefan L Piotrowski (Warsaw), the author and Dr Miroslav Plavec (Ondřejov). (Below) Professor Karol Koziel (Krakow) between Georgiana Kopal and her father, on an excursion to Napa Valley.

the barbarities of World War II was still everywhere in evidence. It took another twenty years before most of this grievous material damage was eventually healed by re-building and re-creation of the historic buildings of the Old Town in their pristine style (the only European city where such damage is still in evidence today remains parts of East Berlin); but the wounds—more grievous—to the population of Warsaw in 1939 and 1944 are even more difficult to heal.

Shortage of time did not allow me to visit also in 1960 the Krakow University, the oldest university in Poland, with which I was to establish such close academic ties in the years to come. The first steps in this direction antedated, however, my visit to Poland in 1960 by several years. The beginning of this acquaintance went back to the Eighth General Assembly of the IAU in Rome in 1952, at which Poland was represented by the late professor Thaddeusz Banachiewicz (1883–1954) of Krakow University, an astronomer of international renown. Neither Dr Piotrowski (then Banachiewicz's assistant in Krakow) nor any other junior astronomer from Poland attended the meetings in Rome; but Dr Piotrowski sent me a message which Professor Banachiewicz was to deliver. And, indeed, a message reached me one day that Banachiewicz would like to meet me at a given place and time.

On the appointed day I set out to meet Professor Banachiewicz on the Campidoglio (under the famous portrait of Copernicus observing the stars in Rome, by Jan Matejko), and found him resting comfortably on a sofa after lunch (no doubt a good one). I knew him (at a distance) from the pre-war IAU meetings in Paris and Stockholm, but had not been introduced to him before; therefore, I stepped into the presence now and introduced myself. Professor Banachiewicz was not asleep—in fact, he was very much alert—but he kept looking at me for several seconds without saying a word, until, at last, he repeated several times, '*si jeune, si jeune!*' I do not know how I was described to him by Piotrowski; but it obviously looked as though he had expected to meet someone more of his own age!

Nevertheless, in spite of this disappointment to his expectations (I was then only 38 years—young), we got along fine, and for the next few days we spent a good deal of time together (I recall, in particular, our joint excursion to Pompeii). But his time was running out; for in less than two years he had departed from this world; and since that time I could only visit him (twice) in the Pantheon of the Monastery 'Na Skalce' in Krakow, where he awaits resurrection to this day.

Professor Banachiewicz's successor and heir, not only to his chair, but also to his interest in the problems of lunar librations, was Professor Karol Koziel (1910–    ). I made my first personal acquaintance with him at the Tenth General Assembly of the IAU in Moscow in 1958;

314  *The Manchester years*

and this acquaintance soon developed into a lasting personal friendship. Our joint interests in selenodesy brought us together again in April 1960 at Bagnères de Bigorre (see figure 5.20); and that conference resulted in a six-month visit by Professor Koziel (accompanied by his assistants Drs Maslowski and Mietelski) to the University of Manchester for further work on problems of joint interest; for our next meeting at Berkeley, USA, in the summer of 1961 see figure 5.41.

It was not the last time that we had the pleasure of welcoming these Polish colleagues to Manchester (for their 1966 visit see figure 5.29); but my first opportunity to visit Krakow did not come till spring 1964, when the Jagiellonian University, Poland's oldest *studium generale* and junior by only twenty years to the Charles University of Prague, celebrated the 600th anniversary of its foundation. I was invited to take part in this academic solemnity; and so was Dr Shapley of Harvard University, then 79 years of age, of whom we heard so much in the preceding chapter. Thus chance brought us together in the academic procession (see figure 5.42) for almost the last time in our lives.

**Figure 5.42** The academic procession to celebrate the 600th anniversary of the foundation of the Jagiellonian University in Krakow, May 1964. The author (extreme right) walks together with Dr Harlow Shapley. This was the last time I appeared with Shapley at a public function; failing health prevented him soon afterwards from undertaking any further travel.

As I marched in the procession on that glorious May afternoon through the streets of Krakow, it did not occur to me that only a little more than ten years later (on 12 December 1974) I would once more become the centre of similar attention, when Krakow University conferred upon me their doctorate of science *honoris causa*. It was on a winter day, without the sunshine which so generously blessed the procession of figure 5.42; and the ceremony took place in the large aula of the University's Collegium Maius.

According to ancient academic custom, the new doctor had to respond to the award with an address on a topic in his professional field, lasting about twenty minutes (though, in deference to changing times, the address was no longer to be delivered in Latin, but in a profane language, for which I chose English). In order to do justice to the occasion, I prepared a few written notes to support my memory, and slipped them into my pocket—just in case of need. These notes were there as I listened to the Public Orator (Professor Koziel) extolling my virtues; but when the time came for me to respond and I reached discretely for them ... oh, horror: what I pulled out of my pocket was not the notes, but our airline tickets! To have fussed for the right piece of paper under my elaborate academic gown would have attracted too much attention. So what else could I do but camouflage the air tickets in the palm of my hand, and deliver my response (on astronomy in the last quarter of the twentieth century) off the cuff, without the aid of any written source. Fortunately, all went well; and nobody seemed to have noticed what I actually held in my hand (or, at least, no-one enquired later about the role of airline tickets in academic graduations). After the ceremony, we were entertained with true Polish hospitality to a sumptuous lunch which lasted till late afternoon; and in the evening I addressed a large professional congregation at the University Observatory on a more technical subject.

This was not my last visit to Krakow or Warsaw: the latest should have taken place in 1983, to take part in the celebrations of the hundredth anniversary of the birth of Professor Banachiewicz at the University of Krakow on May 20-21, in which I was invited to take part and deliver an address. Unfortunately, the Polish Government of General Jaruzelski declined in the last minute to grant me a visa to enter Poland for this purpose; and the 'iron curtain' descended between us again—for how long remains to be seen.

The third country with which our close relations of many years have been the source of particular pleasure and satisfaction is Greece, the cradle of western astronomy, and of many young astronomers who elected to study for their advanced degrees in our science at Manchester. The senior of them, by age as well as service, is Professor Constantinos Leonidas Goudas, PhD (1961) and DSc (1967) of Manchester University. He became in 1967 a professor at the newly-founded University of Patras and was its Rector between 1974 and 1975, the first of my students to attain that academic rank. To the world community Professor Goudas is known not only for his many contributions to science (including his heroic efforts as a shield-bearer to Henri Poincaré in his argument to keep the 'third integral' of celestial mechanics in its proper place), but also as host to the Eighteenth General Assembly of the IAU at Patras in 1982.

Almost a quarter of a century has elapsed since Dr Goudas left his academic home at Manchester; we spent some time together later in Seattle at Boeing (see figure 5.43), and have seen each other many times since in Greece. I always enjoy the company of a young colleague who (unlike myself) can claim (in his more expansive moments) to 'know everything': for such a person can obviously be a useful source of information. Certain it is that the skyline of the modern University of Patras (largely a result of his efforts) testifies to his energy and determination. At the NATO Advanced Study Institute on the Moon at Patras in 1971 (of which Constantine was the local organiser, and at which (see figure 5.44) Professor Urey and I were to receive honorary degrees), he showed that he could do even more—such as inducing the local Metropolitan of the Greek Orthodox Church to bless the lunar rock (on loan from NASA)—an act of doubtful canonical significance, and casting some doubt on whether Greece is really a nation of Greek Christians. But there is one feat which Constantine has not yet managed to pull off: and that is to get elected to the Greek parliament for the Nea Demokratia party. To be sure, he can always rely on my vote (for I am a good democrat myself); but should this not help, he may have to change sides again sometime in the future.

**Figure 5.43** Dr Constantine L Goudas (PhD Manchester, 1961; DSc 1967), left, and the author at the Boeing Scientific Research Laboratories in Seattle, in the spring of 1965.

Greece

**Figure 5.44** At the NATO Advanced Study Institute held at the University of Patras in September 1971, in the course of which Professor Harold C Urey (second from left) and the author received honorary degrees from that University. On the extreme left is the University's Rector, and on the right, the pro-Rector, Professor Constantine L Goudas.

My connections with the University of Patras were closest during its formative years between 1968 and 1973, when that institution (like all others in Greece) was under the military jurisdiction of the 'Government of the Colonels'; the chief administrative officers were (usually) retired generals whose main aim was to keep their faculties as well as students in check. At Patras, this role was given to Lt-Gen Achilles Tagaris (retd)—presumably so honoured because he had published some booklets on the development of strategy in Greece from the battle of Marathon up to the present time.

In my capacity as academic adviser as well as OECD assessor for developments of the new institution I had much to do with General Tagaris at that time, and was decorated by him with the Order of St Andrew (the university's patron saint) for my efforts; but I shall never forget my last dinner with him, at which other faculty members were also present. In an after-dinner speech (delivered in Greek, but translated into English), the General thanked me for having educated so many Greek astronomers at Manchester in the past, but expressed surprise that I had learned so little Greek from them myself.

In order to prove that this was not completely so, and that if I had not learned many Greek words, I knew at least the most important ones, I raised my glass to offer a toast to $Z\eta\tau\omega$ $\sigma\tau\rho\alpha\tau\sigma$ (meaning 'long live the Army'), a battle-cry of the Government under Colonel Papadopoulos. I recall that some of those present (including Michael Moutsoulas) were almost taken aback at my temerity; but General Tagaris did not sense mockery in my zeal and drank the toast with full solemnity.

This was the last time that I saw him; for soon afterwards the military junta in Athens fell out of power, and General Tagaris promptly disappeared from academic life back to the retirement which he should never have left; for all I know, he may still be enjoying a well-earned rest. His successor as Rector of the University of Patras became (who else?) Constantine Goudas. In the tumultous time which followed the downfall of the junta Constantine acquitted himself well; and only a few took exception to the fact that the place held formerly by a Lieutenant-General had devolved on one who by military rank was only an 'untrained soldier'.

The second of my former Greek students in order of seniority is, of course, Dr (now Professor) John Hadjidemetriou of the University of Thessaloniki, who returned to Greece in 1965 immediately after his graduation; while the 'third musketeer' of our Greek Guard, Professor Michael D Moutsoulas, remained at Manchester after his 1967 PhD for several years as a postdoctoral research fellow and lecturer; he took an important part in the last years of our work on the Moon (concerned, in particular, with star-calibrated photography of the apparent lunar disc). It was not until 1972 that our Michalakis left Manchester to return to his home country in triumph as Professor of Astronomy at the University of Athens and (somewhat later) as the founder-director of the Greek National Center for Space Research. He has since been representing his country on many international committees (COSPAR, and others) concerned with space research; and we often see each other at meetings in different parts of the world.

And not only abroad, but even more often in Greece itself. For these first three musketeers were soon followed to Manchester by many other and younger Greek scholars (now likewise occupying academic positions in their country and abroad); and their seniors, headed by Academician John Xanthakis (see figure 5.45), have seen to it that I have often had a chance to visit Greece in the past twenty years—opportunities which I count among the blessings of my life. For if Benjamin Franklin once said that every man of culture has two countries—his own and France—the same could be said even more appropriately of Greece. Thus, in 1969, I had a rare opportunity to visit the excavations of the Minoan civilisation at Thera (Akrotiri) and Knossos with the late Professor S Marinatos, at Mycenae with Professor G E Mylonas, and at Pylos with the late Professor Carl Blegen, the famous excavator of Troy as well as of the palace of Nestor. I was greatly impressed by Blegen and, on being introduced, said how truly honoured I felt to meet so famous an archaeologist. Blegen appeared visibly embarrassed and responded by saying, 'Oh, I am just an old

**Figure 5.45** Among Greek astronomers at the Observatory of Athens in January 1969. From left to right: Dr C J Macris, Dr M D Moutsoulas, the author, Academician J Xanthakis, Mjr Mac Rae, Professor D Kotsakis, Dr C G Banos, Dr G A Antonakopoulos, and Dr G G Banos. The portraits in the background are of former directors of Athens Observatory: from left to right, D Eginitis (1862–1934) and J F J Schmidt (1825–1884), the well-known selenographer.

man', words with which Albert Einstein could have responded to any younger physicist attempting to pay homage to his genius.

At these archaeological excursions I was, to be sure, only an interested spectator; but at others, such as in Delphi (1969) or Samos (1980), I have been called upon to contribute to the proceedings of the repective conferences and have had an opportunity to meet some of the leading men of the country (see figures 5.46 and 5.47(a)) and other guests (figure 5.47(b)). None of these occasions was, however, more pleasant than those which followed my election to foreign membership of the Academy of Athens in 1976. I came to Greece at that time to receive the honour in person. I was received by the President of the Republic in his office; and had the pleasure of escorting his wife, Mme Tsatsos (a distinguished poetess in her own right) to a reception which followed that evening after my address in the large lecture hall of the Eugenides Planetarium on Aristarchos of Samos.

But the most pleasant part of that occassion came at the end of this trip, when I was invited to spend a weekend with my friend Michalakis

**Figure 5.46** At a conference on 'Science and Humanism' in Delphi in September 1969, shaking hands with General Patakos, Greek Minister of the Interior (who came to Delphi to learn some humanism—without much success!).

**Figure 5.47(a)** At the opening session of the International Symposium on Aristarchos of Samos on 17 June 1980, to commemorate the first mortal who, in the third century BC, had conceived the idea of a heliocentric model of the Solar System. This symposium took place on Samos, the island of his birth. From left to right: Academician J Xanthakis, the author (who had just delivered the opening address), Professor M D Moutsoulas, the Hon. G Rallis, Prime Minister of Greece, Professor Hannes Alfvén (Nobel laureate 1970) and his wife Mrs Kerstin Alfvén.

Greece 321

(Professor Moutsoulas) in Thessaloniki, at the official residence of his father-in-law (Hon. N Martis), the Minister for Northern Greece in the Cabinet of Prime Minister (and later President of the Republic) Konstantin Karamanlis. The Martis family awaited us at Thessaloniki airport, and drove us to their official residence (formerly the summer palace of the King), where some two weeks before they had been hosts to the French President Giscard d'Estaing on his official visit to Greece. There must have been some left-overs from that visit still around (at least on the record-player); for when we reached our destination and began to mount the main staircase (Michalakis, for some reason, ran ahead of us), we were surprised all of a sudden to hear the martial tones of the French national anthem: *'Allons enfants de la patrie, le jour de gloire est arrivé ...'* Well, for me, it certainly had!

**Figure 5.47(b)** At the Heraion on the island of Samos, 8 June 1980, in the course of the international conference on Aristarchos (of which the author was co-chairman). On the right is Academician V A Ambartsumian, leading Soviet astronomer and member of the Supreme Soviet. The lady in the background is Mrs Filareti Zafiropoulos of Patras (a graduate of Manchester University).

So many other stories of our adventures with Michalakis (mainly in the times of the Colonels' regime in Greece) come to mind that they could fill a book of their own; but one more should perhaps be preserved for posterity. Thus, in September 1969 in Athens, Michael took me one night for dinner to a taverna in Plaka, called 'Diogenes', on the slopes of the Akropolis. It was not the first time we had gone there; for (according to Plutarch) it is close to the place where Diogenes the philosopher used to live in his barrel, and where (in more modern times) Michael himself was born in 1935, only a stone's throw away.

However, in 1969 we went there not only for old times' sake, but also for Greek bouzouki music; and made bold by at least one bottle of Mavrodaphne wine split between us we asked the musicians to play the well-known 'Song of Piraeus' by Melina Mercouri (then an exiled film-star, and now Minister of Science and Culture) from her famous movie *Never on a Sunday*. To perform that song was strictly forbidden at that time in Greece; and the local musicians were at first reluctant to do so; but eventually they agreed.

However, no sooner had they started than a sleek police car (one of those long Chevrolets supplied by the American Government to bolster the Greek regime of the day) pulled up in front of the taverna and stopped. It was a tense moment; and I was already reaching in my briefcase for my recent photograph with General Patakos (figure 5.46) as proof that I moved in the 'right society'. But Michael (a good observer) set me at ease by whispering 'it's all right, prof, it's all right.'

And so it was; for once the forbidden song was over, the police car pulled away quietly and things returned to normal. 'How did you know that it was all right?' I asked Michael. 'I watched that officer', Michael replied, 'and noted that his hand on the wheel was beating the rhythm of the song.' In other words, the officer had stopped in front of the taverna not to molest anybody, but to listen to music so seldom heard in Athens in those days.

I must only add that, since becoming Minister of Science in 1981 in the Government of Premier Papandreou, Melina Mercouri has grown considerably less popular in Greece (at least, among the astronomers; of whom Constantine Goudas was her particular fan) than she was earlier as a film star; and she reciprocated her feelings (or, rather, lack of them) towards astronomy at the time of the Eighteenth General Assembly of the IAU at Patras in 1982.

Another time when Michael and I had to resort to a ruse to get out of what could have been a ticklish spot was at Heraklion, Crete (I believe in the same month and year that we accompanied Professor Marinatos on an excursion to Knossos). In the evening, encouraged perhaps by a few glasses of a local drink called *chiccuria* on the square,

**Figure 5.48** Participants of the NATO Advanced Study Institute on Close Binary Systems, held on the shores of the Mediterranean at Maratea, Italy 1–14 June 1980.

Standing (from left to right): Mammano, Mantegazza, Milano, Mrs Zafiropoulos, Kopal, Moutsoulas, Bonifazi, Jurkevich, Longo, Lorenzi, Gimenez, Rao, Mrs Carling, Mrs Hack, Rovithis, Mrs Budding, Niarchos, Pastori, Petty, Awadalla, Scalriti, Walter, Miss Cerruti-Sola, Tsouroplis, Burchi, Robertson, Miss Romeo, Etzel, Nelson, Ibanoglu, Kurutac, Mrs Rovithis, Budding and Sedmak. Sitting (in front): Left—Kizilirmak, Güdür and Demircan. Right—Iijima, Hamzaoglu, Miss Balucinska, Mahdy, Miss Otmianowska.

Michael and I began to sing some well-known songs we liked of Mikis Theodorakis—likewise proscribed in Greece at that time. The police were quick on the uptake; and started heading for us. We did not, however, wait to make their personal acquaintance; and betook ourselves away promptly pretending instability. Seeing this, the police lost interest in us; and so we made it safely to our cruise ship to avoid a possibly less comfortable overnight billet on land.

Greece was not the only Mediterranean land with which I was familiar in those days (see figure 5.48); nor did it remain for long the easternmost country with which I was able to establish close relations in the Near East. The lunar work in which we at Manchester got engaged in earnest in the 1960s brought me back to Japan to re-visit Kyoto in 1962, by invitation of Professor Shotaro Miyamoto (who succeeded in 1956 the late Professor Yamamoto, our host in Japan for the observation of the 1936 total eclipse, as Director of the Kwasan Observatory) to collaborate in work on lunar topography (see figure 5.49).

**Figure 5.49** Arrival at Kyoto on 3 October 1962 (my first visit to Japan since 1936) to inaugurate cooperation in lunar research with the Kwasan Observatory of the University of Kyoto. Left to right: Professor Shotaro Miyamoto (Director of the Kwasan Observatory), the author, and Mr Robert Carder of ACIC.

In the preceding section of this chapter I outlined briefly the shadow method which we were engaged in developing for this purpose at that time; and it goes without saying that it is impossible to track the development of such shadows continuously from any single point on the Earth's surface. Japan lies eight hours to the east of Pic-du-Midi; therefore, parallel observations at two localities separated by eight

# Egypt

hours of longitude would enable us to extend the coverage of our records roughly from eight to sixteen hours a day. Professor Miyamoto and his colleagues (among whom his successor, the late Professor Akira Hattori should in particular be mentioned) agreed to collaborate with us in this undertaking; and under the auspices of NASA as well as of the US Air Force (which readily provided, as they had done in France, all logistic support for the undertaking) this cooperation continued for some years.

Moreover, from 1974 we also developed a similar collaborative programme with the Helwan Observatory of Cairo, based on the use of the Observatory's 74 inch reflector at Kottamia in the Egyptian desert (figure 5.50) between the Nile delta and the Red Sea; and with the enthusiastic cooperation of the late Professor A H Samaha (1900–1980) and some of his colleagues (among whom Dr Joseph Sidky Mikhail took a prominent part) the programme was carried out successfully for three years, until the Israeli military attack on Egypt in 1967 (which placed the Kottamia Observatory in the war zone, and caused some damage to its equipment) brought it to an abrupt end. The extension of our work to the Egyptian desert did not increase the geographic span of our observations in longitude, but it considerably augmented (especially at certain parts of the year) the number of clear nights for observations.

**Figure 5.50** The dome housing the 74 inch Grubb Parsons reflector of the Kottamia station of Helwan Observatory in Egypt, where astronomers from Manchester worked for several years from 1964.

Our professional work of several years in the Land of the Pharaohs did not prevent us (mainly during dark-of-the-moon spells) from engaging also in occasional sight-seeing in that enchanted land of the oldest culture of the world; and in the course of time it was my good fortune to see on the ground what I had glimpsed from the plane *en route*

326  *The Manchester years*

to South Africa in January 1963—the valley of the Nile from Alexandria in the north to Abu Simbel in the south; and to admire on the spot all the treasures left behind in the past five millennia: from the (reputed) site of the Museion of Alexandria to Assuan, the baseline over which Eratosthenes once established (very approximately) the size of our Earth; or from the pyramids of Giza to Dashur, the ancient necropolis of the Old Kingdom, whose technical achievements still make us gasp in awe. Not so Tom Rackham or myself; for, as you see in figure 5.51, the height of the Great Pyramid (151 m) did not deter us from climbing to the top (in 1964, before such feats were forbidden by the police to avoid mounting casualties). We descended safely, none the worse for the experience.

**Figure 5.51** On top of the Cheops pyramid (the Great Pyramid of Giza, near Cairo), when it was still permitted to climb it in April 1965, with Dr Thomas W Rackham of Manchester. The Khafren pyramid is in the background.

The Luxor temples and the Valley of the Kings some 600 miles further south belong likewise among the sites you never forget; although most of the treasures found there, including those from the tomb of Tut-ankh-amon, can now be seen in the Museum of Cairo (unless they happen to be, as is now frequently the case, *en route* to far-flung parts of the world). Together with the British Museum and the Louvre in Paris, the Vatican Museum in Rome and the National Museum of Athens, the Cairo Museum belongs to the five-star constellation which no-one interested in the past of the Old World should miss. It has been my good fortune to be able to visit them all, and several times.

South of Luxor, the journey upstream along the Nile becomes all the more enchanting. Tom Rackham and I once rented a car to visit the temples of Esna and Edfu of Ptolemaic days; and the landscape *en route* reminded us of biblical times, but for, now and then, TV antennae sticking out of the roofs of mud huts in the proximity of larger localities. At Assuan, we arrived at the first Cataracts and the Tropic of Cancer: the Nile valley narrows in width, and the desert commences a few hundred yards off the shores covered by tropical vegetation. Do not, however, be tempted to go too far off the road into the desert when visiting some local antiquities (such as the tomb of the Aga Khan). We did it once with Mrs Carling; but Dr Joseph (who accompanied us on that excursion) pointed out on the soft sandy surface only too many cobra trails which could not have been too old—minutes, perhaps—while the serpents which made them may not have been many yards away, hiding from the fierce sun in the shadows of occasional boulders and watching us go by with their evil eyes.

Not far above Assuan is the High Dam on the Nile, a civil engineering marvel of our times, as the pyramids of Giza were of Egypt's Old Kingdom four and a half thousand years before. The man-made Nasser Lake to the south extends for many hundreds of miles into Egypt and the Sudan; and can best be admired from the plane carrying tourists from Assuan to Abu Simbel, another immortal monument from the Ramses era of Egypt's New Kingdom, which rescue action of the United Nations has saved from submersion below the shores of the Nasser Lake, as the High Dam rescued the lovely temples of Philae from submersion by the nineteenth-century low dam just above Assuan.

But our trip to Abu Simbel in 1978 was not motivated solely by the attraction of these historical sites: it was to get acquainted with the tropical station of Helwan Observatory maintained at Abu Simbel for photometry of the zodiacal cloud, which in the morning rises almost vertically above the level of the Nasser Lake; and where the near-perfect atmospheric transparency permits photoelectric measurements to be carried out almost down to the horizon. This station, originally set up by Soviet astronomers there in the 1960s, is still in use; but the quality of the site would merit its being equipped with a much larger instrument than it possesses at the present time.

But to return to the Moon, in the years which followed the conclusion of our work at Kottamia, the Manchester-trained astronomical community began to establish itself also in other Muslim countries of the Near and Middle East; and the occasional calls of our former students for professional assistance from their former academic home gave me the chance to visit other enchanted places about which I had heard from childhood, but never hoped to see. Thus when the Government of Iraq decided in the early 1970s to provide their country with a modern astronomical observatory, I was the first astronomer from abroad to be invited (in the spring of 1973) for consultations and a preliminary search for a site. The stay in Iraq at that time was too short to do full justice to the cause; but my hosts made it possible for me to visit also the excavations of Babylon and Ctesiphon south of Baghdad, as well as the remnants of Ninevah (near Mosul) and of its great palaces from the times of the Assyrian Empire.

In visiting the universities of Mosul and of Sulaymanyia (where we travelled by car via Arbil and Kirkuk) I was suprised to learn how many of their staff were former Manchester graduates (not only of astronomy); even in these distant localities they remained patriotic Mancunians and 'supporters' of our local football clubs! In the course of a preliminary site testing in the mountains east of Sulaymanyia we happened to run into an armed band of Kurdish guards; and (although I did not understand their language) I gathered that we had temporarily become their captives. Eventually, however, they allowed us to go about our business, but each member of our party, including my young colleague Dr Abdul Al Sabti (PhD Manchester, 1972), was accompanied by a huge soldier carrying a gun heavy enough to pierce a tank. But they were really good men, these soldiers; and apparently had nothing against astronomy. My guard (as I gathered from the pidgin French in which we conversed) was once a French Foreign Legionnaire; and when they saw how cold we were (the temperature dropped rapidly below freezing after sunset); they supplied us with hot tea. Never had I drunk so much of it before at Manchester!

After this auspicious beginning, I watched the further development of astronomy in Iraq mainly at a distance; and did not visit the country again until January 1982 when (under a new President) a former student of mine, Dr Hamid M K Al Naimiy, became the Director-General of the Iraq Space and Astronomy Research Centre; and his ranking colleagues, Drs Aziz Sadik and Talib Kadouri, together with several more junior astronomers, are all former Manchester students and graduates of our department. Their new observatory will be located at Mt Korek in northern Iraq, not far from Arbil (in the proximity of which Alexander the Great destroyed in 331 BC the might of Darius III, the last Great King of the Achamenid dynasty of Iran), 2127 m above sea-level. Its principal instruments (a 3.5 m reflector made by Zeiss-Oberkochen, and a 30 m radio telescope working at millimetre wavelengths, supplied by Krupp-Essen) should make it the second largest observatory on the Eurasian continent; and (by aperture of its optical equipment) the ninth largest in the world. The Mt Korek

Observatory, now in an advanced stage of completion, should be in full service by the spring of 1987; and its astronomers certainly face a bright future.

From Iraq it is, of course, not far to Iran (the jet flight from Baghdad to Teheran lasts only a little more than one hour) and it was that country which I was invited to visit in May 1977 by the Government of the late Shah of Iran to deliver a series of Pahlavi Lectures at different universities of the country on different aspects of space research in general, and on the Moon in particular. The invitation came to me in the autumn of 1976 from the Minister of Education (at that time Dr Abdul Hossein Samii; but when I eventually arrived, the office was held by Professor G Motamedi, former Vice-Chancellor of the University of Isfahan); and involved also consultations with the officers of the Government on the foundation of a national astronomical observatory in Iran—similar to (in fact, larger than) that contemplated for Iraq at about the same time.

Tragically futile as this exercise turned out to be in the light of subsequent events, I met nothing but kindness at every stage of my three-week journey, on the part of officials, university colleagues and students alike. At the time of my arrival, the latter were, to be sure, on the cool side, especially at Teheran (the starting point of my tour), no doubt on account of the official sponsorship of my journey. But word must soon have got around among the students of the other universities I was to visit—Isfahan, Shiraz and Ahwaz—that I was no reactionary; and the local students came to listen to what I had to say in large numbers.

My colleagues at different ports of call—Dr Arkani Hamed of Arya-Mehr University in Teheran (a graduate of MIT, with whom I became well acquainted during my sabbatical year of 1971/72 in Houston), Professor A Kiasatpoor at Isfahan (a graduate of the University of Pennsylvania); Dr J Sobouti in Shiraz, a graduate of the University of Chicago and a keen student of binary stars (brother, I believe, of the Rector of the University of Teheran); or the Dean of Science at the Jondishapour University of Ahwaz—all received me very kindly and provided me with an ample opportunity to visit some of the famous sites of their ancient country: the rose fields of Shiraz in full bloom, the archaeological treasures of Persepolis (see figure 5.52), or the excavations at Sus, the ancient capital of Elam, within reaching distance of Ahwaz, a site whose history goes back to the dawn of human civilisation (the famous stele with the laws of Hammurabi, now in the Louvre Museum in Paris, was unearthed there by French archaeologists in the year 1901).

Yet, in spite of all this, the general atmosphere I encountered almost everywhere in Iran at that time was already such that the events which shook that country to its foundations so soon after my visit did not come as very much of a surprise to me. The overthrow (and subsequent murder) of the Shah's popular Prime Minister Dr Mossadegh in 1950, arranged by the Western Powers with oil interests in Iran, deprived the government of the young ruler of support at the grass roots; and marked, in effect, the 'point of no return' on the road which made the advent, in due course, of the Islamic Republic with Ayatollah Khomeini almost inevitable. For, in the second half of our century, it is no longer possible for any government, sitting on the bayonets (and motivated largely by tribal greed) to maintain indefinitely an absolute rule of 200 families over a nation of 30 million dispossessed people.

**Figure 5.52** (left) In the palace of Darius (fifth century BC) at Persepolis and (right) at Sus, once (in the second millennium BC) the capital of ancient Babylonia, not far from the spot where French archaeologists excavated the stele with the laws of Hammurabi.

What happened in Iran should serve as almost a textbook example of what is eventually bound to happen in every country under such circumstances; and it was only the Shah's timely escape abroad which saved him from the fate of the less fortunate French King Louis XVI in 1793. To mention only a few portents of it evident already at the time of my visit in the academic community, almost half of the country's universities (including the prestigious Arya-Mehr in Teheran, the Iranian version of MIT) were already closed because of student unrest (broken windows and other damage testified to the ferocity of the pitched battles fought between students and police on the campus); and the University of Tabriz, not far from the site of the ancient Marâgha Observatory of the thirteenth century (see p. 439) which I also wanted to visit, but could not, experienced the same fate. In spite of the enormous mineral wealth of the country (most of which drifted abroad, never to come back), inflation was rampant, and the cost of living such that although the salaries of university teachers were (on the official exchange rate) two to three times as large as those prevalent in Britain, they were (I was told) barely sufficient to cover the necessities of life; very few of our colleagues in the cities could afford to own a house.

But before we conclude our tour of astronomical institutions in the Near East, I must mention some which were not on the beaten track of many astronomers in the 1970s; and this track led to Libya. My first visit to that country took place in 1975, in response to an invitation by the Vice-Chancellor of the University of Benghazi, Professor A Hasan (a graduate of Durham University in England, who once heard me lecture when I came to Durham to address a student society) to advise them on the creation of an astronomy department, and the establishment of an observatory in the proximity of that city. My second visit to Benghazi (on the same mission) took place in spring two years later; the new Vice-Chancellor, Dr Al Zelatny, was a Manchester graduate (of economics); he visited me at Manchester in 1977, before my second trip to Benghazi; and has since become the Minister of Education in his country. He saw to it that postgraduate students from Libya came to study astronomy with us as well, and was favourably inclined to approve of the proposal to equip the University of Benghazi Observatory with a 2 m reflector from Zeiss-Jena.

Over the course of the years, we have provided for Benghazi University astronomy staff from the ranks of our former graduates from Muslim countries (Pakistan, Turkey), and also educated two young Libyan scholars to PhD level. Since, however, they returned to their home country, I lost any direct contact with the Benghazi project, and do not know how far it may have progressed. However, before leaving Libya from the second (longer) visit, I had one unforgettable experience: an opportunity to visit the site of the Hellenistic city of Cyrene, capital of Cyrenaica (ancient Apollonia and once the home of Eratosthenes), largely excavated by the Italians before World War I. By its extent and degree of preservation, there is nothing comparable with it in Greece or elsewhere in Africa: and if the site were more easily accessible, it should certainly attract large crowds of tourists.

But all these astronomical involvements in the Near East did not prevent us from pushing further eastward. Our connections with Japan (where I returned in 1962, after sixteen years, in the service of the Moon, and have kept returning ever since) have already been mentioned. Since 1964, these activities extended also to India; and a few words may perhaps not go amiss to explain how this came about.

In post-war years, shortly after India recovered its independence, the government of Pandit Nehru decided to set up a certain number of National Centres for Research in different sciences; the one for astronomy was assigned to Osmania University in Hyderabad. As a gift to this nascent institution, the US Department of State presented to it a 48 inch Cassegrain telescope (made by Fecker of Pittsburgh); and Dr A G Wilson, then of Lowell Observatory (and later of the Rand Corporation in Santa Monica), was sent to India for site-testing to determine its location. This work (carried out in the late 1950s) settled on an optimum site at Rangapur hill, about forty miles south of Hyderabad, overlooking the village of Gopal (on the outskirts of the jungle) in countryside still teeming with cobras. The expedition almost cost Al his life (not from snake-bite, but from an amoebic disease contracted in the course of his work, which was identified only in the nick of time to save him); and nothing further was done on the project, largely for lack of local staff, until Dr R V Karandikar agreed to return from the United States to take up the directorship of the new institution in 1963.

The Chairman of the Indian University Grants Commission (under whose jurisdiction the Hyderabad Centre belonged) was then Professor D S Kothari, formerly of the University of New Delhi and a distinguished astronomer in his own right (who, years before, had laid down the theory of the internal structure of white dwarfs). Naturally enough, he had a full appreciation of the difficulties facing the new institution; and one of his first acts in this connection was to invite me to act as its astronomical adviser. Of course I accepted; and so it happened that between 1964 and 1969 I was called upon to visit India almost each year.

On my arrival for the first visit, we found the 48 inch telescope with all its accessories still packed in the boxes in which it had been delivered several years before; and although the road to the site had already been built and buildings erected, they were still without electric power. In order to install the latter, by great good chance we could avail ourselves of the volunteer help of a group of highly competent Czech engineers from the Škoda Works of Plzeň, Czechoslovakia, who were then in the neighbourhood of Hyderabad erecting a large steel plant for the Government of Andhra Pradesh. Their contract called for a six-day work week; but on the seventh day they were free to give us a helpful hand at Rangapur; and since electrical installations were their profession, we soon made some headway with their help.

The work during my early visits to Hyderabad consisted partly of logistic help, but mainly teaching; for, in the meantime, Dr Karandikar had succeeded in assembling a sizeable body of staff and students (figure 5.53), many of whom, led by Professor Abhyankar (a student of Otto Struve in Berkeley), had a distinct interest in close binary systems.

In due course, a photoelectric photometer for observations of southern eclipsing variables was secured for the 48 inch reflector; and especially after the return of Dr Sanwal from Kitt Peak in the United

States, the new facility has accomplished a large amount of very useful work. Moreover, photographic work on the Moon (which also formed a part of our original programme) opened an additional source of logistic support for the new observatory; and also the possibility of bringing some of the young Indian scholars (Messrs Jamsheddi and Swaminathan) to Europe to work with us for a time at Manchester and Pic-du-Midi. Dr Swaminathan eventually returned to India where he now holds a senior position at Hyderabad; but Naushir Jamsheddi eventually remained in Europe for good.

Figure 5.53 At Osmania University in Hyderabad, in front of the new building of the National Centre for Advanced Studies in Astronomy (to which the author then served as scientific adviser) in April 1965. In the middle row, reading from third left to right, are Professors Alladin, Sharma, Karandikar (then Director of the Centre), the author, Professor Abhyankar and Drs Peraiah and Rao. Sitting in front of Drs Karandikar and Kopal are Messrs (later Drs) Swaminathan and Jamsheddi, who both spent several months at Manchester.

One near-accident occurred during my first visit to Hyderabad which could have cost me my life. I was then accommodated in the University Guest House on the outskirts of the city (erected on land but recently reclaimed from the wilderness); sharing its facilities with Dr Hopgood, a young geologist from the University of St Andrews in Scotland. We were cautioned, however, not to go out too much after sunset, not because of the two-legged nuisances which infest our Western cities these days, but because of snakes which were still plentiful in the vicinity. In a lack of artificial illumination one could easily step on one, with incalculable consequences. The university car brought us back for dinner as a rule; and called again on us next morning to take us to our daily work.

One evening, after we had had a rest in the lounge and gone to the dining room to eat, a commotion arose in the lounge (not too much excitement, perhaps: for the Indians are not demonstrative in the face of everyday events). On enquiry we learned, however, that they had just killed a little snake, called a krait, which must have been hiding under one of the easy chairs in which Hopgood and I had been resting after our daily exertions only a short while ago. Now the krait, like the American coral snake (which I met once on the steps of the Lunar Science Institute in Houston), is an inconspicuous and timid little reptile, which does not bite unless scared: but if it does, its bite is deadly, and no serum against it is of any avail.

To this day I am not sure what would have happened if either Hopgood or I had wiggled too much in our easy chairs while waiting in the lounge for dinner. The day was hot, and the snake had no doubt crawled inside (the lounge was on the ground floor, and its door always open for a bit of breeze) to hide from the Sun under one of the two chairs: we shall never know now which one, and how close its occupant may have been to the end of his terrestrial pilgrimage. But no-one should blame us if, next morning, we both quietly moved out in quest of new quarters closer to the city centre; we have never seen each other again. I should add that Indian large cities are full of life, and snakes are known to be allergic to commotion; so one feels safer from them there. This was true not only for my sake, but also for that of John Meaburn, whom I brought with me to Hyderabad in 1969 to teach for some weeks at the International School for Young Astronomers there; and with whom I played chess (with varying fortunes) in the evenings, but under more secure circumstances.

Incidentally, *en route* to that school (by air) from Delhi, I had an experience at the airport not at all unusual in India (at least in those days). In full sunshine, I am in the habit of protecting myself from sunstroke by wearing a white 'Gandhi cap'—very functional for the purpose, but susceptible also of alternative interpretations. Thus as I was pacing to and fro at Delhi airport in my Gandhi cap awaiting the departure of the plane for Hyderabad, a sturdy Russian (obviously a diplomat) came in with his interpreter; and very soon turned to him, pointing to me, with a question: 'Who is that man?' 'I do not know,' responded the guide; 'he must be some politician of the Congress Party.' My Russian is good enough to have understood what was said; and for the rest of the time I maintained a fierce silence, to uphold the dignity of the role so gratuitously assigned to me for the occasion.

Ten years later, when I was at Bangalore for the COSPAR meetings in 1979, Professor Abhyankar (who had succeeded Professor Karandikar as Director of the Rangapur Observatory since my last visit) invited me to visit Hyderabad once more; and I agreed to fly there when my presence at COSPAR would be most easily missed, at the ceremonial opening of the occasion, when COSPAR was to be addressed by Mr Desai, then Prime Minister of India. In order to catch the Hyderabad plane, I had to miss Mr Desai's address, but I did not miss him at Bangalore airport; for our flight to Hyderabad was delayed until the Prime Minister's private plane (a Soviet Tupolev jet) had taken off.

This delayed our flight for about an hour; but I saw Mr Desai at close range taking the salute of a military Guard of Honour that was lined up on the tarmac, as he walked to his plane. I started after him almost immediately (not to miss mine), carrying an official-looking briefcase (containing only astronomical material), and wearing a Gandhi cap similar to that worn by the Prime Minister. Great, however, was my astonishment when the Guard of Honour suddenly snapped to attention once more to present arms as I walked by. I must obviously have been taken for someone of Mr Desai's entourage. This experience convinced me very quickly that countries exist in which it is of advantage to be a politician!

Since that memorable occasion, I was back in Bangalore three years later (see pp 341–343); but I never visited Hyderabad again, though I have continued to watch from a distance with great interest the further growth of the Observatory of Rangapur, well content that the seeds which I helped to plant there more than twenty years ago are now bearing ample fruit.

But in the midst of all this the time was inexorably approaching when my thirty years of service as Professor of Astronomy at the University of Manchester was bound to come to a close; and it did so at the end of September 1981, when I was $67\frac{1}{2}$ years of age. At the beginning of that month I presided, however, over the IAU Colloquium (number 69) on 'Binary Stars as Tracers of Stellar Evolution' at Bamberg (see figure 5.54); and in the second half I travelled on my first visit to Spain, from Madrid to Granada and almost to the Mediterranean shores of the Iberian peninsula.

Astronomically speaking, southern Spain is as favoured a part of the world as southern California or Arizona. What a thrill it was to see from La Veletta (some 3400 m above sea-level) the peaks of the Atlas Mountains in Morocco, some 300 km across the Mediterranean! At the Calar Alto Observatory (within reach of Granada) the German Max-Planck Gesellschaft is putting up optical telescopes, the larger of which will be a twin of the 3.5 m Zeiss reflector of the Iraqi National Observatory at Mt Korek; and the same is true of the 30 m radio-telescope working at millimetre wavelengths, and operated at La Veletta by a Franco-German consortium.

336  *The Manchester years*

On the scenic side, the beauties of Granada (Alhambra!), the seat of the Astrophysical Institute of Andalucia, made a visit to this monument (and last foothold on the Iberian peninsula) of the Moorish culture unforgettable also for non-astronomers. And when, before leaving Spain, I had an opportunity to visit with Dr Alvaro Gimenez the Escorial (north of Madrid), it was not without emotion that I passed by the tombs of so many Spanish monarchs who, at the time of the Renaissance, had so much to do with my native country (Ferdinand I, King of Bohemia between 1526 and 1564, was the younger brother of the Emperor Charles V who is buried at the Escorial; while Ferdinand himself, Emperor from 1556 in succession to his brother, is buried in St Vitus Cathedral of Prague).

**Figure 5.54** After a convivial beer party in Bamberg, October 1981, offered by the Local Organising Committee of IAU Colloquium number 69 on 'Binary Stars as Tracers of Stellar Evolution'. Members of the scientific organising committee (from left to right): Peter van de Kamp (Holland), the author (the Colloquium chairman), Rudolf Kippenhahn (Munich) and Remo Ruffini (Rome).

Finally, at the end of September 1981, the time came for me to hand over the astronomy department at Manchester to younger hands. My successor was nobody else than my old colleague and friend Franz D Kahn (see figure 5.55), holder of a personal Chair of our subject since 1966: the first colleague to join my staff in 1951, and the only one who has remained faithful to Manchester ever since.

# Retirement

**Figure 5.55** Professor Franz D Kahn, presiding at a Farewell Dinner at the University of Manchester on 28 September 1981, on the occasion of the author's retirement from his Chair, to which he had been elected in the autumn of 1950. At right (partially eclipsed) is Dr John Meaburn.

On 25 September, he (together with Dr Meaburn and my faithful secretary Mrs Carling, the second oldest member of our staff by service) arranged for a farewell dinner in my honour at which over a hundred of my former students and friends, now scattered in several countries, congregated at the University's Staff House to listen to my last lecture, give me a 5 inch Celestron telescope (which has been much in service since) and—the greatest surprise, which was as touching as it was unexpected—present me with a *Festschrift* of papers contributed to this occasion and edited by Franz Kahn under the title of *Investigating the Universe* (Kahn 1981), published by the Reidel Publishing Company of Dordrecht. The first (leather-bound) copy of this was presented to me by my old friend (if this is the right term to apply to a young lady) Mrs Nel Pols-van der Heijden (figure 5.56), who had made a special trip from Holland for this purpose.

In order to keep the size of the book within practicable limits (even so it contains over 460 pages), only our former students were invited to joint the roster of its contributors, led by my first student of eclipsing variables (while still at Harvard), Dr Vainu Bappu (at that time President of the International Astronomical Union) who introduced the volume with a very gracious Foreword. Last but not least, my gratitude for this surprise went out to my secretarial staff headed by

Mrs Carling (see figure 5.57) for the preparation of a camera-ready copy of the entire volume in their own time—an effort which they had managed to conceal from me to make the surprise complete.

**Figure 5.56** Mrs Nel M Pols-van der Heijden of the Reidel Publishing Company presents the retiring professor (to his complete surprise) with the first copy of a *Festschrift* collection of papers by his former students.

Much indeed went through my mind as I rose to my feet that evening to thank all those who were present, and to listen to messages from those whom distance or other reasons prevented from being with us otherwise than in spirit—thinking of the human and professional fellowship which would bind us together for the rest of our lives. Since the commencement of my teaching career at Harvard more than forty years ago, it had been my good fortune to be able to educate well over 100 PhD's, now practising our subject in fourteen countries all over the world; and if each of them produced a further ten or more students of their own, I should be leaving behind me between (say) 1000 and 2000 spiritual grandchildren (not counting those I may have influenced at a distance through my books). With all these we may not know each other in person; and much of that offspring may be engaged in work other than that in which I guided their teachers in PhD research. Yet something of me may have descended to them all. In this sense, I may perhaps echo the words of the ancient poet, 'Non omnis moriar'; but will continue to live in them after I am no longer in this world to watch their further progress.

**Figure 5.57** In the author's office at the University of Manchester in May 1981. From left to right: Mrs Ellen B Carling (Finlay), senior secretary (since 1955) and Personal Assistant to the Head of Department (standing on her left), Mrs Barbara Barlow and Miss Joan ('Juanita') M Suthers (junior secretaries). Juanita was the only person we all really had to look up to!

But, to return to the present, when I came to Manchester in 1951, one of my first steps had been to establish a series of publications through which the original work of my colleagues and students could be shared with our sister institutions. To this end Series III of *Astronomical Contributions from the University of Manchester* was launched in 1952; while Series I (the *Annals*) were sponsored jointly with Jodrell Bank. About half of the material which appeared in the latter was contributed by us before the *Annals* became extinct in the late 1950s; but by the time I retired the number of contributions to Series III already exceeded 500, and the series is still going strong. My own personal participation in these totals has, perhaps, not been entirely negligible; for the *Astronomische Jahresberichte* (together with *Astronomy and Astrophysics Abstracts*, which took over the role of the *Jahresberichte* in 1969) disclose that (including my pre-Manchester years) I published almost 350 individual papers, and wrote (or edited) 51 volumes of professional literature (if repeated editions or foreign translations are separately counted)—quite apart from just about 150 volumes of professional journals which appeared since 1962 under my editorship.

In more ways than one, at that Farewell Dinner, I felt like old Mr Chips must have on a similar occasion. In my recollections I still saw all those present as they had been in their formative years in the (sometimes) distant past, and not as sedate gentlemen who had come to join us that night for old time's sake. How could I recognise at first sight among them (say) John Hazlehurst, who, 25 years before (see figure 5.12), had been a starry-eyed innocent, coming back now from Hamburg, not only with a beard which would have done credit to many an astronomer of bygone days, but (*incredibile dictu* for one who had so long remained a confirmed bachelor) with a wife!

To switch to the more dignified aspects of the occasion, the University was represented by Professor Mark Richmond, the fourth Vice-Chancellor under whom I had served the University in the past three decades (and who, although a distinguished bacteriologist, does not believe in life in interstellar space). Sir Bernard and Lady Lovell came from Jodrell to reminisce about the past; while I had the quiet satisfaction of observing that Chairman Kahn and John Meaburn (perhaps the best student I helped to educate in the past twenty years, and who presented me with the Celestron) acquired eloquence in presentation that may serve them in good stead should they decide to stand for Parliament.

This was, however, by no means the end. By conferring upon me the rank of Emeritus, the University gave me an opportunity to carry on with purely academic work while being relieved of all administration; and to be able to travel where I wanted without the need of official leave. I availed myself of this opportunity and continued to lecture to students (albeit with interruptions caused by many trips abroad) until the spring term of 1983 when I was 69 years old; and I educated in retirement five new PhD's from four countries (Egypt, Iraq, Libya and Pakistan) before reaching the age of 70.

The principal reason why my systematic lecturing at Manchester gradually came to a halt at that time was the fact that an increasing part of my time was being claimed abroad, either by my former students now settled in many countries, or by other professional colleagues who wanted to get better acquainted with our recent work at Manchester on the Fourier analysis of the light changes of eclipsing variables (see the next chapter)—not to speak of other topics of possibly wider interest.

Thus already on the second day of my retirement I weighed anchor for a trip to Greece to participate in a symposium on 'Space Technology for Mankind', held on 14–16 October at the European Cultural Centre at Delphi, and chaired by Professor Moutsoulas (then President of the Centre's Executive Committee). Our return journey (which took me for the first time to Thermopylae) led from Athens (which we saw then at the height of election fever, leading to the replacement of the Rallis Government of Nea Demokratia by that of Andreas Papandreou) to Dubrovnik, Yugoslavia, where the sixth IAU European Regional Meeting on 'The Sun and its Planets' was to take place on October 19–23. As a member of its Organising Committee,

*Further travels* 341

I was invited to open its proceedings on 19 October by an address on 'The Solar System—Known Fact and Unsolved Problems' (cf Fricke and Teleki 1982); and to chair some of its sessions.

On 4 January 1982, I was due to deliver the Appleton Lecture on 'Space Exploration of Major Planets' before the British Institution of Electrical Engineers in London; and the day after I flew to Baghdad for a reunion with my former students, and to address an international conference on 'Astronomy in Mesopotamia', inaugurated by H E Taha Yassin Ramadhan, Deputy President of the Government of Iraq (see figure 5.58). Soon after my return, however, a much longer journey loomed ahead of me, to Bangalore where I had agreed to spend several weeks in February and March 1982 giving an intensive course of lectures on Fourier analysis of the light changes of eclipsing binary systems to an all-India audience at the Indian Institute for Astrophysics, by invitation of its director, and my old student and friend, Vainu Bappu.

**Figure 5.58** Addressing the conference on 'Astronomy and Space Research in Mesopotamia', Baghdad, Iraq, January 1982.

It was not the first time I visited Bangalore on professional business (cf pp 334–5); but it was the first time I could properly renew my acquaintance with Vainu who, since we had lost sight of each other after my departure from Cambridge to Manchester in 1951, had also travelled a long way in his line of duty. After his return to India in 1954, he established a new observatory equipped with modern instruments at Manora Peak near Naini Tal (within sight of the Himalayas). Six years later, he took charge of the Kodaikanal Observatory (which he equipped with modern instruments); and in the 1970s he created at Bangalore the National Institute for Astrophysics, where I visited him first (briefly) in 1979, and again (for a longer period) in spring 1982.

By that time, Bappu was successfully building a new observatory, his third, at Kavalur (some 150 km south-east of Bangalore; see figure 5.59) equipped with a Zeiss 1 m telescope for photometric and spectroscopic work, to which a 2 m reflector (with Indian-built optics as well as mounting) was soon to be added. We spent one weekend together with the Bappus at Kavalur, and I had the chance to admire also the wonderful landscaping of the Observatory's site (which, I was told, was the work of Mrs Bappu). I was, incidentally, not the only one who was attracted by it: so were the monkeys which loitered in the trees of the surrounding jungle in great numbers (and with even greater noise); and the observatory's fence offered only a small obstacle to their desire for closer acquaintance with the astronomers inside.

**Figure 5.59** In the course of his short life, Vainu Bappu (1927–1982) made many important contributions to astronomy in his home country. After his return to India from his student years in the US in 1954, Bappu established first a new observatory at Manora Peak of Naini Tal in Uttar Pradesh (on the outskirts of the Himalayas) in 1955; then modernised and re-equipped the solar observatory in Kodaikanal (1961) before coming to Bangalore to establish a new Indian Institute of Astrophysics as a national centre of research in this field. Associated with this institute is a new observatory at Kavalur, shown here. It is the largest in India to date, and its principal instruments are a Zeiss 1 m reflector, housed in the dome at left, and a 2 m reflector in the right-hand dome, built in-house and completed a few months after Bappu's untimely death.

However, my real place during this 1982 visit was not in the field, but in the classroom at Bangalore; but there, in spite of Vainu's kind words at the conclusion of the course that 'As a result of it, double-star astronomy in India will never be the same again', I felt that I had not conveyed to my audience as much of the subject as was expected (but I have remained in correspondence with some of the students ever since).

However, all good things are bound to come to an end; and the end of this came when Vainu and other friends took me to Bangalore airport on the day of my departure. The connecting flight to Bombay was late, but astronomers never waste their time; we immediately started to discuss the problems of the evolution of binary systems, and continued until my flight was called for boarding about one hour later. As we clasped hands (as it turned out, for the last time in our lives) at the gate, we promised each other to continue our discussion in August at Patras, where Vainu was to preside over the Eighteenth General Assembly of the IAU. Alas, it was not to be; for only a few days after the start of the meetings in Greece, the grievous news reached us from Munich of Vainu's sudden and premature passing, at the age of only 55 years. *Requiescat in pace*!

The next astronomical occasion that year which I wish to recall was indeed the General Assembly held that August in Patras—the first I had attended since the Fourteenth Assembly in 1970 in Brighton (England). Several reasons attracted me to Greece at that time: not only to see Vainu again after our spring meeting in Bangalore, but also to get together with my many students and friends in Greece who would congregate at Patras. They did, and prepared for us an unforgettable occasion.

As local host to the IAU, Professor Goudas was at that time a very busy man; but not too busy to arrange for a 'Manchester evening' in a pleasant hotel at the sea-shore of the Gulf of Corinth (not far from the place where our NATO Advanced Study Institute had been held in 1971), which was like a 'class reunion' to many of us—and not only Greeks. Several members of the IAU Executive came to join us on that occasion. Unfortunately, Vainu Bappu (who was to have presided over the General Assembly) had left this world some days before; but, in his place, we had the pleasure of having with us the new IAU President-Elect—my old friend Robert Hanbury Brown with his charming wife. This made the family reunion even more complete, and so did the presence of Patrick Wayman (the IAU Secretary General, who had already for some time held jurisdiction at Dublin over a part of my former astronomical library—at least until it is burnt down again by the IRA!).

Incidentally, Patrick came to our reunion at Patras with one more piece of happy news: the IAU Executive had approved of a proposal that one newly discovered asteroid (number 2628) was henceforward to carry my family name. Strictly speaking, by that time this was no secret for me; for the discoverer of that asteroid, Dr Eleanor ('Glo') Helin of the California Institute of Technology, had informed me of this some days before at a convivial occasion (see figure 5.60) at which more than one toast was drunk to everyone concerned—including all asteroids still to be discovered by Glo in the future.

After my return from the Mediterranean area by the end of the summer I was expecting to remain home in Wilmslow at least for the rest of the year, but it was not to be. In December 1983, the Government of Sri Lanka (Ceylon) was to inaugurate a new Institute of Fundamental Studies at Colombo; and how could I resist the invitation of that country's President, J R Jayewardene (prompted, no doubt, by my friend Chandra Wickramasinghe, a scion of the family of ancient kings of his country; as well as by Dr D Asoka Mendis, one of our best students (PhD, Manchester 1967; DSc, 1978)) to attend the occasion?

And so I did, and have never regretted the trip: for apart from a very interesting international conference convoked between 1 and 14 December for this occasion (in a palace erected for the purpose by the People's Republic of China as a gift to Sri Lanka), we were privileged to take part in a three-day expedition to see the sights of that enchanted island, at Kandy, Sigiriya, and Anuradhapura, with interesting examples of cultural and economic life. It was for the first time at Ceylon that I saw motorcars sharing the roads with 'working' elephants; while, in cities and towns, two ubiquitous sights continued to remind me constantly of my native country: advertisements for 'Pilsner beer' and 'Bata shoes'. Apparently it is not only the Czechs, but also the Sri Lankans, who seem unable to get along without either of these commodities! Yet it was with genuine regret that we had to leave this enchanted island before Christmas, back to foggy Manchester; and it was with deep sorrow that we learned, less than two years later, of the outburst of civil intolerance between the Sinhalese and Tamil populations of Sri Lanka which (like similar events elsewhere) can entail only tragic consequences.

The next year (1983) was to prove one of the busiest and most exciting of my life. After a brief visit to Egypt in the spring (during which I took farewell not only of the pharaohs of old, and of so many monuments which they had left behind them, but also of my old friends Professors Asaad and Fahim, both scheduled soon to retire), later that spring I was called upon to undertake a long trip to another country which I had never visited before, Indonesia, on three separate missions: to take part in the IAU–UNESCO International School on the Solar System, held in 1983 at the Lembang Observatory, to participate in the

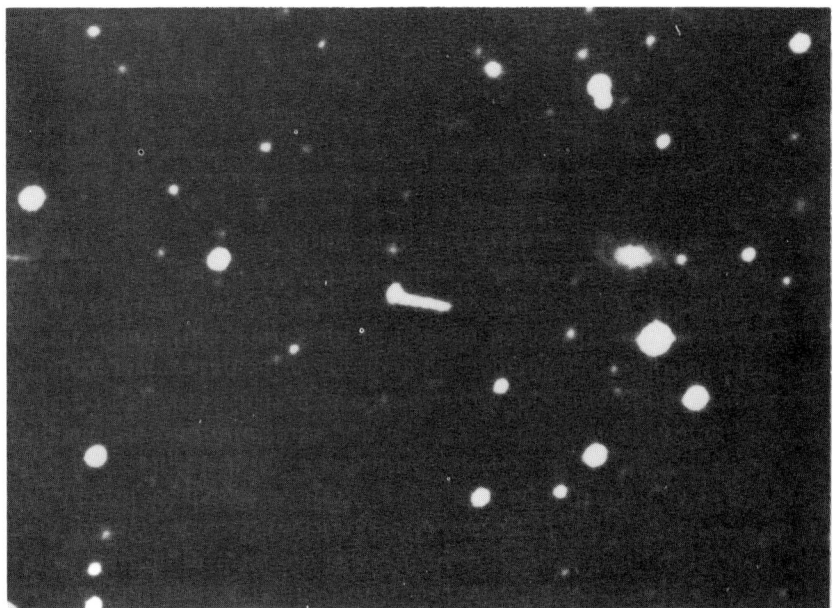

**Figure 5.60** (above) Asteroid 1979 MS$_8$(2628) 'Kopal', discovered on 25 June 1979 by E F Helin and J S Bus with the 48 inch Schmidt telescope of the Siding Spring Observatory in Australia. It revolves around the Sun in a mildly eccentric orbit, with a period of 4.96 years at a mean distance of 2.906 astronomical units. (Below) Dr Eleanor ('Glo') Helin of the Jet Propulsion Laboratory with the author at the Eighteenth General Assembly of the IAU at Patras.

IAU Colloquium number 80 on 'Double Stars—Physical Properties and Generic Relations' at Bandung on 3-7 June (see figure 5.61); and afterwards to set out to witness a total eclipse of the Sun, the shadow cone of which crossed central Java in the morning hours of 11 June.

Figure 5.61 (left) At the Lembang Observatory in Java, June 1983, which the author visited to deliver a series of lectures on the Solar System at the IAU–UNESCO School for Young Astronomers of South-East Asia, held at Bandung at that time; and also to introduce (on 3 June) IAU Colloquium number 80 on the physical properties of double stars. (Photo Dr Josip Kleczek, Director of the IAU–UNESCO School.) (Right) Dr Bart J Bok (left), Dr Edwin Budding (centre) and the author at IAU Colloquium number 80. This is probably the last photograph of Bart Bok (see also figure 4.9) taken before he passed away two months later (on 5 August 1983) aged 77 years.

After the end of these conferences in Bandung, and before we embark on this memorable journey two days before the eclipse, it is high time for me to introduce to the stage of our narrative Dr Josip Kleczek of Ondřejov Observatory and Charles University of Prague—a veritable *secretaire perpétuel* of the IAU–UNESCO Schools for Young Astronomers, and a very dear friend, not only of those who passed through these schools, but of myself as well. The 1983 school in Bandung (the second to be held in that city; the first one was held there ten years before) was already the thirteenth to be organised by Dr Kleczek, in twelve countries so far; and while others may have been 'fronting' for them in the IAU, Kleczek did all the work. Several hundred young astronomers all over the world (mainly from countries left 'off the beaten track' by the progress of our science) thus gained access to higher education which would otherwise have been outside their reach; many of these schools laid down the foundations of their professional careers (and not a few met on such occasions their future spouses!). As the director of the first such school (in 1967) and teacher in three others (1969, 1975 and 1983) I can attest from my personal knowledge that these were indeed invaluable.

Let us wish the best to future schools of this type which, as long as Jožka Kleczek remains in charge of them, will, I am sure, be as successful as they have been in the past; and return to the total eclipse of 11 June 1983. The reader may recall from Chapter 3 the story of the first total eclipse of the Sun which I had occasion to observe in a cloudless sky on 19 June 1936 at Hokkaido. At the Japanese eclipse of 1936 totality lasted only just under two minutes; whereas at Java in 1983 its duration exceeded five minutes, thus rendering it one of the longest total eclipses observable from any part of the Earth (the theoretical maximum duration of totality anywhere on the terrestrial surface amounts to 7 minutes and 40 seconds).

In order to try our luck with the weather, a few of us (Dr Edwin Budding, formerly from Manchester and now of New Zealand, Dr and Mrs Duerbeck of Bonn, Dr Tibor Herczeg of USA, Dr Ernst Zinnecker of Munich and I) rented a car at Bandung, and by the evening of 9 June had reached Yogjakarta in central Java, some 300 km to the east. Most of the next day we spent sightseeing (there are several memorable sites in that region, including the ancient temples of Borobudur) under cloudy skies (and occasional rain made our chances of witnessing the eclipse next day doubtful). Only towards evening did the sky begin to clear up somewhat; and a few stars appeared between the clouds as we returned to Yogjakarta for our second night. When, however, we woke up next morning the skies were clearing up; and only isolated clouds floating here and there were tempering our optimism in the success of our long journey.

The centre of the belt of totality was to the north-east of us; and we reached pretty close to it about an hour later, near the town of Magelang, where we decided to make camp and await our chances. However, no sooner had we set out to do so than we became the centre of attention for the local inhabitants, and received an official welcome from the mayor as well as the chief of police, who saw to it that we could carry out our programme (on the slopes of Tidar Hill) without any hindrance from the local population.

The more the eclipse progressed, the fewer clouds remained in the sky; and during totality we were privileged to observe the Sun with its corona, high in a cloudless sky of rare transparency (see figure 5.62). When, after more than five minutes, the first rays of the Sun emerged from behind the dark disc of the Moon, the happiness of the observers knew no limits (especially my own, for having witnessed the second total eclipse of my life under near-perfect conditions). In 1936 at Nakatombetsu I had been the youngest astronomer on the campus;

while 47 years later at Magelang I was the senior of the group, without much chance of seeing another total eclipse within the remaining years of my lifetime; and all that remained for me to do was to return home.

This, of course, was easier said than done. It was around noon when we bade farewell to Magelang whose skies had treated us so generously on that occasion; and it took an additional fifteen hours before we returned by car (via Semarang) along the coasts of the Java Sea to

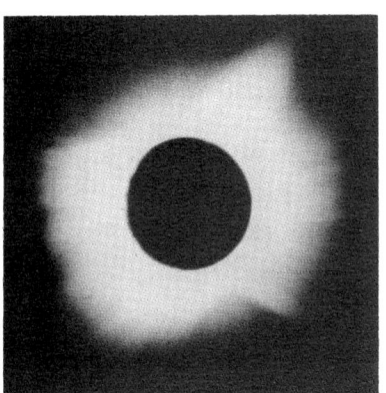

**Figure 5.62** (above) Darkness before noon: the total eclipse of the Sun as photographed on 11 June 1983 at 11.29 AM at Magelang in central Java. (Below) Jubilant observers of the event, both local and visiting. The gentleman at extreme left is the local chief of police, charged with enforcement of the government ruling that no-one should look at the eclipsed Sun except through a telescope (an edict not too closely observed by the chief himself!). Of the observers from distant countries, the third from left is Dr H Zinnecker of Munich (now at Edinburgh); sixth from left is Dr H W Duerbeck of Bonn, with his wife (Professor W C Seitter of Münster) ninth from left. The tall astronomer behind her is Dr Edwin Budding of New Zealand; fourteenth from left is Professor T Herczeg from the United States. (Photo by the author.)

Jakarta. After a good night's rest, I departed from Jakarta airport by an overnight flight back to Manchester, in less time than it had taken us to drive from Magelang to Jakarta by car the day before.

If I thought that this trip to Java was to be my longest hop by air in 1983, I was mistaken; for in the autumn of that year another trip was to take me to the Far East, via Hong Kong to China, Japan, and Korea; before I returned (through Hong Kong again) back home by the end of the year. My first port of call, Hong Kong, was no longer new to me: for I had already stopped there several times in the 1960s *en route* from India to the United States via the Pacific basin; but my visit to China was the first since 1936 (see Chapter 3); and the country I entered in Shanghai in October 1983 was wholly different from the one which I had left via Siberia 47 years ago.

It would take a whole chapter—if not a book—to describe fully the changes which stared one in the face wherever one looked. As in 1936, my astronomical visit to Shanghai started at Zi-ka-wei. In looking around the horizon (all built-up) there, I identified at least the tops of the spires of the cathedral which the Jesuits erected there in the nineteenth century, and which I remembered well from the previous visit. But (my Chinese friends told me) if I had re-visited China only a few years before, the spires would not have been visible at all; for they were demolished during the 'Cultural Revolution', and only re-built since. And the same was apparently true at Zô-Se: the cathedral built by the French on top of the Sheshan hill was still raising its spires to the heavens, but encased in scaffolding (used to repair the damage inflicted probably at the same time as in Zi-ka-wei).

The first thing that must strike any visitor now is the tremendous, teeming crowds met wherever one goes: with its population of over twelve million Shanghai is now the largest city on the Eurasian continent (the largest city of the world, with its fifteen million inhabitants is, of course, the conurbation of Tokyo–Yokohama in Japan). Long gone are the days when (as in the years of my youth) the largest cities, like London or New York, were on the shores of the Atlantic. Both of them together would not add up nowadays to Tokyo–Yokohama! And hand-in-hand with this veritable explosion of population goes a tremendous building activity to house this population and the industrial activities supporting them. To give one example (of many that could be quoted) at the beginning of this century, about 80% of the entire population of (say) the United States were still living in the country; and only 20% in the cities. Today, 80 years later, this proportion seems just about to have reversed itself in the US; while in China it is still about the same as it was in the US around 1900. Is this situation extrapolable to the future? And if so, will 80% of the Chinese population by the year 2050 AD (by then some 1500 to 2000 million

350  *The Manchester years*

**Figure 5.63**   A 10 km long bridge across the Yang-tse river in Nanjing, for road and rail traffic, erected in the last ten years of Chairman Mao's rule.

**Figure 5.64**   Chairman Mao-tse-tung welcoming users of his bridge across the Yang-tse river.

people) also become city-dwellers with all that would mean for the industrial potential of the nation? A disturbing perspective for nations which are likely to be demographically left far behind!

To give another example, when in 1936 we travelled from Nanking to Peking by train (a journey of two and a half days at that time), the Shanghai express had to be transported across the Yang-tse river on a raft; now it can be driven across a modern bridge (see figure 5.63) several kilometres long, which speeds up transport (by train or road) ten times—far faster than Chairman Mao, whose statue welcomes you at the entrance to the bridge (figure 5.64), took to swim across in the late years of his life. And in Peking (and its neighbourhood) you see more of the same: tens of thousands of people come each day to visit the Great Wall of China (see figure 5.65)—about the only handiwork created by human hands on the Earth which could be seen from the Moon with the telescopes now at our disposal (under special conditions; cf pp 278–9). Its magnitude is indeed awesome (sufficient to build a thousand Cheops pyramids from its stone masonry) but also pathetic: it was erected as a 'Maginot line' to protect the Heavenly Kingdom from Western invaders, who indeed took it many times in the past—but not with weapons in their hands: for it was far more economic to bribe its defenders.

Figure 5.65  A section of the Great Wall of China west of Beijing.

From China my autumn journey of 1983 led to Japan, in the reverse direction from that of 1936, and by air rather than ship, to visit my astronomical friends at the Mitaka Observatory (see figure 5.66). To return to Japan after so many visits in the past was now almost like a home-coming (figure 5.67); and this time my friend Professor Kitamura had arranged a pleasant surprise: namely, a visit to Nakatombetsu where, 47 years before, my Japanese colleagues and I had observed my first total eclipse in 1936. This part of my journey was already described in Chapter 3. Suffice it to add here that, in the afternoon of that memorable day, my Japanese friends drove me to Esashi, a village some 10 km from Nakatombetsu, on the shores of the Okhotsk Sea, where in 1900 a French expedition under Professor Henri Deslandres (1853–1948) had observed another total eclipse of the Sun (two Saros cycles before 1936!). It was with emotion that I read, in the local memorial book, an inscription entered there by Deslandres in his own hand: 'The nation which chose the Rising Sun for its national emblem must love astronomy.' Rarely had this prophecy been better fulfilled; for a short time afterwards the Japanese Government announced their intention of underwriting a reflecting telescope of 7.5 m aperture (by far the largest in the world) to be erected for the use of their astronomers at Mauna Kea in the Hawaiian Islands!

**Figure 5.66** 'Mancunians in exile' at the Mitaka Observatory of the University of Tokyo in October 1983. Standing (left to right): Dr Okazaki, Dr H Ando, Dr N Awadalla, Professor M Kitamura, Dr A Yamasaki and Professor N Sekiguchi. Sitting: Mrs Yamasaki, Mrs Kitamura and Mrs Sekiguchi.

Incidentally, on our return journey from Esashi back to Asahigawa I was taken to a local observatory, on a hill overlooking the Okhotsk Sea, from which the tragedy of the Korean 747 plane from New York to Seoul (shot down by a Soviet military plane in the spring of 1983) would have been observable well above the horizon. The Japanese radio tracking-station at Wakanai (not far from Nakatombetsu, which we drove by on our return) could provide more details on that tragic event—should it wish to disclose them!

I should, however, record gratefully that, next morning, the Chief Political Officer of the Prefecture of Asahigawa presented me, in his office, with a Gold Medal to commemorate the 1936 eclipse at Hokkaido, and my participation in its observations. The Prefect took a personal interest in the occasion; for his family (he told me) came from Nakatombetsu (his sister was still living there); and he received his primary education in the school in which our expeditions were housed. Of all the observers of those days, only two were still alive in the autumn of 1983: Mr Sigemaro Kibe and myself (figure 3.7); so that the medal I received was, perhaps, intended to commemorate longevity rather than any particular scientific feats which we achieved on that occasion.

**Figure 5.67** On the roof of the Astronomy Department building at the University of Tokyo on 3 October 1978. From left to right: Dr H Yoshimura, Dr H Ando, Professor M Kitamura, Emeritus Professor Y Hagihara, the author, Professor G Hori and Mr (since Dr) T Kim (now Director of the National Observatory of Korea). This was the last occasion Professor Hagihara visited his old department; for he died less than four months later (20 January 1979), aged 82 years.

On the return from the Far East my itinerary took me to South Korea, where I spent a week at Yonsei University in Seoul, a very active centre of double-star work (see figure 5.68). During my first transit through Korea in 1936 by train, the country was still under Japanese domination; but now, divided as it was between North and South (the astronomical observatory of Yonsei was located very close to the dividing line!) it offered another eye-opener to astronomical visitors. Not far from the observatory, one can see a very large and

354  *The Manchester years*

elaborate system of antennae for radio transmission; and when I asked my colleagues playfully if these are there to provide the observers with TV programmes for cloudy nights, I was told no; the antennae are there to jam North Korean broadcasts to the South (just as South Korean broadcasts are probably being jammed somewhere else north of the 38th parallel).

**Figure 5.68**  Graduate students and staff of the Astronomy Department of Yonsei University in Seoul, at the time of the author's visit to Korea in October 1983.

South Korea (the only part I had a chance to visit) itself is almost as large in area as the whole of Britain; and Seoul, its capital city, has a population comparable to that of London. Its downtown (business) part is as lively and glittering with goods as that of any similar city in the US. But it is refreshing to see, in the midst of it, a tall tower exhibiting the Foucault experiment, demonstrating the axial rotation of the Earth, for the benefit of the shopping throngs (figure 5.69). What I found less edifying, I confess, were the military vehicles (manned with submachine-gun-carrying soldiers) parked in force in front of every school of advanced learning which I had an opportunity to visit in the Seoul area—obviously to keep the students in check. Moreover (as I was told by my colleagues) this was no anomalous situation, but a permanent feature of South Korean life. I tacitly recalled what I had seen in Teheran in 1977, and kept my peace.

When 1984 came, George Orwell notwithstanding, I had a pleasant occasion to celebrate in April: my 70th birthday, which happened to coincide with the appearance of the 100th volume of the journal

*Astrophysics and Space Science*, founded in 1968, of which since that time I have been Editor-in-Chief. The circumstances under which this journal was born will be set forth in greater detail later. The completion of its 100th volume (the first such anniversary of any journal published by the Reidel Publishing Company of Dordrecht) was (perhaps appropriately) celebrated in Dordrecht; and I was glad to have an opportunity to express my appreciation to all my collaborators of those years: Mrs Ellen Carling-Finlay of Manchester, Mrs Nel Pols-van der Heijden, Mr Jan Hattink and Mr Patrick Wharton of Dordrecht, for their cooperation and fellowship in work on our chosen part of the vineyard. The true originator of the journal, Mr Anton Reidel, was no doubt thinking of us in the United States, where he had transferred his residence several years before.

**Figure 5.69** A tower in the downtown (business) part of Seoul, housing a Foucault pendulum to demonstrate the rotation of the Earth to shoppers who drop in.

But I was not destined to spend the next year at home either. In the autumn of 1984 I took off once more (for the eighth time) to Japan to attend the Third Regional Asian–Pacific Meeting of the IAU in Kyoto (which we were committed to publish in our journal in due course). On a Sunday excursion to Nara I recalled my first opportunity to feed the deer there 48 years before. I doubt if any astronomer I met in Japan at that time was still alive, save, perhaps, Professor Miyamoto, Emeritus Director of the Kwasan Observatory in Kyoto (who in 1936 was Professor Yamamoto's student); but he no longer attended meetings.

After the Kyoto meetings, I took once more to the air to fly (with only one stop-over in Fiji) to New Zealand, for my first visit to that country. Never before have I been 'down under' so far south, to renew my acquaintance with the southern stars. This time I was, however, among only friends: and, indeed, in the family. For already from Japan I had braved the vast expanses of the Pacific Ocean in the genial company of Dr Edwin Budding (see figure 5.61), a former student of mine and later colleague from the University of Manchester, to whose first-born son I became godfather. It was also in Ed's company (and in a rented car) that I had a chance to explore the National Parks of North Island, and gave colloquia at the Universities of Auckland, Wellington and Christchurch, as well as addressing the local astronomical societies (one of which made me an 'honorary Kiwi').

Since 1982, Ed Budding and his family have been settled in Wellington, where Ed is Deputy Director of the National Observatory of New Zealand. But most professional astronomy in New Zealand is being carried out at South Island, at Blackbirch Mountain (which houses also the southern station of the US Naval Observatory) and Mt John in the New Zealand Alps, equipped with instruments up to 1 m aperture at altitudes of 1450 m and 980 m respectively. At the first of these, during my visit we experienced winds almost strong enough to launch us into space. But at Mt John (which I visited in the company of Dr Hearnshaw of the University of Canterbury in Christchurch) the weather was wonderful—never before had I seen the southern sky so clear and the Magellanic Clouds so high above the horizon—and so was the daytime visibility of the main chain of the New Zealand Alps, with Mt Cook as its highest summit (3768 m; climbed by my son-in-law, Dr Smith, in 1973 after the IAU meetings in Sydney).

But as summer was coming to the south, it was time to go north. The first stop on my return journey was Tahiti, where I searched (vainly, as it transpired) for the remnants of an observatory set up to observe Halley's comet in 1910 by Dr (later General) Štefánik, at whose observatory in Prague (see Chapter 2) I had started my astronomy career. With a short stop-over in California I returned home by the north polar route, thereby completing my twelfth circumnavigation of our planet.

The last of my long trips before this book went to press was again to the Far East: to India, to attend the Nineteenth General Assembly of the International Astronomical Union, held between 19 and 28 November 1985 in New Delhi. For me it was a special occasion: the fiftieth anniversary of my accession to the IAU (when I was elected a member of Commission 27 on Variable Stars at the Fifth General Assembly in Paris).

That occasion was described in Chapter 3 (pp 106–109); but many things have changed since those times. The membership of the Union, which in 1935 fell just short of 500, has by now outgrown 6000; of the 'old guard' who have remained members since 1935 only 23 are still with us in this world—and only three of us (W J Luyten, K Aa Strand and myself) made it to Delhi in 1985! Moreover, the IAU Commission 42 (on close binary systems), whose birth I attended in 1948 (and over whose formative years I presided till 1955) has since grown to become one of the largest commissions of the Union—almost as large as the whole Union was at the time of its foundation in 1922. It was a source of genuine pleasure for me to see it presided over at Delhi by Dr Alan Batten, a former student of mine at Manchester whom we have already met in figure 5.12 (and whose closer acquaintance we shall make in the next chapter; see figure 6.2).

After so much travel mentioned throughout this section, the reader may scarcely believe me if I say that since that time I have remained home at Manchester, writing up all these reminiscences. I have, however, gradually grown tired of dwelling so much on the past; and if I have nevertheless persevered in this task, it was mainly because of the awareness that unless I do so now, who knows if another opportunity would present itself later? Yet these reminiscences should be regarded only as an interruption of more serious work to which I itch to return as soon as I can; and if, in writing this book I had perforce to dwell too much in the past, I kept consoling myself, 'Do not worry; your best books are yet to be written. You are still so young!'

## Summing up

By the time I retired from the Manchester Chair of Astronomy at the end of September 1981, my academic service (commencing with my earlier years at Harvard University and MIT) extended over 41 years—long enough perhaps to justify an attempt to draw some more general conclusions from the experience spanning now (as this book goes to press) a lifetime of over seventy years. In summing-up the lessons of this experience, what have I learned in the course of this time that could be of interest to the younger generation?

In making an attempt at such a summary, it is perhaps proper to start with a recount of the blessings for which one should be grateful to one's fate. In my case, these blessings have indeed not been insignificant. The first is one of mere survival. The twentieth century, which commenced not long before my young eyes opened up on this old Universe of ours, has been infested with strife unparalleled in the history of mankind so far. Its first half alone saw two world wars, more devastating (and claiming greater loss of life) than any fought before, which I managed to survive; at least I stand a good chance of joining my ancestors before the third, which will probably be much more devastating than any of its predecessors, unleashes its full fury on this troubled world. With the possible exception of certain times during World War I, I have never suffered from hunger or the other deprivations which have since become endemic among an increasing fraction of the human population; and was able to spend most years of my life in relative comfort.

When I grew into manhood, and World War II descended on this world while I was in my twenties, I was not called upon to perform military service in the barracks. To be sure, in 1936 I was conscripted in the Signal Corps of the Czechoslovak Army; but that army ceased to exist before the time of my call-up; and when war came to the United States in 1942, it was as a technical expert, rather than a man in uniform, that I was summoned to serve my adopted country (for the work assigned to me was normally the responsibility of an officer of fairly high rank which, as technically still an alien, I could not have formally received till after the war).

Moreover—and this I consider to have been the greatest blessing of all—wherever my fate placed me, and whatever I was called upon to do in the course of my life, I did so gladly. I may have been born with an aptitude to do certain things better than many of my contemporaries, and a very few of them perhaps better than anyone else. Since early adulthood nobody had to tell me what these were; and I was fortunate that most of my teachers (and early superiors) saw it in the same way. When I eventually reached the level of achievement where I got ahead of others and may have (as is only natural) generated a certain amount of jealousy among my contemporaries or even seniors, I was already in the fortunate position that I could continue going my own way and no longer had to mind the consequences. It was only later in life that I learned how increasingly rare such a fortunate position was becoming in our midst; and that before this century is over, academic freedom with all that entails may vanish from this world as snow in the spring. In the face of such a situation, should I apologise for 'corrupting the youth' in my lectures by a confession that I was glad not to be any younger (according to the calendar) than I was?

Fate willed it that I was born at the time of decline of a civilisation—not far from its summit, perhaps; and still far from the abysmal depths to which it may decline in the future—but the gradient of decline is already sufficient to make one contemplate it with apprehension; and be aware of the fact that the blessings we still enjoyed in our lifetime may not last. This explains perhaps that (throughout most of my life) I always seem to have been in a hurry 'to make hay while the Sun shines'. At any rate, anyone who is pregnant with thought is bound to feel an irresistible urge to deliver his message, even if this will keep him awake at night, or away from the more trivial pursuits of ordinary life.

A waste of time has always been repugnant to me for this reason, and perhaps never more than at present, when by all the laws of nature my time must be inexorably running out; and the Latin proverb '*carpe diem*' must constantly be, not in the back, but at the front of my mind.

Favourable circumstances made this easier for me by sparing me the need to move too much from place to place, and thus to avoid the distractions bound to arise from such actions. As a matter of fact, since I began to show any signs of intellectual stirrings before the age of ten, my entire life has been spent essentially in only four places: Prague (1923–1937), both Cambridges (1938–1951), and Manchester (from 1951 on); and I made, therefore, only two *major* moves in my lifetime. After returning from the US to the UK in the autumn of 1951 (to be sure, at a time when life in post-war Britain was still far from easy), it took me the better part of three years to get properly settled and organised to resume and expand my former activities; and a reduced number of papers I published in those years bears it out in a rather convincing manner.

It is true that I became unfaithful to Manchester at times (for instance, during my frequent travels abroad); but these were only temporary interruptions; and my house at Wilmslow (where I wrote this book) has been my veritable home for almost half my life. It was only in this way that one could accumulate, in the course of years, a professional library (as well as other *Handapparat*, as the Germans call it), the sheer weight of which alone would make it very difficult to move about (and without which systematic work becomes difficult to accomplish).

Since 1940, when I joined the faculty of Harvard University as its very junior member, my research activities have always been connected with teaching, mainly at postgraduate level, but sometimes also to undergraduates and to a more general public on subjects of wider interest. This I consider to have been another blessing for which I am duly grateful to my fate: as a third-generation teacher in our family, how could I have been content with anything else? In particular, during my Manchester years I could lecture to distinguished classes of postgraduates for a third of a century on any topic of current interest to me; and (with the exception of my years at MIT) no course I ever gave was repeated later with the same contents. In fact, my performance consisted mostly of thinking aloud in front of the class; and although I often came to the lecture room equipped with some notes, my students know only too well how soon I departed from these in the course of a session, with the result that my initial monologue became a discussion, lasting (sometimes) twice as long as the lecture originally scheduled. The ways of academic life at British universities during my years of active service still made it possible to do this at postgraduate level; and the results seem to have borne out the advantage of such informality.

In the past thirty years and more, my main personal contact with students has been, then, at postgraduate level; and when they came to me in quest of a suitable topic for research, as a rule I readily suggested some; though I was apt (perhaps at some risk of injustice) to mark off the individual in quest of a topic as one of second-class mind. For persons of truly first-class mind already know what to do when they come to join a postgraduate course, and need no-one to tell them in which direction their principal talents lie. In such cases—and these are the happiest discoveries which members of the teaching profession can make—the most that professors of the subject can do is to bring about this process of crystallisation, and help to develop it.

The same was also true when students came to enquire about the effects of this choice on the success of their careers at postdoctoral level. To these I usually answered (quite truthfully) that in the course of my whole life I never met a good astronomer without a satisfactory job (it is only about second-raters that any doubt may arise). I was, however, apt to add that while it may be fortunate to have a job, it is more important to have a mission; and, as Carlyle put it once, 'Blessed is he who has found his mission; he should ask for no other blessing.'

To discover it is a matter of will. All that teachers can do is to assist you on your way towards attaining self-expression. Where there is a will, a way will eventually be found; before the strong will of a dedicated personality a carpet of opportunity will unroll like before a king. And if not in this life, it will be so later. 'What would you prefer', I sometimes teasingly asked my young friends: 'to be well-known in your lifetime, or a hundred years later?' For should you become famous in your own day (that is, if what you did could be readily appreciated by your contemporaries), the chances are that you 'sailed too close to the wind'; and that your contributions to science will soon be absorbed in the substrate which supports the onward march of science. On the other hand, if you are really outstanding (that is, ahead of your time) only posterity will fully appreciate what you have done; and while your work may live, your name may become a mere label, your person submerged in anonymity. Which kind of fame would you opt for?

The head (until recently) of the astronomy division of the National Science Foundation of the United States expressed to me his opinion that 'Whatever may have been the case in the days of G W Hill, American astronomers today do not yearn for his kind of posthumous glory; they prefer to collect their kudos still in this world'; and many of their colleagues in Europe seem to share the same sentiments as well. But is such a stance merely a vote of distrust for the future, and of uncertainty as to whether there will be any? Is it another weather-vane pointing towards a decline of our civilisation, and a symptom of its fall?

Besides, are we really free agents in such matters? For whichever other motives may enter into his decision, no truly great scholar can betray his mission without cowardice (Isaac Newton providing perhaps the most conspicuous example in history of such a betrayal); and a conscious quest of goals which are beyond the horizon of one's contemporaries requires a greater degree of heroism than to seek easy laurels within one's lifetime. It is perhaps just as well that the choice is made for us at the subconscious level; for the greater the eventual prize, the longer it may take to attain it.

Needless to say, it is not for me to express in this place even a personal opinion as to the category to which any one of the hundred or more students whom I helped to educate to PhD level would qualify in such an assessment; for only the future can tell, and a teacher is bound to be biased in favour of his former charges (or should keep to himself the extent to which they may have fallen short of his expectations). But one maxim should be respected above everything else: notwithstanding what may have happened in the past, the sins of the predecessors should never be held against their successors. The latter deserve at least the benefit of the doubt!

But looking back over the past, I can note with some satisfaction that most of the hundred-odd students whom I conducted towards their doctorates subsequently had reasonably successful professional careers; and many of them I shall recall with genuine affection till the end of my life. If their own affection to their chosen subject perhaps left (in individual cases) something to be desired, one should keep in mind that no two men are the same; and, as the great German poet Friedrich Schiller once described our muse Urania, 'Dem einen ist sie eine holde, himmlische Göttin; dem anderen tüchtige Kuh, die ihn mit Butter versorgt.'† Some retain a productive relationship with their subject, unaffected by the passage of time; while for others their initial

---

† 'To one she is a noble heavenly goddess, but to another a stalwart cow which keeps him supplied with butter.'

romantic attachment may change to a mere platonic love, of the kind which produces no children.

What can I add about the problems of teaching astronomy at postgraduate level at the present juncture? Perhaps the most important point to get across to our young charges is to induce them, not only to think, but above all to *question* everything they hear in the classroom or read in the literature, by asking themselves (or their teachers): 'Is that so?'; or (even more to the point): 'How do we know it; and with what degree of assurance?' And the true hallmark of a scientist worthy of that name is to know when to say: 'I do not know; nobody does as yet for sure; it is for you, the scientists of the future, to find out!' This is the way to bring the best out of good students; and if others may be perplexed it does not matter so much; the sky is always darkest before dawn! Pierre Simon de Laplace, a great French astronomer from the turn of the nineteenth century (and, incidentally, a man for whom modesty was not the principal virtue), caused an inscription to be engraved at the portal to the Paris Observatory, conceding that 'What we know is very little; what we do not know is immense.' And who of his successors would dare to dissent?

Do not—I should stress above all—lay before the students of the natural sciences (outside mathematics) any piece of knowledge as an article of faith. Such faith (even if it were generated in less critical minds) could act only as a crutch for those who may want to know without having to search for the truth themselves. Moreover, I am convinced that it is not possible to teach science at all effectively (certainly at a more advanced level) unless teachers themselves give an example in advancing our knowledge by contributions of their own. To do otherwise would be like expecting parents to succeed in teaching virtues which their children do not see them practise.

And what is true of classroom teaching is even more so of the reading of 'popular' books, filling our markets in the wake of recent advances in science in general, and of astronomy in particular. In reviewing recent literature on these subjects, we can only observe that one can indeed write an awful lot on what one does not know; what is clearly understood can usually be expressed quite briefly. I am convinced that even very complicated subjects can be presented to the public, inside or outside a university, in a language intelligible to anybody willing to listen with an open mind; and any obscurity in presentation usually (though not necessarily) corresponds to obscurities in the mind of the speaker. To be sure (as was once remarked rather cynically by Talleyrand) 'Sometimes it is sufficient to pretend ignorance to acquire a reputation of knowing what one does not know'; but it does not always work that way—at least in schools.

As regards scholars, it has been my observation of many years that few people (scientists included) improve much with advancing age. Only the greatest are truly ageless; while many others actually deteriorate in the latter part of their lives—or, rather, return to what they had always been; but what later in life they no longer consider it necessary to hide.

The contents of this section are essentially personal comments on his profession by a single individual. Let us, however, say a few words concerning the *organisations* in which individual scholars of the present have been constrained to work throughout their lives; in my case, these were predominantly the institutions of tertiary education commonly known as universities. Their history goes back to the Middle Ages (indeed, Charles University, my own *alma mater*, was founded in 1348; and at least some of its original constitution survived for the next 600 years).

It cannot, of course, be the aim of this chapter even to attempt an outline of the evolution of universities in Western Europe, and of the contributions they made to our civilisation; but only to touch upon a few points of their more modern history, in so far as these can help us to anticipate their future. Those regarding the Charles University (and other schools of its vintage) should be almost a part of our family history; for my father and I were associated (as students and professors) with it for almost eighty years. Its essential features (which I found still in force at the time of my matriculation in 1933) were laid down in 1848, and survived for 100 years. By 1848, as one permanent gain of that 'spring of the nations' which swept through Europe in that memorable year, all institutions of higher learning in central Europe were secularised (emancipated, whenever necessary, from the influence of any Church) and their upkeep taken over by the State.

However, so high was the esteem in which universities were held at that time, that appointments to chairs in them were the prerogatives of the Head of State (acting no doubt on the advice of the cognisant Minister, but reserving for himself the performance of the actual act). Between 1848 and 1916 this was, for Prague, the Emperor Francis Josef I, who also received each new professor in private audience, to which the incumbent (not only the Emperor, who was very fond of uniforms) had to present himself suitably caparisoned. Like everyone employed by the state, university professors too had to have a uniform—and very resplendent indeed was it (consisting as it did of a three-cornered hat trimmed with feathers; and, of course, a sword). Alas, this was worn as a rule only once in a lifetime (at the inaugural audience with the Emperor). Albert Einstein had to wear it too on his one visit to the Emperor in this connection; but, as he mentioned to his eight-year old son, he looked in his uniform like a 'Brazilian admiral' (cf Frank 1947, p. 100).

After World War I (the aftermath of which swept Francis Josef and all the uniforms under the carpet—or to Hollywood), the universities too became more informal places. All decrees appointing my father to his posts (see Chapter 1) still carried, to be sure, the signatures of the Head of State, President Thomas G Masaryk (who held that office from 1918 till 1935), but an audience was no longer required (nor would it have been necessary, as father was Masaryk's student). As to myself, none of my appointments was signed by anybody but Registrars (or Secretaries of Corporations), as a reflection of the extent to which the prestige of the academic profession has come down in the eyes of our contemporaries.

There is, of course, more than one way in which to assess the standing of any profession in society, and its impact on the contemporary world; another is the extent to which society is prepared to support it in financial terms. In this respect, World War II will be regarded as the Great Divide, across which significant accelerations have taken place; and by no means all for the better.

To give one example from my early experiences in Great Britain, in the years up to World War II, the intellectual leadership of the country in theoretical astrophysics was beyond dispute; with people like Eddington, Jeans, Chapman, Milne and many others in the prime of their age, how could it have been otherwise? It was another one of my blessings in life that I caught a last glimpse of these times before the deepening shadow of the gathering storm brought them to a premature end (when I returned to England in 1951, only Chapman of the above-mentioned quartet was still alive); the former 'scenario' had changed almost beyond recognition.

Consider Eddington († 1944), at whose feet I had a chance to sit for some time in 1938 (see Chapter 3). Eddington lived at the Observatory; and if one wanted to see him, it was not necessary to make an appointment: in fact, it was impossible to do so, for he had no secretary to keep track of his affairs! There was also no typewriter at the observatory; Eddington wrote all his books, papers and correspondence long-hand (and papers going out to press were 'manuscripts' in the true sense of the word). The only computing machine which I found at the Cambridge Observatory in January 1938 was a crank-driven Brunswiga-Dupla (and just as well; for I was already familiar with it from Prague). That Eddington never had a car (bicycle served all his needs for locomotion) has already been mentioned; and the telephone (of very old-fashioned type) did not ring too often to disturb the concentration of astronomers working in the library or elsewhere.

There is another feature in which contemporary astronomical life differs greatly (and to the better) from the times I still remember of the past. Throughout the greater part of the nineteenth century, most astronomers at Cambridge (and elsewhere) were trained as mathematicians, and had little contact with physics (let alone engineering). Edward C Pickering of Harvard, or George Ellery Hale of Yerkes and Mt Wilson Observatories in the United States, who were physicists rather than mathematicians by training as well as inclination, remained for a long time voices crying in the wilderness, who met with only limited sympathy among their professional contemporaries†.

Perhaps nowhere else did this trend survive longer than in England, as readers may recall from the relatively recent past. What was the greatest event, which influenced more deeply than anything else the development of astronomy in the past quarter of a century? Without doubt the sudden onset of the Space Age in 1957, which so greatly increased our horizons and permitted us to lift our telescopes above the terrestrial atmosphere, and some to the close proximity of our nearest celestial neighbours, an effort culminating in 1969 in the first manned landings on the Moon. And yet only a few years before, the Astronomer Royal of Great Britain, Dr (later Sir) Richard Woolley, let himself be heard to pronounce *ex cathedra* (on his assumption of office) that the idea of space travel was 'utter bilge'! This unfortunate statement, which even then received much publicity and some adverse comment (see the London *Times* for 3 January 1956), in the wake of what followed certainly could not have created much confidence in the foresight of at least official British astronomers at that time. A gaffe of this calibre was by no means an isolated instance at the Royal Greenwich Observatory of that time; for only a few years before the (then) Director of HM Nautical Almanac had expressed doubts that artificial satellites could one day revolve around the Earth with impunity inside the terrestrial 'Roche limit' (cf Sadler 1949).

The roots of so profound a lack of insight into the shape of things to come went, of course, back to the education of astronomers of the old school who were brought up as mathematicians rather than physicists; and so the fact that spacecraft do not generally behave as fluid bodies held together by self-attraction, but possess a considerable rigidity of their own, did not register as firmly in their minds as it would have done with a physicist. Yet the fact that their ill-fated predictions were so completely refuted by actual events soon filling the front pages of newspapers all over the world could not but have given

---

† I once heard the late Professor Russell recall that when E C Pickering came in 1876 from MIT to Harvard and launched the well-known Harvard Photometry Catalogues, he was roundly upbraided by his senior colleague, Benjamin O Peirce (Hollis Professor of Natural Philosophy at Harvard, of *Short Table of Integrals* fame), for wasting his time on photometry of the stars, quantities which one could (then) measure with an accuracy of (at best) 1 part in 100, when he could be measuring the positions of the stars with a precision better than one part in a million!

rise to doubts in the public mind as to whether astronomers (or, at least, their official representatives) knew what they were talking about; and the Government kept the purse-strings tight even when they voiced more legitimate proposals.

Fortunately, this is no longer the case at the present time: most UK astronomers today possess a physical, rather than purely mathematical, background. But in the first ten years after my return to England from the United States, this was not yet the case. Indeed, apart from special branches (like radio astronomy), the war did not seem to have made much difference to the way in which astronomers went about their business. The real change commenced to make itself felt only in the 1960s; and has gone on gathering momentum through the 1970s to the present time. But there are also *caveats* to keep in mind! The total budget of the same observatory these days is perhaps a hundred times as large as it used to be by 1938; and although inflation may have reduced the real value of this income (say) ten times, its real income today should be at least ten times as large as it was in Eddington's days. But could one sustain a claim that, as a result, its current contributions to our science are ten times as valuable or extensive as they were in the 1930s? And if not, where has the rest gone?

But astronomy is by no means the most expensive branch of natural sciences. Physics—especially nuclear physics—is another kettle of fish; and as it originated (very largely) during the 'Rutherford decade' between 1907 and 1919 at Manchester, what had its *magnus parens*, Ernest Rutherford, to say on the way in which this branch of science should be properly administered and financed? The financing of nuclear physics in its 'string and sealing-wax' days at Manchester was no problem (the Schuster family fortune saw to that). However, looking into the future, Rutherford foresaw that it was 'essential for men of science to take an interest in the administration of their own affairs or else the professional civil servants would step in—and then, the Lord help you!' (cf Eve 1939, p. 238).

That this could happen was foreseen at the beginning of this century by Rutherford as clearly as the nucleus of the atom; and today we can only wonder at his prescience. The lever that pried out this particular Djinn from his bottle was, of course, of our own making: namely, greed coupled with lust for power. After the end of World War II, in the course of which it transpired that men of science can contribute at least as much to victory as men handling the weapons or wily diplomats, the claim of science (then still largely centred in the universities) to a larger share of the national budget became undeniable, and was eventually met by the creation of the National Science Foundation in the United States, and of various Research Councils in Great Britain.

The original aim of these organisations, to channel public funds to institutions of tertiary education to enable these to survive, largely miscarried; for, in practice, very small fractions of the funds so earmarked reached their legitimate end. By far the largest portion has been withheld 'in house', under the direct control of the bureaucrats, in support of various national (self-styled) 'centres of excellence', to which talents from the universities were siphoned off by liberal application of the well-known principle of 'the carrot and the stick': carrots mainly so far, but the stick may yet come.

Preceding generations of astronomers had it easier (or, at least, simpler): throughout the nineteenth century (and, in fact, up to the middle of the twentieth) most of their extraordinary financial requirements used to be provided from the private means of rich individuals who were interested in astronomy as amateurs (or who, like Charles T Yerkes, had ulterior motives in mind). Certainly the largest refractors of the nineteenth century (Lick, Yerkes) and the great reflectors of Mt Wilson and Palomar Observatory were financed in the first half of this century by private foundations (Carnegie, Rockefeller). However, since the end of World War II the gifts from private sources rapidly dried up (income or inheritance taxes saw to that); and the only 'capitalists' capable of raising money (by taxation) for scientific purposes in adequate amounts became the governments; and the disbursing agent of their largesse became the Civil Service.

'When a physicist runs out of money'—another one of Rutherford's caustic comments going back to the beginning of this century—'he must begin to think.' Half a century later, this advice was given an ominous twist. 'Don't think—scheme!' is the advice today given by seductive voices of the modern dispensers of Government funds. 'For it is up to us now to determine, not only what will be done, but also who will get a chance to do it. It is true that, some time in the past, the University senate and faculty committees possessed sufficient autonomy to conduct their own affairs not only by name, but also in fact. In the second half of this century this has, however, become largely a matter of the past; for with Government money behind us we have already reduced the University senates to rubber-stamps by overcrowding their numbers; and as regards its committees where matters are cooked up, we have already seeded enough of our "hatchet men" there to do our bidding. And if, on occasion, some of their operations may cause a certain amount of smell, remember that the ends can sanctify the means; and that—in politics—a week is a long time. If you are smart enough to catch the drift, in order to attain your ends you need only make some friends on the disbursing committees to scratch backs with; they will be glad to cooperate if they can trust you that, in due course, you will return the favours.'

The rest is easy: when reports are needed, you should write them cleverly, and dress them up impressively enough to impress those who have little time to read. Or, another and more effective way of cover-up (not as much in use among astronomers as it is in physics) is to have the contents of an intellectually empty report 'classified'—i.e., made accessible only to fellow-augurs for the rest of your life. Such a strategy requires connivance on the part of the guardians of 'official secrets'; but, if granted, represents a sure way to endow even mediocre work with an aura of importance ('for'—the gimmick goes—'if it did not contain something important, they would not have classified it').

In retrospect, it is perhaps not difficult to grasp why such a situation was eventually bound to arise. The principal cause underlying all this is, of course, the rapid growth of world population. It need not take place only where you may live; for its explosion anywhere is bound to affect all other parts of the world as well, as more and more hands are reaching for dwindling supplies of essential raw materials; and more and more mouths open for food whose production does not keep pace with growing needs. All this is bound to cause inflation with which not individuals, but only national governments can cope—not only as far as food or raw materials are concerned, but with the entire superstructure of civilised life as well.

However, in the course of this process, the servants of the governments—the Civil Service—are only too tempted to desert their original mission as well as name: their authority to handle other people's money makes them over-reach themselves; and former servants rush to assume the role of the masters. In our days this is already affecting the field of education. The rush of academics to join their new masters has already begun; and nothing in sight is likely to stop it in the foreseeable future.

To give another example of how far we have progressed down this road (by tricks unbeknown to our ancestors), I recall that not long ago a former Astronomer Royal of one of the kingdoms constituting the UK noted that a relatively junior Civil Service employee (without any particular scientific qualification) at the headquarters of the Science Research Council received a salary much higher than he; and was given the following explanation: 'That particular officer is authorised to sign cheques up to the sum of £50 000, while you can do so only up to £5000.' I do not know if the dispenser of this particular bit of wisdom realised its horrible implications (do civil servants have to be bribed to remain honest?); but I do know that the astronomer so instructed grasped the lesson, and quit his office not long thereafter for private enterprise which (hopefully) will prosper well enough to enable him to sign cheques for even larger amounts than that civil servant.

This is, in brief, the situation which men of science in our time have to live with. The fact that it is not germane only to the countries of the Atlantic community, but seems (as far as I know) equally at home in the socialist countries of the East, indicates that it represents a common response to a situation facing us all at the present juncture of the history of the world; and that its cause is deeply rooted in human nature rather than in any particular political or economic system.

But, in the meantime, the institutions of tertiary education have to survive as well as they can; and nowhere did this new situation make itself felt more clearly than in the post-war years in the United States. When General Dwight D Eisenhower (before he became the President of the United States) was elected in the late 1940s to the presidency of Columbia University in New York (in the hope that his name might attract financial contributions from wealthy alumni), on one of his first occasions he addressed its staff as 'his faculty'. 'Mr President', responded one veteran scholar on the Senate, 'we are not your faculty; you are our President.'

Now, forty years later, this is still true of some, but not many, private universities in the United States, and of next to no universities supported by state funds. Instead, professional administrators (and money-raisers) have increasingly taken over institutions in which they were originally brought in to assist academic administrators recruited from the Senate; but having rapidly become 'the tail that wags the dog', an increasing fraction (more than half) of the university's income is now deployed to maintain them.

Indeed, the schools for tertiary education in the United States are rapidly becoming 'big business' with all that this entails, and the roles traditionally performed by the faculty are on the defensive. 'Which President or Trustee' (I recently heard a professional 'educator' of one of the US Universities in the Mid-West proclaim with a voice which did not lack conviction) 'would be willing to pay the faculty—which cannot do anything else but teach, and so are a drain on the University budget anyway—more than the school's athletes or their coaches? For it is they who make money for the school from profits on ball-games even if they never saw the classroom.'

Why do I include such pearls of human wisdom in these reminiscences? Because the same stream of thought has already reached the universities in Europe as well, and may make further headway in the future. It is not dominant yet in all its aspects: in particular, professional sport (in the style of circus games) is still in Europe being kept out of academe (for reasons which may retain their validity also in the future). But, in other aspects, much of it is already among us today. If you hear today in academic circles promoters of 'student questionnaires' in which the novices on the scene are invited to 'classify' (anonymously) the veteran scholars who teach them, you cannot escape a suspicion that 'something is rotten in the state of Denmark': for if student scrutiny is necessary to weed out deficient scholars, the latter must have been wrongly selected by their seniors (whose responsibility it was to make the right choice to begin with; they should not be shifting a quest for remedy on to shoulders too immature for the purpose).

In writing this I do not wish to deprecate in the least the quality of teaching; or to by-pass the obvious fact that a good teacher should be capable of expressing himself clearly (although he should not necessarily 'play to the gallery'). However, it is much more important for the future of our civilisation that our academic teachers provide inspiration for their students, and are looked up to by them as examples worth while to emulate and follow. Many great scientists of the recent past who did so—for instance (to mention only a few) Lord Kelvin, Sir Arthur Eddington (or, among those still living, Sir Harold Jeffreys)—happened also to be rather atrocious performers in the classroom. Should (in our own days) 'student questionnaires' be used as a tool of denigration of such men by their colleagues of much smaller stature (who can be expected to preponderate by numbers in any school)?

Another significant development of our lives has been the fact that, as time goes on, university halls or classrooms are ceasing to be the only place in which communities of scholars are congregating to transmit knowledge to posterity; an increasing burden of this role is being taken over, certainly in the second part of this century, by symposia or conferences at which scholars often from distant parts of the world congregate to present the latest results and discuss them with their professional *confrères*. The idea itself is not new; and learned societies inaugurated a tradition of regular meetings of this kind more than a hundred years ago. However, the novelty which came with increasing ease of travel was to make such occasions the rule rather than the exception.

In the nineteenth century most astronomers still knew each other only by name and personal visits of members from different places were relatively rare; though when a man of the stature of J U Leverrier had to call off an announced visit to Pulkovo he sent instead a life-size oil portrait of himself to his colleagues to be remembered by, in the manner in which medieval kings used to woo prospective brides. When, after World War I, astronomers felt a growing urge to see each other more often and founded their International Astronomical Union, the first General Assembly (held in 1922 in Rome) brought together a little more than 200 *confrères*, a membership which, in the sixty years which has elapsed since that time, increased more than thirty-fold and now comprises over 6200 members from 49 countries.

Moreover, in the second half of this century (and especially since the onset of the 'Space Age'), astronomers congregated in the Union felt an urge to meet more often, and for more specific purposes. As a result, up to the present time the IAU has sponsored not less than 125 topical 'symposia' (and almost as many 'colloquia') held in different parts of the world, thus converting the world-wide astronomical community to truly one family of students of our subject.

That this is so should only be welcomed; but the trend entails also disadvantages which may tend to make such conferences actually counter-productive. The organisation of conferences can be useful to science only up to a certain point; and beyond it the opposite may actually be the case. I recall, in particular, the early 1960s when men were heading for the Moon, and nothing seemed impossible on Earth: a symposium on the Moon held at Pulkovo Observatory (see figure 5.28) was followed only six weeks later by one in Dallas, with the participation of much the same cast of speakers who barely had an occasion to adjust to the change in local time, and who repeated more or less the same texts.

In those years I too used to spend more time than I should have on this merry-go-round. However, after a while I could not escape the feeling that this was not the work for which I was born; but that I was running away from what I should have been doing in the first place. I learned from observation that earnest scholars dislike, in general, conferences to which they cannot usefully contribute; and are apt to regard them as a socially acceptable waste of money of the sponsoring organisations, but a waste of time for the doubtful benefit of ambitious busybodies. Readers who wish to recapture some of the spirit of the 'roaring sixties' of this century are advised to reach for a delightful novel, by Arthur Koestler, entitled *The Call-Girls*!

## Publish or Perish

But it was not only the role of *teaching* at university level which underwent a gradual but relentless change in the course of my life; an equally profound change has taken place in the way in which the results of current *research* should be made available to the scientific community at large through the medium of *publications*. The change, one may say, represents another import across the Atlantic to Europe under the label 'publish or perish'—implying that the intellectual calibre of a member of the academic community can be assessed from the number of publications which the latter has produced per unit time.

There is some merit in such an assessment; for it is indeed essential for creative scholars to combine teaching with research, and thus strive to expand the frontiers of our knowledge into realms hitherto unknown. The strongest motive behind this drive propelling each scholar worthy of his salt is (I think) to satisfy his own curiosity as to what lies beyond the horizon. Next in importance stands, however, the moral obligation to share the results of this quest with others, for only if the carriers of the sacred fire, ignited in the human brain by the divine spark of curiosity, find hands at the end of their journey ready to grasp the fruits of their labours, and add to them in due course, can the further progress of science be forever assured. No single man can match the accumulated wisdom or experience of his ancestors. Even Isaac Newton, a very secretive man when it came to science, confessed once (in a moment of weakness) that 'he stood on the shoulders of the giants.' How did this come to pass?

As long as accumulating wisdom of the past could be transmitted only by human memory (a process replete with 'information noise'), progress was bound to be slow. Indeed, thousands of years had to come and go before language became articulate enough to transmit knowledge, and before writing became a durable and useful tool; but when this happened some 5000 years ago (at different times in different parts of the world) the budding sciences of mathematics and astronomy were among its first beneficiaries. Whether this was so on papyri of Egypt, or in more primitive ways in which celestial pictographs were utilised to facilitate navigation, is for historians of our subject to decipher; our more immediate aim will be to trace the flow of this knowledge in the form of astronomical literature to our own days.

The sources of this flow can be traced with some assurance to the first millennium BC; and their locations were in the Near East and the area of the Mediterranean. It represented a very meandering river, with many twists and turns which are scarcely intelligible to us today. The continuity of transmission in contemporary science is ensured by so many agencies as to be almost automatic, and the individual scientist nowadays need not spend much effort in gaining access to any source. On the other hand, in the ancient world, and even in the Middle Ages, the transmission of information was a very uncertain process. If, in particular, one bears in mind the untold number of wars and other calamities that have occurred in the Mediterranean world since the days of Hellenistic civilisation, how did the writings of men like Aristarchos or Archimedes escape destruction, and in what form did they eventually reach us?

When Archimedes composed one of them—say the *Psammites*, which preserved for us the first reference to a heliocentric model of the solar system—the number of readers who could have been interested in it must have been severely limited from the outset, and have remained so throughout the centuries. It is unlikely that the 'first edition' issued by the author consisted of more than a dozen handwritten copies. Some of them may have found their way to the libraries of Alexandria or Pergamon; but those libraries were in time destroyed. Other copies may have been for a time preserved in private libraries, such as those of Archimedes himself, of the king Hieron or his son Gelon; but how much of a chance of survival did these possess? Did any one of them escape destruction at the sack of Syracuse by the Romans in 212 BC? Did any palace of antiquity, anywhere in the world, come down to us with its contents intact? How did the books they once contained ever reach us, especially those which could never have enjoyed any degree of 'popularity'?

It is probable that the same questions must also have arisen in the minds of their authors, and that, as a result, men of science in bygone days did not always write up their discoveries as they would do today, because they may have thought 'What's the good of it? Whom will the text reach, and who will preserve it?' Inhibitions arising from such misgivings were probably a contributing cause to the slowness of progress in ancient times. Thus the relationship of Ptolemy to Hipparchos appears to us like that of a younger contemporary to his senior, and yet they were separated by almost three centuries in time! Moreover, the doubts which ancient authors must have had about the fate of their contributions to humanity were, unfortunately, only too often well-founded. Modern scholars agree that at least nine-tenths of all ancient scientific production has never reached us in any form (except, perhaps, through cryptic allusions or references) and are totally lost, and how much of this lost knowledge concerned astronomy we shall never know.

The survival of the rest may indeed appear miraculous; and yet again it may not have been as rare an event as one might think. Ancient people had an almost superstitious respect for the written word, and esoteric writing in particular. This explains why Greek manuscripts, even though of no use to the average person, were often jealously kept and transmitted from generation to generation—from owner to robber to looter, from looter to a new owner—until, from time to time, they fell into the hands of people who were sufficiently appreciative or enthusiastic to prepare new copies or translations, or add commentaries.

The Archimedean and other manuscripts which have finally reached us have thus very likely escaped not one catastrophe, but many. To be more specific, the oldest extant manuscript of the tract *On the Size and Distances of the Sun and the Moon* by Aristarchos survived as a part of the Codex Vaticanus Graecus 204, written sometime in the tenth century AD—a thousand years before our time, but thirteen centuries after the composition of the original. Moreover, the oldest and most trustworthy extant manuscript of the *Psammites* of Archimedes, contained in Codex 28 of the Laurentian Library in Florence, hails only from the end of the fifteenth century (though it appears to be a direct copy of another manuscript (now lost) going back to the ninth or tenth century of our era).

A decisive turn for the better occurred only in the past 300 years, when the centre of gravity of Western civilisation began to shift from the Mediterranean area to the shores of the Atlantic. The invention of printing (and its introduction to practical use) in the second half of the fifteenth century did not, at first, mean much for astronomy: Copernicus's masterpiece *De Revolutionibus Orbium Coelestium* was printed (at Nürnberg) in 1543 in only about 150 copies (114 of which are still extant); and the famous *Astronomia Nova* by Johannes Kepler, that veritable watershed between ancient astronomy and modern times, appeared in 1609 at Heidelberg in a still smaller number of copies (of which only a few can be traced today); to make this feat possible at all, its distinguished author had to take part in setting his text in print (a task from which his more fortunate contemporary successors are debarred by union rules!). Moreover, to print astronomical books in those days was not a very safe task even for the professionals, as Landini (the printer of Galilei's *Dialogues on the Two Great World Systems* in Florence in 1632) found to his sorrow; and although this printer was not burned at the stake for his misdeed (some of his books were!), it was the opinion generally held by the Establishment of that time that the author as well as the printer of such unorthodox works would he accorded a similar treatment on the day of the Last Judgment.

Largely as a result of this regrettable *imbroglio* between progressive astronomers and a conservative establishment, the further flow of our river of knowledge had to go underground for several decades; but the flow was irrepressible; and its further propagation owes much to private correspondence by such men as Père Marin Mersenne (1588–1648) in France, and many others who augmented the scanty periodicals then coming into existence by private letters (subject less effectively to public censorship), thus shepherding astronomy well into the Age of Enlightenment.

The first specifically astronomical journals (some of which are still with us today) made their appearance only in the first half of the nineteenth century; and they too had their origin in scientific correspondence—such as that conducted from Gotha by Franz Xaver von Zach (1754–1832) in the form of a *Monatliche Correspondenz zur Beförderung der Erd- und Himmelskunde* (1800–1813); and later (between 1818 and 1826) published (from Genoa) in French under the title of *Correspondence astronomique* . . .

Of more permanent nature were the efforts of Heinrich Christian

Schumacher (1780–1850), who in 1823 founded in Altona the journal *Astronomische Nachrichten*, now in its 306th volume. This was the first international journal devoted to our science, and throughout its first hundred years remained the most important link between astronomers of the world (and it was on the pages of this journal that the present writer embarked on his own astronomical career in 1930).

One cannot indeed browse through the early volumes of *Astronomische Nachrichten* without emotion in realising how close we still are to our past as one encounters papers reporting the first measured parallax of a star (by Bessel in 1838—incidentally, styled Geheim-rath Ritter Bessel—the greatest astronomer of the nineteenth century); or to come across numerous contributions from the pen of that prince of mathematicians, Carl Friedrich Gauss (editorially styled Hofrath on the pages of the journal!).

But *Astronomische Nachrichten* did not for long remain alone in fostering the interests of astronomy in the world; for soon afterwards the Royal Astronomical Society in London commenced (in 1827) to publish their *Monthly Notices* (now in their 217th volume). In the second half of the nineteenth century, these two veterans were joined by the American *Astronomical Journal* (1851), the French *Bulletin Astronomique* (1884), the *Publications of the Astronomical Society of the Pacific* (1889) and, in the last years of the century, by the *Astrophysical Journal* (1895). Several of these journals, with more than a hundred volumes behind them, have witnessed on their pages an expansion of our science which would have astounded their founders. They also represent three different types of journal (those sponsored by a national academy, scientific society and private publisher) which are still with us today.

Throughout the nineteenth century, astronomical journals did not constitute the only—or even the main—vehicle for the dissemination of scientific results; most results customarily appeared in special publications issued by most observatories, and exchanged freely among them—a practice which economic exigencies have brought almost to a standstill in our own time; and only series of volumes of different observatory annals on library shelves bear witness to a more genteel way of scientific exchange in bygone days.

During the first two-thirds of the present century, the evolution of astronomical periodicals took a turn whose consequences are still with us today: namely, a gradual transition from international to national publications sponsored by Governments or local scientific societies. Thus the *Journal of the Royal Astronomical Society of Canada* commenced publication in 1907, and the French *Journal des Observateurs* in 1915; to be followed by *Annales d'Astrophysique* in 1938. The *Memorie della Società Astronomica Italiana* were initiated in 1920; the Russian *Astronomicheski Zhurnal* opened its editorial pages in 1924; and the same year saw the foundation of the *Japanese Journal on Astronomy and Geophysics* (since 1949 continued as *Publications of the Astronomical Society of Japan*).

Of other regional periodicals which were founded in the past fifty years, the Dutch *Bulletin of the Astronomical Institutes of the Netherlands* (1921) and the Polish *Acta Astronomica* (1926) were followed by the *Bulletin of the Astronomical Institutes of Czechoslovakia* (1947), the *Acta Astronomica Sinica* (1953) and the Scandinavian *Arkiv för Astronomi* in 1955.

In spite of many differences in age and form, all these periodicals possess many features in common. They all represent publications founded to serve mainly national communities or society membership; and although not monolingual in principle, several of them became so in the course of time. Moreover, several formerly independent journals became affiliated with national scientific societies to secure firmer financial backing in their membership. Thus the *Astrophysical Journal*, one of the most distinguished periodicals in our field, removed in 1963 from its masthead the subtitle of 'An International Review of Spectroscopy and Astronomical Physics', to underline its affiliation with the American Astronomical Society—as the *Astronomical Journal* had already done in 1942.

This trend towards national publication in a field of science so international by its subject was not, to be sure, without exception. Thus the *Zeitschrift für Astrophysik*, founded in 1930 by Robert Emden and published since by Julius Springer Verlag, opened its pages to the whole international scientific community regardless of national or regional affiliation. In more recent years, the same has been true of *Icarus*, an international journal of solar-system studies, published since 1962 by the Academic Press of New York; and, still more recently, of *Solar Physics* (1967) published by the Reidel Publishing Company of Dordrecht; followed by *Astrophysics and Space Science* in 1968 and *The Moon* a year later; while several of the European journals mentioned above have coalesced in *Astronomy and Astrophysics*, supported by the governmental funds of participating countries.

The names of some of the journals founded in the past twenty years bring to my mind recollections which I should like to share with the reader in this place. In a way, being my father's son I was almost predestined to follow in his footsteps as I grew up; and to this day I recall a modest bit of pocket money which I earned for spotting misprints in proofs of the journals edited by my father before I was ten years old; and more substantial money in my teens for typing father's manuscripts for the press (father never learned to type properly, and I was one of the few who could read his handwriting with reasonable accuracy).

My part in early efforts to edit a class journal in my secondary-school days (not at all unusual at that time) have already been mentioned in Chapter 2; but for the appearance of my name on the cover of a scientific journal I had to wait till 1950, when I was invited to become a Consulting Editor for the *Journal of Mathematics and Physics*, published by the Massachusetts Institute of Technology (on the faculty of which I then served as associate professor). The job of the Managing Editor was in the competent hands of Professor Eric Reissner, who did not have to consult his colleagues too often (to no harm of the journal); and my connection with it lapsed in 1953 when I definitely settled in England.

My re-entry (this time permanent) to the field of scientific publications occurred in 1962, in the early years of the Space Age. Shortly before, the Academic Press of New York engaged me in writing for them a major book on the Moon (Kopal 1962); and the late Mr Kurt Jacoby, Vice-President of Academic Press at that time, asked if I would be willing to launch a new journal for them devoted to solar-system studies. I was willing—nay, eager to do so; and not without the support of a respectable editorial board listed in the first issue of the first volume; with Albert G Wilson (formerly Director of the Lowell Observatory, and then on the staff of the Rand Corporation in Santa Monica) as co-editor. The first issue of this journal appeared in April 1962 under the name of *Icarus*, and it has been with us ever since that time.

The name *Icarus* for the new journal was my choice, and gave rise to a certain amount of apprehension. I recall that my late friend Pol Swings (1906–1983), who joined the editorial board of the new journal from its inception, noted that the saying 'C'est un Icare' in French is tantamount to the English phrase 'up like a rocket, down like a stick' which, in Pol's mind, could augur ill for the future; and old Mr Jacoby was heard murmuring that their new journal on nuclear physics should then be called 'Prometheus'. But I stuck to my guns, and convinced wavering colleagues by a resort to the following quotation from the introduction to *Stars and Atoms* by Sir Arthur Eddington (1927), which stated in delightful language that

> In ancient days two aviators procured to themselves wings. Daedalus flew safely through the middle air and was duly honoured on his landing. Icarus soared upwards to the sun till the wax melted which bound his wings and his flight ended in fiasco ... The Classical authorities tell us, of course, that he was only 'doing a stunt'; but I prefer to think of him as the man who brought to light a serious constructional defect in the flying-machines of his day.
> So, too, in science. Cautious Daedalus will apply his theories where he feels confident they will safely go; but by his excess of caution their

hidden weaknesses remain undiscovered. Icarus will strain his theories to the breaking-point till the weak joints gape. For the mere adventure? Perhaps partly, that is human nature. But if he is destined not yet to reach the sun and solve finally the riddle of its constitution, we may at least hope to learn from his journey some hints to build a better machine.

This was the battle-flag which the journal carried on its mast for many years; and the forebodings of Pol Swings proved fortunately ill-founded: the journal is now in its 64th volume, and will no doubt see many more. Not, however, under the original editorship: in 1968, in view of my increasing commitments in Europe, we invited Carl Sagan to join us as a third co-editor; he became sole editor in 1969, when Hannes Alfvén, Harold Urey and I started a new journal on *The Moon*, published by Reidel—a journal which under its expanded name (*The Earth, Moon and Planets*) and expanded editorship (A G W Cameron, M D Moutsoulas and the present writer) is now in its 34th volume. On assuming the managing editorship of this journal in 1969, I relinquished all administrative connections with *Icarus*; and Carl Sagan gave it up too after some time, with no harm to a good cause; for the journal has continued to flourish under his successors to this day.

My life's principal venture in the field of journal publication has, however, been intimately connected with *Astrophysics and Space Science*, a journal which by the end of 1985 had rounded off its first 117 volumes, and still keeps gathering momentum. The invitation to start this journal as its Editor-in-Chief came to me in the spring of 1967 from Mr Anton Reidel (a direct descendant of Dirk Reidel, who founded the venerable firm of printers in Dordrecht, Holland, in the nineteenth century; though Anton Reidel was the first of its owners to extend its scope also to book publishing), in a letter of 17 March which reached me in Seattle. After consultation with my friend Al Cameron (who was visiting the Boeing Scientific Research Laboratories at that time), I replied, expressing readiness to discuss the matter further. On my return from the US to Europe by the end of that month, this discussion took place in Brussels, where Mr Reidel (with his assistant Dr Edelman) came to meet me from Holland.

It was my first meeting with Mr Reidel; and a beautiful spring day, the morning of which we spent in the Parc de Bruxelles (not far from the Royal Palace); and there it was, on one of the benches of that park, that the new journal was born. I accepted the challenge; and by the time the Thirteenth General Assembly of the IAU met in Prague in August of that year, we had enlisted a total of 68 colleagues from 26 countries to serve on its Editorial Board, to underline its character of an international journal, open to contributors from all countries in the newly opening field of cosmic physics, with emphasis on space physics and its theoretical back-up.

The editorial policies which guided us in this mission were to maintain an open forum, encouraging neither undue conformism, nor cultivation of theories too far removed from contact with observations. Such a policy is by no means new; it should reflect only a part of our common cultural heritage, developed since modern science was born. In fact, nobody formulated its basic tenets more succinctly than did Michael Faraday (1791-1867), one of the greatest physicists of the nineteenth century, in 1818 (then 27 years of age) in a paper which only recently came to light (cf Faraday 1818). We have adopted his message as our own:

> I know of no right, [wrote Faraday] nor can I conceive of any, by which a person is entitled to call that foolish, false, or improper, which he cannot answer. It is easy to say *Yes*, and *No*, particularly when the speaker will assign no reasons to support his monosyllables; and it is easy to assume an appearance of sagacity and experience, and give unjust censure the dress which belongs only to correct criticism. It is also easy for a man to imagine that his own views of things, however partial and imperfect they may be, are right; but these facilities do not by any means make such modes and feelings just, *nor is a man's wisdom increased in proportion to the good opinion he has of himself.*
>
> That *every* one has a right to enjoy his own opinion is allowed; but that *any* one may condemn the opinion of another is denied. Why should I borrow the reason of another, or place my own in subjection to it? Why should I go elsewhere for what nature has planted in my own breast? I should be the very meanest of all beings, so to degrade my nature—but I should also be the most tyrannical of the human race, if I wished, and endeavoured, to make my opinion govern all others, For I am not entitled, 'though I have a mind within myself, to make other men, who also are supplied within, of my mind too. I am made, as to reason, independant; each other man is also independant, and therefore the opinion of one is worth just as much as that of another; and if it should happen to be unanswerable, it certainly ought not to be contemned. I claim no privilege for individuals, except that of Liberty; no exemption, but from impertinence: but I would have that elevation destroyed which some are pleased to assume to themselves—An elevation, which though in the fancy of the possessor, it has an imposing dignity, in reality only serves to lift him up to contemptuous observation.
>
> It is also an injustice to an argument, to quote contrary opinions without supporting them; or to compare the opposite opinions of the *same* person. Nothing of this kind proves that, what has last been said, is either right or wrong. It only shows that other persons, or perhaps the

same person, thought differently at other times. Because I said yesterday that St. Paul's church was larger than St. Peter's, and today that St. Peter's is the largest of the two, no one can advance the contradiction as a proof that *neither* is largest: the only thing is, that it diminishes confidence, but does not affect argument.

To conclude, man *is* independant of others for his opinions, and *ought* to be so. Every man has a right to form his own judgement; and, if he chooses, to decline hearing the judgement on others, nor even a title to have it hear, unless he can support it by argument. Conviction frequently attends upon falsehood; Truth is often accompanied by scepticism; and, as no one can assert of others that their conviction is well founded, each should be suspicious of that he has harboured in his own breast.

If I have quoted Faraday's words so extensively, it is because Faraday, a great observer of Nature (not only inanimate, but also human), expressed in them what should dwell in our hearts as well; and human nature certainly has not changed in the past 170 years to any appreciable extent. Many points which Faraday made were probably addressed *ad hominem*, with specific contemporaries in mind; their identity may be inferred by the historians, but for most of us the results of such detective work would be of only antiquarian interest. However, their species is by no means extinct even today; and every reader has probably met some during his lifetime (or at least heard about them). But young students of science who may frequently share Faraday's feelings towards unsympathetic 'anonymous referees' may take consolation in the fact that they have in him a distinguished predecessor.

Why, in recent years, has there been such growing apprehension about the role of 'referees' in assessing other people's work, and why are the editors (and others) so willing to surrender their judgment to others? Is it because the editors have no time to read themselves; or is it because the front lines of current research are expanding so fast (and, in consequence, any particular sector of the front so thinly manned) that even colleagues working in related fields of the vineyard become only nodding intellectual acquaintances? Much of it is indeed true; yet it does not follow that even outstanding scientists must always be right. History provides indeed glaring counter-examples which counsel caution: remember the way in which the great Carl Friedrich Gauss dismissed in 1824 the work of Niels Henrik Abel on the irreducibility of quintics; or the cavalier way in which Augustine Cauchy treated Evariste Galois and his theory of groups!

To mention another example, when Lord Rayleigh (1842–1919) became Secretary of the Royal Society and (out of sheer curiosity) took the trouble to inspect the file of papers rejected for publication, he came across one by a certain J J Waterston on the kinetic theory of gases which (as he told his son, and the latter published on pp 169–171 of his father's biography; cf Strutt 1968) anticipated by 10–15 years much of the work of Joule, Clausius and Maxwell. Yet this work, which in more recent times would have been rewarded with a Nobel Prize, was rejected with contempt by the Royal Society as being 'nothing but nonsense, unfit even for reading before the Society' (*op.cit.*, p. 170) on the basis of a report of one of the referees—nobody less than Sir John William Lubbock, Bart, by profession a banker who spent his leisure time dabbling in astronomy. But as he had been educated at Eton and Trinity College, Cambridge, this did not prevent him from joining the Royal Society and eventually (1838–1847) becoming its Vice-President. It was in this capacity that he confined Waterston's paper to limbo; from which it was not rescued till almost half a century later (when its author was long dead) by Lord Rayleigh.

Why am I rescuing these incidents of the past from oblivion? Because they are by no means extinct in our own days. It is only natural that they can happen precisely to the greatest; for their minds are mostly so preoccupied with their own work that they may have no time to spend on thoughts of others. Besides, the Gausses or Cauchys are rare; but the Lubbocks are still very much among us; and it is against these that any editor worthy of his salt should be on his guard.

The role of 'referees' (or 'censors', as they were called) in scientific literature has a long history, which can be traced at least to the time of the Renaissance. When Galileo in 1616 submitted his views on Copernican astronomy to Cardinal Bellarmine for his opinion, the latter, acting as a 'referee', gave Galileo some good advice (which he failed to heed); but when Galileo in 1636 submitted the manuscript of his *Dialogues on the Two Great World Systems* to Father Riccardi to obtain the *imprimatur*, the roles of 'referee' and 'censor' got inextricably mixed up (as they often have been ever since), and science was the loser. Of course, no one would advocate today the usefulness of the Roman 'Holy Office' in regulating scientific thought. However, the spirit (if not the method) of 'thought-police' in scientific literature is by no means extinct to this day—it has only shifted position from spiritual to economic grounds; and because it continues to exert a significant brake on the further growth of science, it may not be out of place to throw light on some of its contemporary manifestations.

The motive power behind the tendency to thought control in our own days is, in general, practised under the guise of the trend of 'publish or perish', which has now become endemic in many parts of the world (and it is most virulent precisely in countries which consider themselves the most developed ones). Its most conspicuous external manifestation is the growing tendency towards a monopoly of the channels of information (in the widest sense), and consequent rationing of access to its sources.

From my own experiences, extending now over more than half a century, I can think of many examples of such situations, as well as of the problems and difficulties one can run into when one happens to run against the vested interests of such monopolies, which may be worth while recording in these reminiscences. Perhaps the most illuminating ones I experienced go back to the spring months of 1967, when (as already mentioned) I was invited by Mr Anton Reidel, then owner and director of the D Reidel Publishing Company of Dordrecht, Holland, to accept the editorship of a proposed new international journal of the Reidel family, concerned with astrophysics and space science. Extensive consultations, during the spring and summer, with professional colleagues in many parts of the world convinced me that the proposed new journal could indeed fulfil a useful purpose, especially as the publishers agreed not to exact any page charges from authors of articles accepted for publication (so that editorial work would not be connected with any financial transactions); and, in fact, offered to supply each author with a certain number of free reprints of his work for personal use. By the summer of 1967, four co-editors (A G W Cameron from the US, C de Jager from Holland, M Kitamura from Japan and M Plavec from Czechoslovakia) had agreed to join us in this venture, supported by an international editorial board of 63 well-known scientists from 26 countries.

One would have thought that a new journal so well supported, and launched under such auspices, would have been welcomed by the rest of the astronomical community; but if anyone thought that its support would become unanimous, he would have been mistaken. For, in the meantime, parallel negotiations were going on behind closed doors about the merger of certain pre-existing west European journals into what eventually became (1969) *Astronomy and Astrophysics*, the plans for which were disclosed to the world astronomical community during the Thirteenth IAU General Assembly, at the time when *Astrophysics and Space Science* was openly launched.

It was interesting to observe the reaction (or, rather, interaction) of these plans and ideas in different people's minds. The offensive against the foundation of the new Reidel journal was started with a letter (dated 12 July 1967) which Mr Reidel received from Professor S R Pottasch from Groningen, then Editor of the *Bulletin of the Astronomical Institutes of the Netherlands* (and one of the two editors-designate of the European journal), trying to dissuade him from sponsoring the new journal by various innuendos. 'We do not believe that you have asked advice of the astronomical community at large'—'we' referring presumably to 'other Dutch astronomers, a.o. Blaauw, van de Hulst and Oort', with whom, Pottasch wrote, he had discussed his letter.

But this was only the beginning. Not more than two weeks later Mr Reidel received another letter in this matter (dated 26 July) from Professor Jean-Claude Pecker of Paris, at the time Secretary General of the IAU. The text of this letter (put at my disposal by Mr Reidel) asserts (after a homily on the virtues of organisation journals, and against private publishers) that

> I am aware of the letter sent to you by Dr. Pottasch. I must say that I share completely his firm and far-reaching views; any creation of a new journal should be, for a pure matter of courtesy, submitted to the commission 5 of the IAU, of which the name has been in the past 'Commission pour l'Economie des Publications'.
>
> I urge you not to take steps any further in that kind of dangerous direction. We have seen too many empires of Science publishers having a complete set of productions duplicating each other ... At the contrary, I would favour any effort in favour of an European Journal, obtained by the fusion of existing reviews.

Disturbed by Claude Pecker's reference to the IAU, I contacted Professor Pol Swings, who was then President of the Union, in quest of explanation; and Pol responded very promptly. He made it clear that, in the matter of astronomical publication, the International Astronomical Union must remain entirely neutral, and that none of its commissions can take official note of any plans currently under discussion. None of the functionaries of the IAU as such is authorised by the Union's charter to undertake any action in such matters; and if Pecker had done so, he had clearly overstepped the limits of his competence. Unofficially, however, and to let me know where he stood as an individual, Pol Swings accepted my invitation to joint the Editorial Board of *Astrophysics and Space Science*; and remained a valued member of it until his death in 1983.

Shortly thereafter, however, in the course of the Thirteenth General Assembly, a preliminary announcement was circulated, under the signatures of Drs F Denisse, J H Oort, S H Pottasch, J-C Pecker, D H Sadler, F G Smith and J L Steinberg, to the effect that these signatories planned to recommend a fusion of the national journals of their countries into a joint European one. This led, in turn, to further attempts to stifle the birth of our journal on the part of Professor Evry Schatzman from Paris, who sent a letter to me dated September 14, but with copies to all members of the Editorial Board (as Schatzman had himself agreed to be one, he had obtained previous information from me about their identity). In this letter he 'supported strongly'

Pecker as well as Pottasch; and urged me to desist from our own plans; or at least to declare a willingness to merge with the European journal as soon as the latter could commence to appear.

To this I felt impelled to reply (on 18 September) that I did not possess any official knowledge of the proposal for the creation of a European journal prior to the receipt of Sadler's circular letter in Prague.

> Moreover, [I went on], I note that their proposal constitutes so far private action of 8 signatories of the Steinberg and Pottasch group, to whom I send my best wishes. In Great Britain, the proposal is now being considered by a special sub-committee of the Royal Astronomical Society; but I understand it is too early yet to pre-judge the issue—or even the time at which the Society may arrive at any kind of decision in this matter.
>
> Under these circumstances I am sure you would agree that it is premature to consider any further mergers of astronomical journals—or even the question whether or not a merger of the *Astrophysics and Space Science* and the proposed European Journal at some time in the future would serve the best interests of our science. Many other parties would be concerned in such deliberations besides myself; and I think we can safely defer the problems for future consideration. A copy of my reply to Sadler's letter is enclosed for your information.
>
> There is only one aspect of this matter which I consider unfortunate: namely, the fact that Sadler's group, as well as yourself, thought fit to circulate copies of the letters to me to all members of the Editorial Board of *Astrophysics and Space Science* at this time; for such a move could obviously be misinterpreted as an effort, on the part of the group behind the European Journal, to interfere with a sister periodical at its formative stage. I am sure this was far from your minds; but any further effort to use for the purposes of circulation information which anyone receives from me in confidence as a member of our Editorial Board would inevitably be so interpreted.

In due course, I received comments on the subject matter of this correspondence from almost all members of our own Editorial Board; and realised how 'counterproductive' the action on behalf of the nascent European journal had turned out to be. Suffice it to stress that not a single member of our board (including as it did three Nobel laureates, and three Presidents, past and future, of the IAU) changed his mind about collaboration with the Reidel journal. None was convinced that new astronomical journals are superfluous—how could they, when Evry Schatzman with Alan Maxwell started publishing another new journal (*Astrophysical Letters*) for another private publisher (Gordon and Breach) in the same year; of which I was invited to become (and still am!) a member of the Editorial Board myself? At any rate, not one of our board members was convinced by the Pecker-Schatzman arguments to withdraw their cooperation—not even Schatzman himself, who continued to serve on the board of our journal from its foundation for 88 volumes till 1982.

Moreover, when the Council of the Royal Astronomical Society came to consider the Sadler-Smith proposal for a merger of their *Monthly Notices* with the European journal at their October 1967 meeting, the proposal was unanimously defeated (even the proposers abandoned it in the end). So it came to pass that issue 1 of the first volume of *Astrophysics and Space Science* appeared on schedule in January 1968; and *Astronomy and Astrophysics* (the European journal) followed suit a year later, its subtitle having lost some of its justification (unless one adopts the view of General de Gaulle that Britain is not yet fit to become a part of Europe).

Whatever politicians may believe, however, we are all part of one world; and so one might think that the 1967 episode about astronomical journals could be legitimately confined to oblivion. Not so, however, Stuart Pottasch; for when, in the spring of 1968, the Reidel Company made a proposal (solicited by the Editors of *Astronomy and Astrophysics*) for the printing of their journal, this is the answer (dated 15 April 1968) which Mr Anton Reidel received from Professor Pottasch of Groningen:

> The financial offer you made for publishing the new journal was the lowest we received and we agreed that the quality of your other publications, i.e. the printing and esthetic impression, were completely satisfactory to us. However, the fact that you also publish *Astrophysics and Space Science* has created such a distrust among my colleagues for your publishing house that this was felt to be an insurmountable difficulty. Personally I share their concern. That *Astrophysics and Space Science* was established without adequate consultation among astronomers testifies to inadequate concern of your publishing house to the needs of astronomy and astrophysics. We require a publisher who will subordinate his interests to those of astronomy in general. The 'inadequate concern' was confirmed when, after hearing of your plans to publish *Astrophysics and Space Science*, I myself, and several other European astronomers wrote to you and asked you to wait one year in publishing this journal so that the astronomical world could be consulted. You refused to consider this possibility.
>
> While no definite decision has yet been reached concerning a printer for *Astronomy and Astrophysics*, I'm sure that you can appreciate our reasons for not considering your publishing house further as long as you publish *Astrophysics and Space Science*.

It is, perhaps, unnecessary to quote further sources to put the spotlight on the operators, behind the scenes, in actions so contrary to

their openly professed intentions: their methods certainly differ but
little from those used in practice to scare away unwanted business
competition from the 'High Street'. For although Mr Reidel's invited
bid for the job in question was admittedly the lowest, the operators
behind the European journal were quite willing to pay more (from
public money) to attempt to penalise the lowest (and otherwise wholly
satisfactory) bidder for lack of subservience to local pressure-groups.

Needless to say, Mr Reidel declined to have anything to do with
Pottasch and his committee; and in forwarding it to me he termed their
proposal as 'ridiculous'; my own opinion was no different. Yet, in
fairness to everyone concerned, I do not think that it was Drs Pottasch,
Pecker, or Schatzman who actually originated the campaign against
*Astrophysics and Space Science* in 1967–68: they only allowed themselves
(perhaps by suitable encouragement) to be used by others; while the
'prime movers' were lurking sufficiently far behind the lines not to 'lose
face' if their advance guard had to beat a retreat without having
accomplished its purpose.

It is betraying no particular confidence to disclose (the reader may
have guessed it already) that the focus of resistance was located in
Leiden—like so many other strings which were pulled in the IAU for
more than forty years. Because of this, it is with all the greater
gratitude that I hereby acknowledge the loyal and unstinted collaboration of our colleagues from the University of Utrecht, three of whom
served on the Editorial Board of our journal from its inception:
Professor M G J Minnaert up to his death in 1970, Professor Ann
Underhill until her return to the United States in 1971; and, above all,
Professor Cornelis de Jager, who has served as co-editor of the journal
since its inception in 1968.

Why have I included excerpts from the above-quoted correspondence
in these reminiscences? Although their subject matter belongs indeed
to the past, their spirit is by no means dead. However, *Astronomy
and Astrophysics* (the 'European' journal—for Europe stopping short
at the Straits of Dover!), and *Astrophysics and Space Science* (a truly
international journal), launched in 1968 and 1969, respectively, have
both now more than a hundred volumes behind them; and since
Dr Pottasch ceased to have anything more to do with the 'European
journal', friendly editorial cooperation (helpful, I hope, to both sides)
has been established between both journals—as should indeed be
expected. The success of both in their respective missions has more
than sufficiently answered the 'whispering campaigns' of 1967 that new
astronomical journals are superfluous (unless they are under the thumb
of 'vested interests' claiming the right to pontificate on behalf of our
science).

The reader may ask: what was the reason for all this ado to keep the number of astronomical journals to a minimum, and to establish control of the remaining ones? The reasons are several; and the roots of most are financial. Not that publishers make much money on them; for the world astronomical community is still too small to make journal publication a lucrative preoccupation; the revenues from well-established ones are steady, but not munificent. The journals published by scientific societies for their consortia can however improve their situation by levying 'page charges', or making subscriptions the condition of membership of their 'trade unions'. Private publishers (Academic Press, Gordon and Breach, Pergamon, Reidel, Springer) have no means of resorting to such gimmicks; and the fact that they remain in business is sufficient proof of the fact that page charges are not really necessary, but rather are optional measures intended to increase the revenues of the sponsor organisations (and defray their 'overheads').

I sometimes came across specious arguments being advanced to make payments of page charges more palatable to their hapless victims by an assertion that papers published in journals requiring them will be more widely read—an assertion made very openly by David Lazarus (Editor-in-Chief of the American Physical Society), and published in *Physics Today* (1982). I confess I have seldom encountered in the literature addressed to scientists a statement covering so shabbily the mercenary spirit lurking behind it: does he want us to believe that original work published by *bona fide* scientists will not be noted by their colleagues unless served to them in the right kind of package? Such techniques have, to be sure, been developed on New York's Madison Avenue to fool the gullible public by experts of the calibre of the late Dr Dichter *et consortes*; and have indeed proven their worth in inducing men in the street 'to buy products which they do not need, with money they do not have, to impress friends they do not like'; but is it not an insult to place astronomers (and scientists in general) in the same category?

Speaking from my own experience, I vividly recall from my young years a few instances of sudden inspiration which reached me from media which nobody would call—now or then—popular. This was certainly the case with the fundamental papers by S Takeda (1934, 1937) which appeared in the *Memoirs of the College of Science of Kyoto Imperial University*; or a paper by V P Tsesevich (1936) from the *Publications of the University Observatory of Leningrad* (and its subsequent continuation on which the late Professor Tsesevich worked while in relegation in Kazakhstan during the years of the Stalin purges). Others had, no doubt, the same experience: witness the effects which the papers by G C Wick (1943) or V A Ambartsumian (1944) had on the subsequent creative productivity of S Chandrasekhar during the years of World

War II which brought a normal exchange of information almost to a complete standstill. 'The wind blows for him who listens' is an old proverb which has not lost its worth in the past, nor will it do so in the future; and it has nothing to do with packaging or page charges.

There are, to be sure, also other hidden factors by which financial interests (albeit indirect) can be connected with the 'publish-or-perish' policy; namely, by acquisition of control of the editorial boards of 'prestige' journals by faculty members of (usually a small group of) 'elite' schools, who then have it in their power to make their colleagues or students enjoy special editorial privileges at the expense (and sometimes to the detriment) of those 'on the outside'. As a result, for instance, high tuition charges now levied on the students of schools with established vested interests in existing channels of communication are being regarded by many (and not without reason) as representing a sound investment in their future, bringing returns on the science marts which other and less favoured institutions are ill-prepared to match.

I confess I have always viewed the concept of 'elite' schools, and the various reasons advanced from time to time in different countries to justify special privileges for them to carry the torch of science, with a considerable amount of distrust. It is, to be sure, true that not all schools or institutions are (or can be) on the same level, any more than individual scholars working or teaching in them are; but their performances are functions of time, describing transients rather than the steady state.

Indeed, the general level of the universities, or of individual departments within them, cannot avoid being in a constant state of flux—the decisive factor being who is there at any particular time, and even his age. For most scholars in the prime of their life will no longer be the same twenty years later; and when the need arises to replace any aging scholar, his successor will invariably be better or worse—but very seldom the same! To give some recent examples from the field of astronomy within my lifetime: the Yerkes Observatory of the University of Chicago after Otto Struve was never the same; nor was Harvard after Shapley; though Princeton under Lyman Spitzer was no doubt an improvement on what it used to be during the long term of office of Henry Norris Russell. Other similar examples could be multiplied almost without limit.

Exceptions exist, of course, when a man can retain his creative fires burning much longer than his contemporaries (within the circle of people I knew well, Harold Urey was a shining example) but these are rare, and recognisable as such only *ex post*. There are, of course, ways of camouflaging this very natural situation; and perhaps the best description of their *modus operandi* I have read about was in one of the

Cambridge novels of C P Snow, in which that distinguished author (himself a Cantabrigian) described a conversation between an old don and a younger one in the safe haven of the combination room of their college. 'There have always been one or two really distinguished men at Cambridge,' said the wise old man; 'the rest become distinguished by bestowing distinctions on each other.'

The amount of back-scratching in academic communities is probably much greater than among other groups of different occupation (because it can be practised with greater impunity from detection); but this is exactly why members of academe should also exercise a greater degree of self-control. And nowhere are the consequences of 'tribal warfare' between different groups more evident than in the field of scientific publications. Some examples of such warfare have already been mentioned in this chapter; and we can perhaps close the subject with a statement by which my old friend Lawrence Aller commented on Schatzman's letter to me of 14 September 1967: 'As the volume of astronomical work increases, we approach a situation like that which exists in physics where huge quantities of material have to be processed. I think,' went on Lawrence, 'this could be better accomplished by a number of smaller teams operating smaller journals rather than by one huge publication factory.'

And who of us, on the receiving line of new issues of monthly (or fortnightly) journals of jumbo size (not necessarily astronomical) containing sometimes more than a hundred papers in a single issue, would not agree? Even looking over the titles of individual papers (let alone over their 'key words') one simply cannot register all that the incoming issues contain; and much of their contents may slip through immediate scrutiny, and be mentally passed over to await accidental discovery at a later date.

If I look back over more than twenty years of my editorial work, what do I see as the principal lessons of this experience? As far as the duties of the Editor are concerned, his principal task towards the publisher as well as the scientific public should be to protect the journals from three kinds of paper: those which (on closer inspection) are found to contain demonstrable errors; those whose information content is too thin to justify the place which they would occupy; and those in which the authors may seek a forum to quarrel with other colleagues†.

The first and second cases (by the far the most frequent encountered) are dealt with as a matter of course, in collaboration with the outside

---

† The fourth duty should be, of course, to protect the journal in question from deliberate plagiarisms. In these days of prolific production this is no longer as easy to fulfil as it was in the past. However, among some four thousand papers published in *Astrophysics and Space Science* between 1968 and 1985, only one such case slipped through our editorial net.

advisers generally known as 'referees'. In journals which I have had to deal with, this has been the primary function of our co-editors or other members of the editorial boards; only for papers on very specialised topics does it become necessary to seek the opinions of outside referees. Our referees do not, however, function anonymously; and the opinion of no single one was ever taken as a sufficient basis for rejection. Above all, no paper is rejected merely because its subject may be controversial; for in too many cases of this nature the final judgment must rest with the future.

If I were to give an example of an Editor to look up to from the early days of my practice, it would no doubt be the late Otto Struve (1897–1963) when he used to edit the *Astrophysical Journal*. An outstanding astrophysicist himself, he was prepared to assess personally the quality of all incoming contributions. On occasion, he had no doubt to seek outside advice; but if he agreed with it, he never hid behind the authority of anonymous 'competent referees' but presented it to the author as an opinion of his own. To this end he also never used any forms, and did most of his correspondence long-hand. This correspondence with the authors of papers submitted for publication contained not only criticisms which Struve held as valid, but also (especially in correspondence with young authors) praise when he thought the latter was deserved—a combination which produced the confidence in the author's mind that the opinion received came not only from the pen of an outstanding scientist interested in the subject, but also from one who had the author's interests at heart. Several such letters received from Otto Struve are still on my file as treasured memories; but how far we have departed from this spirit since that time!

Or, to recall another outstanding editorial example of past days, when the late Dr Daniel Barbier (1907–1965) used to edit the French *Annales d'Astrophysique*, and when some of his authors visited Paris from abroad, he considered it part of his duty to entertain them for lunch or dinner. Barbier (a confirmed bachelor) was a real gourmet in the best French tradition; and meals with him used to be a treat which no-one would forget for the rest of his life. How glad I would have been to uphold the same tradition—except that, unfortunately, the culinary arts of Manchester cannot compete with those of Paris!

To turn to other changes which the tides of time have brought to the field of journal production and which may merit a comment, another innovation in the style of most journals of this century has been the introduction (in the 1930s) of 'abstracts' at the head of a paper (with an optional summary at the end), followed (since the 1970s) by 'key words' which specify at least the field dealt with in the text.

I am old enough to have seen the advent of both with mixed feelings, and have avoided inclusions of 'key words' in any journal with which

I have had anything to do, for the following reasons. An abstract in front of the main body of the text seems (to me) to invite the user to stop reading afterwards—or at least offers the reader an absolution to do so—a process which a statement of 'key words' merely carries one stage further. It is indeed true that scientists (especially senior scientists) of our time have less time to read (or even think) than we had in the past; and the availability of abstracts may seem to offer an easy way out. Yet the way is not without its perils; for a credulous reader may easily become the victim of 'gamesmanship' on the part of unscrupulous authors, who may raise high hopes in the abstract with attractive statements; but their justification can be vainly sought for in the text.

With 'key words' it is usually more of the same. From decade to decade, certain terms become more popular than others—such as 'black holes' these days. What wonder then that an author ambitious to be read slips in such terms among his 'key words', although they may only perfunctorily (if at all) be referred to in the text. Or, in the field of close binary systems, every author who 'knows the ropes' would now use the term 'RS CVn-type' among his key words; while the only reference to them in the text may be a surmise that '... the system in question may (but need not) be of this type' (though, incidentally, the author may not have bothered to define such stars at all). The information content of such sentences in the text may be zero; but among 'key words' the symbol RS CVn makes you sit up—and this is precisely what it was meant to accomplish!

In inspecting 'key words' (the abstract of an abstract) in many journals these days, one wonders where we go from here, and what the next stage of this process of simplification (absolving us even further from the need to read) is going to be—encoding the message in a numerical system of a very large base? The latter would certainly be very convenient for storage purposes; but is not the aim of our efforts to *use* the information acquired rather than merely to store it? Only the future can provide the answer.

My recollections of the years in which I was active extend not only to the field of journal production, but also that of astronomical books; and the present text would not be complete without at least a brief mention of my participation in the publication of four series of monographs in which I took part in different times of my life. Thus (after the premature passing of Dr M E Ellison in 1963), I joined Bernard Lovell as General Editor of 'The International Astrophysics Series' published by Chapman and Hall of London; but after the death of Geoffrey Parr (Director of that publishing house) the series eventually passed into oblivion.

Since 1962 I accepted, however, the editorship of a series of annual volumes on *Advances in Astronomy and Astrophysics* published by Academic Press of New York. In the decade 1962–1972 nine volumes of this series made their yearly appearance; but the series was discontinued thereafter by my increasingly closer involvement with the Reidel Publishing Company of Dordrecht. Since 1972, I joined the Editorial Board of the Reidel 'Astrophysics and Space Science Library', now numbering more than 120 volumes, and the most successful venture of this type in modern times; and, in addition, in 1974, I also accepted responsibility (with A G W Cameron) for a series of monographs on 'Developments in Solar System and Space Science', published by the Elsevier Publishing Company of Amsterdam.

From the technical point of view, the editing of books confronts one with problems no different from those encountered in work on the journals, except that the number of contributors is smaller; and their contributions are larger in size. Common to both is, however, a problem created by the economic facts of contemporary life: the ever-increasing costs of production, reflected in the price of the 'finished products', have effectively phased out the 'private sector' of the market from existence as a significant economic factor—relatively few individuals (especially students!) can afford to buy books, or subscribe to journals, in their professional fields any more. In the second half of the twentieth century the main customers for the literature dealt with in this section have become public libraries—university as well as institutional—whose budgets can better keep up with inflation. However, an extrapolation of current trends gives rise to uneasy thoughts about the future, and the ways in which our descendants in successive generations will have to cope with the consequences of the 'explosion of information' and its dissemination through the scientific literature: it may have to be done in a way very different from the one to which we have been accustomed so far.

## References

Ambartsumian V A 1944 *J. Phys. (USSR)* **8** 65–75
Baade W 1945 *Astrophys. J.* **102** 309–317
Beer A and Kopal Z 1954 *Annales d'Astrophysique* **17** 443–455
Blackett P M S 1947 *Nature* **159** 658
Carson D, Davidson M, Goudas C L, Kopal Z and Stoddard L G 1966 *Icarus* **5** 334–359
Chandrasekhar S 1933 *Mon. Not. R. Astron. Soc.* **93** 390–404, 449–461, 462–471, 539–574
—— 1944 *Astrophys. J.* **100** 76–86

Clairaut A C 1743 *Théorie de la Figure de la Terre, tirée des Principes de l'Hydrostatique* (Paris)
Crawford J A 1955 *Astrophys. J.* **121** 71–76
Darwin G H 1900 *Mon. Not. R. Astron. Soc.* **60** 82–124
Davidson M, Goudas C L and Kopal Z 1967 in *Measure of the Moon* ed Z Kopal and C L Goudas (Dordrecht:Reidel) pp 140–175
Eddington A S 1927 *Stars and Atoms* (Oxford: Oxford University Press)
Eve A S 1939 *Rutherford* (Cambridge: Cambridge University Press)
Faraday M 1818 'On Argument', reprinted in *Isis* **52** 89–90 (1961)
Frank P 1947 *Einstein, his Life and Times* (New York: Knopf)
Fricke W and Teleki G 1982 *The Sun and Planetary System* (Dordrecht: Reidel) pp 13–21
Goudas C L 1967 in *Measure of the Moon* ed Z Kopal and C L Goudas (Dordrecht:Reidel) pp 237–281 and other papers referred to therein
Hale G E 1935 *Astrophys. J.* **81** 97–106
Hanbury Brown R and Hazard C 1952 *Nature* **170** 364
Jansky K G 1933a *Nature* **132** 66
—— 1933b *Proc. Inst. Radio Eng.* **21** 1387–1398
Kahn F D 1981 *Investigating the Universe* (Dordrecht: Reidel)
Koestler A 1959 *The Sleepwalkers* (London: Hutchinson)
—— 1972 *The Call-Girls* (London: Hutchinson)
Kopal Z 1950 *Computation of the Elements of Eclipsing Binary Systems* (*Harvard Obs. Monogr.* no. **8**) (Cambridge, Mass.: Harvard University Press)
—— 1953 *Mon. Not. R. Astron. Soc.* **113** 769–775
—— 1954a 'A Study of the Roche Model' *Jodrell Bank Ann.* **1** 37–57
—— 1954b in 'Les particules solides dans les Astres' *Mém. Soc. R. Sci. Liège* (4) **15** pp 684–686
—— 1955a *Ann. d'Astrophysique* **18** 379–430
—— 1955b *Numerical Analysis* (London: Chapman and Hall and New York: Wiley; 2nd ed. 1960)
—— 1956a (ed) *Astronomical Optics* (Proceedings of a Symposium on Astronomical Optics, held at the University of Manchester Apr 19–22 1955) (Amsterdam: North-Holland)
—— 1956b *Mon. Not. R. Astron. Soc.* **116** 154–156
—— 1956c *Ann. d'Astrophysique* **19** 298–335
—— 1958 in *Numerical Approximations* (ed R Langer) (Madison: Univ. Wisconsin Press) pp 25–43
—— 1959 *Close Binary Systems* (London: Chapman and Hall and New York: Wiley)
—— 1960 *Figures of Equilibrium of Celestial Bodies* (Madison: Univ. Wisconsin Press)
—— 1962 (ed) *Physics and Astronomy of the Moon* (New York: Academic Press) (2nd ed. 1971, Russian transl. 1972)

—— 1965 *Space Phys. Reviews* **4** 737–855 (cf in particular, pp 773–774)
—— 1966 *An Introduction to the Study of the Moon* (Dordrecht: Reidel)
—— 1967 in *Measure of the Moon* (ed Z Kopal and C L Goudas) (Dordrecht: Reidel) pp 407–413
—— 1968a *Exploration of the Moon by Spacecraft* (Edinburgh: Oliver & Boyd) (Swedish translation 1970)
—— 1968b *Telescopes in Space* (London: Faber & Faber; New York: Hart 1970)
—— 1969 *The Moon* (Dordrecht: Reidel)
—— 1970 *Widening Horizons* (London: Kahn & Averill; New York: Taplinger 1971)
—— 1971 *New Photographic Atlas of the Moon* (New York: Taplinger)
—— 1972a *Man and his Universe* (London and New York: Rupt Hart-Davis and Wm Morrow) (Czech transl. 1976)
—— 1972b *The Solar System* (London: Oxford University Press) (Japanese transl. 1974)
—— 1973 *The Moon* **7** 1–2
—— 1974 *The Moon in the post-Apollo Era* (Dordrecht: Reidel)
—— 1978 *Dynamics of Close Binary Systems* (Dordrecht: Reidel)
—— 1979a *Language of the Stars* (Dordrecht: Reidel)
—— 1979b *The Realm of the Terrestrial Planets* (London: Institute of Physics) (Bulgarian and Czech translations in 1985)
—— 1984 'Vainu Bappu Memorial Lecture' *Astrophys. Space Sci.* **99** 3–21
Kopal Z and Carder R W 1974 *Mapping of the Moon* (Dordrecht: Reidel)
Kopal Z and Carling E B 1981 (eds) *Photometric and Spectroscopic Binary Systems* (NATO Adv. Study Inst. Ser. C) (Dordrecht: Reidel)
Kopal Z and Finlay E B 1961 'Problems of Lunar Topography' *Astron. Contr. Univ. Manchester* Ser. III, no. **90**
Kopal Z and Goudas C L 1967 (eds) *Measure of the Moon* (Dordrecht: Reidel)
Kopal Z, Hidayat B and Rahe J 1984 (eds) *Double Stars—Physical Properties and Generic Realtions* (IAU Coll. no. **80**) (Dordrecht: Reidel)
Kopal Z, Klepešta J and Rackham T W 1965 *Photographic Atlas of the Moon* (London: Academic Press)
Kopal Z and Mikhailov Z K 1962 *The Moon* (IAU Symp. no. **14**; New York: Academic Press; Russian transl. Leningrad 1963)
Kopal Z and Millns P Y 1956 in *Astronomical Optics* (Proceedings of a Symposium on Astronomical Optics) (Amsterdam: North-Holland) pp 389–401
Kopal Z and Rahe J 1982 (eds) *Binary and Multiple Stars as Tracers of Stellar Evolution* (IAU Coll. no. **69**) (Dordrecht: Reidel)
Kuiper G P 1938 *Astrophys. J.* **88** 497–507

Lazarus D 1982 *Physics Today* **35** (no. 4, April) pp 9ff
Lovell A C B 1968 *The Story of Jodrell Bank* (Oxford: Oxford University Press)
—— 1975 in *Bibliographical Memoirs of the Royal Society* **21** 1–115
McMath R R, Petrie R M and Sawyer H E 1937 *Publ. Univ. Obs. Michigan* **6** 67ff
Meaburn J 1976 *Detection and Spectrometry of Faint Light* (Dordrecht: Reidel)
Nasmyth J and Carpenter J 1874 *The Moon, considered as a Planet, a World, and a Satellite* (London; 4th ed. London 1903)
Newell H E 1973 *The Moon* **7** 1–5
Oort J H and Walraven T 1956 *Bull. Astron. Inst. Netherlands* **12** 285–308
Ring J 1955 *Observatory* **75** 249
Roche E A 1849 *Mem. Acad. Sci. Montpellier* **1** 243
Sadler J H 1949 *Occasional Notes R. Astron. Soc.* **2** no. 13 (p. 126)
Saward D 1984 *Bernard Lovell—A Biography* (London: Hale)
Schuster A 1883 in *British Assoc. Rep. for 1883* pp 427ff
—— 1905 *Astrophys. J.* **21** 1–22
Schwarzschild K 1906 *Göttinger Nachr.* p. 41
de Sitter W 1924 *Bull. Astron. Inst. Netherlands* **2** 97–108
Strutt R J 1968 *Life of Rayleigh* (Madison: Univ. Wisconsin Press)
Struve O 1957 *Sky and Telescope* **17** 70–72 and 93
Sullivan W T 1984 *The Early Years of Radio Astronomy* (Cambridge: Cambridge University Press)
Takeda S I 1934 *Mem. Coll. Sci. Kyoto Univ.* (Ser. A) **17** 197–217
—— 1937 *Mem. Coll. Sci. Kyoto Univ.* (Ser. A) **20** 47–86
Tomkin J and Lambert D L 1978 *Astrophys. J. Lett.* **222** L119
Tsesevich V P 1936 *Publ. Univ. Obs. Leningrad* **6** 48–60
Vashakidze M A 1954 *Astr. Tsirk. USSR Acad. Sci.* no. **147**
Walter K 1931 *Königsberg Veröff.* no. **2**
Wick G C 1943 *Z. Phys.* **121** 702–718

## Chapter 6

## The Binary Stars

> *Twinkle twinkle little star,*
> *I don't wonder what you are,*
> *What you are I know quite well,*
> *For your light changes will tell.*

The principal contributions which I have been able to make towards the advance of astronomy in the past fifty years and more of my professional life have been devoted to a study of binary stars—particularly of the objects which we call close binary systems. Ever since this field was opened up more than a hundred years ago by the discovery of the first such system, continuing advances have been achieved by the combined efforts of astronomers whose basic talents and training have been in very different fields of theory and observations. No-one can be equally proficient in both, except down to a very shallow depth; but each branch of specialists should at least be aware of the limitations imposed by Nature on the efforts of their colleagues working in different parts of the vineyard.

In retrospect, it would appear that my mind must have been 'pre-programmed' for work on close binary systems since the early years of my life; for my first modest contributions to the subject were published in 1932 while I was still a grammar-school boy; and although during the 54 years which have elapsed since that time my mind has wandered away also to other subjects (the most notable gap occurred between 1959 and 1969, when I was occupied with work on the solar system and in particular the Moon, as described in Chapter 5), double stars have really never been too far to the back of my mind. It is, therefore, only appropriate that, before giving any account of my contributions to this subject in the past half century, I should share with the reader some of the lure by which these objects have held me spellbound since the time of my adolescence, and will no doubt continue to do so till the end of my days; and to embark on our story let us first introduce to the stage its *dramatis personae*.

The term *binary star* was apparently used by William Herschel to designate 'a real double star—the union of two stars that are formed together in one system, by the laws of attraction' (Herschel 1802). The term *double star*, on the other hand, is of much earlier origin: at least its Greek equivalent was already used by Ptolemy to describe the appearance of $\nu$ Sagittarii, two fifth-magnitude stars whose angular separation is about $14'$ (i.e., a little less than the apparent radius of the Moon); and it has been used ever since to describe close pairs of stars resolvable with the aid of a telescope. Not every 'double star' defined in this sense constitutes, to be sure, a 'binary system'; for a large majority of them may be *optical* pairs, owing the accidental proximity of their projections on the celestial sphere to the laws of chance. The first double star which was know to form a binary system was $\zeta$ Ursae Maioris (Mizar), discovered around 1650 by Father Giovanni Baptista Riccioli in Bologna (and whose principal component was recognised by E C Pickering as the first spectroscopic binary in 1889). In 1656, Christian Huygens saw $\theta$ Orionis resolved into the principal stars of the Trapezium; and in 1664 Robert Hooke noted that $\gamma$ Arietis consisted of two stars. At least two additional pairs (one of which proved to be of more than ordinary interest) were discovered before the end of the seventeenth century—namely, $\alpha$ Crucis, discovered in 1685 by Father Fontenay, a Jesuit missionary at the Cape of Good Hope; and $\alpha$ Centauri, discovered by his *confrère*, Father Richaud, while observing a comet at Pondicherry, India, in December 1689.

All these discoveries were accidental, and made in the course of observations taken for other purposes. No suspicion seems to have been entertained by all these observers or their contemporaries that the proximity of the two stars in such pairs was due to other reason than chance; but although they were regarded as mere curiosities and no special effort was made to augment their list, their number grew from decade to decade until, around 1750, several dozen such pairs had been noted and recorded. As their number gradually became sufficiently large to lend itself to rudimentary statistical analysis, the cosmic significance of double stars appeared to conceal more than met the eye of their discoverers.

The first scientific argument in favour of the view that at least some, and probably many, double stars then known were the result of physical rather than optical association we owe to another faithful servant of Christ, the Reverend John Michell. On May 7th and 14th of the year 1767, Michell read a remarkable paper before the Royal

Society in London, published subsequently in the Society's *Transactions* (Michell 1767), in which he pointed out that the frequency-distribution of the angular separations of double stars known in his time deviated grossly from one that would be expected for a chance association of stars uniformly distributed in space. There appeared to be far too many close pairs among them and, according to Michell, 'the natural conclusion from hence is, that it is highly probable, and next to a certainty in general, that such double stars as appear to consist of two or more stars placed very near together, do really consist of stars placed nearly together, and under the influence of some general law... to whatever cause this may be owing, whether to their mutual gravitation, or to some other law or appointment of the Creator' (*op. cit.*, p. 249).

Who was Michell? We do not know, unfortunately, where he was born and when; but the records of Queens' College of Cambridge University disclose that he received his MA degree there in 1752, and his BD in 1761. Moreover, between 1762 and 1764 he held the Woodwardian Chair of Geology; but in 1767 he was appointed rector of St Michael's Church in Thornhill near Leeds, a prebend which he held for the rest of his life. He died there on 21 April 1783, and is buried at Thornhill; the parish registers describe him as having attained the age of 68 years. Accordingly, he was born in 1724; but the day and month of his birth are unknown; and so is his birthplace. No effigy of this remarkable man seems to exist at the present time; but a photograph of his memorial stone in the Church of St Michael in Thornhill is shown in figure 6.1.

When I stood before this stone many years ago I could not help wondering what led a man of his indubitable gifts to spend a major part of his life in a God-forsaken place like Thornhill. Why did he leave Cambridge where he—a Fellow of the Royal Society since 1761—had previously held a professorship of some significance? The usual reason in those days, marriage (Cambridge dons then still had to observe celibacy), does not seem to apply in his case; for there is no record that he ever contracted holy matrimony. Were the incumbents of Cambridge chairs so poorly remunerated in those days that Michell went to the parish of Thornhill to better himself; and if so, why—as he had no family to look after (apart from his brother mentioned rather prominently on the memorial tablet)?

We do not know; but whatever may have been the case, a transfer to Thornhill did not bring Michell's scientific work to a standstill. In 1784, he discovered (independently of Coulomb in France) and constructed a torsion balance and attempted to establish with its aid the mean density of the Earth by an experiment repeated later by Cavendish with the same instrument, purchased from Michell's estate (the two men were acquainted; and after Michell's death Cavendish merely

# John Michell of Thornhill

refined his experimental results). While in Thornhill, Michell also built small telescopes of his own; and tradition has it that William Herschel, who spent seven years (between 1759 and 1766) at various places in Yorkshire, visited Michell more than once, and may have acquired from him the rudiments of the art of telescope making—an art which, in the fullness of time, led the great astronomer he later became to 'break the barriers of the Heavens'. The story (related by Barbara H Nuttall on p. 62 of her *History of Thornhill*) is plausible; and although I was unable to authenticate it from any other source, it does not lack verisimilitude.

**Figure 6.1** Memorial tablet in the parish church of Thornhill, Yorkshire, to the Reverend John Michell BD FRS (1724–1793), who served for 26 years as incumbent of that parish. It was Michell who predicted the existence of physical double stars in the sky.

But even if this story were apocryphal, Michell's renown in the annals of our science should be forever secure on the grounds of two indubitable accomplishments: one we have already mentioned; and the second (although not connected directly with the subject of this chapter) was the first realistic estimate of the distance to the stars, made more than fifty years before the first reliable parallax of any fixed star was actually triangulated.

In order to do so, Michell took note of the fact that the planet Saturn at its opposition to the Sun appears equally bright as the star Vega ($\alpha$ Lyrae), and exhibits to us an apparent disc about 20 arcseconds in diameter, which, viewed from the Sun, would be reduced to 17 seconds. Therefore, Saturn's illuminated hemisphere should intercept $(17/3600)^2 \times (\pi/720)^2 = 4.245 \times 10^{-10}$ of the light sent out by the Sun. Now if Vega's apparent brightness in the sky is equal to that of Saturn, it follows from the inverse-square law of the attenuation of light with distance (established already by Kepler in 1604, and verified experimentally by Bouguer before Michell's time) that Vega must be $(3600/17) \times (720/\pi) = 48\,500$ times farther from the Sun than Saturn. Since, moreover, (by Kepler's third law) Saturn is known to be 9.548 times farther from the Sun than the Earth, it follows that the distance to Vega should amount to $9.548 \times 48\,500 = 463\,000$ 'astronomical units'; or 7.32 light years!

This value represents, to be sure, only about one-quarter (or, more accurately, 27%) of the actual distance of Vega, first triangulated by F G W Struve in 1837 (the disparity going back to the fact that Vega proved to be intrinsically much brighter than the Sun). It is also true that Michell's idea of computing the distance of cosmic light sources from their apparent brightness on the assumption of the complete transparency of space was anticipated, in principle, by Christian Huygens in his *Cosmotheoros* (published posthumously in 1698) in Holland; and later Loys de Chéseaux in Switzerland (in a memoir 'Sur la force de la lumière et sa propagation dans l'Ether, et sur la distance des Etoiles fixes', published in Paris as an appendix to his book on the comet of 1743) put forward similar ideas. Nevertheless, the fact remains that Michell applied this method of computation of 'photometric parallaxes' (anticipating the 'spectroscopic parallaxes' of more modern times) to a specific star and obtained the first realistic estimate of its distance; while Huygens (in his application of the same method to Sirius) was still more than eighteen times short of his goal. Besides, there is no indication that Michell would have been acquainted with Huygens' *Cosmotheoros*, and still less that he would ever have heard of Chéseaux; so that his work represents an important and independent milestone in our growing acquaintance with the Universe around us.

But let us return to Michell's first great contribution to stellar astronomy, of having proved from the data at his disposal, by a method of unimpeachable logic, the existence of physical double stars in the sky. The directness of Michell's expression already quoted leaves perhaps something to be desired; but his argument by which he convinced himself of the message of his data is equally convincing to us as it must have been to him. Unfortunately, Michell's contemporaries did not see the matter in quite the same light. Consider, for instance, the reaction of his younger contemporary William Herschel, who may have known Michell in person. In a paper entitled 'On the Parallaxes of Fixed Stars' (Herschel 1782), Herschel held to an opinion that the components of double stars which are very unequal in brightness must be at very different distances from us and, therefore, particularly suitable for measurements of the relative parallax of their brighter (i.e. nearer) components. This prompted him to embark upon a systematic search for such pairs, and the results of his efforts were gathered in the *First* and *Second Catalogues of Double Stars* published in 1782 and 1785. Michell was, to be sure, quick to point out that Herschel's new discoveries greatly strengthened his earlier probabilistic argument (Michell 1784), but Herschel still remained unimpressed. It was not until 1803 that the aging astronomer admitted, in a paper entitled 'Account of the Changes that have happened during the last Twenty-five Years, in the relative Situation of Double-stars: with an Investigation of the Cause to Which they are owing' (Herschel 1803) that they must indeed be true binary systems.

And more; for Herschel's sustained observations of several such systems (of which Castor was one) for over a third of a century demonstrated that the relative motion of the fainter component around the brighter represents (in projection) an *ellipse* on the celestial sphere, with the brighter component situated at one of its *foci*. The geometry of the orbits of the two stars around their common centre of mass is a direct consequence of the characteristics of the field of force binding them together to form a binary system. The elliptical form of orbital trajectories by itself is consistent with *two* distinct types of field, in which the force varies either with the direct first, or inverse second, power of the distance between the two stars. In the former case, the centre of mass would be located at the centre of the apparent ellipse (i.e. at the intersection of its principal axes); in the latter, at its focus. The fact that Herschel's observations demonstrated the latter alternative to be true was of great cosmogonical importance in his day; for it supplied the first observational proof that Newton's law of gravitation is universally valid also outside the confines of our solar system.

Since Herschel's time, double-star astronomy by visual (and, later, photographic) means has made great strides, and has been enriched by the discovery of thousands of new pairs. Moreover, those binaries (less than a hundred in number) for which absolute orbits as well as parallaxes could be determined have become an important source of our knowledge on the masses of the stars. Were we, however, to be limited only to the data forthcoming from this source, our knowledge of the masses (and absolute dimensions) of the stars would be severely limited, and confined largely to those with masses smaller than that of the Sun. Fortunately, Nature has provided other sources to this end in the sky, in the form of *close binaries*, which are (mostly) beyond the means of actual optical resolution on account of the close proximity of their components, but which can be recognised as such by other means. Thus, with diminishing separation, the velocity of their orbital motions is bound to increase, and may eventually become sufficiently large for its radial component to produce measurable Doppler shifts in the line spectra of such systems. Many hundreds of close binaries of this nature (we call them *spectroscopic binaries*) have been discovered in this way since the first such object was recognised by Vogel in 1889; and observations of them have provided invaluable data on the physical properties of a much wider class of stars.

While spectroscopic observations by their very nature are capable of providing information on absolute properties of the stars regardless of their distance, they cannot furnish actual masses or absolute dimensions unless the inclination of the orbits of such pairs of stars to the celestial sphere can be established by other methods. This can indeed be done if this inclination is large enough for the two stars to eclipse (partially or totally) each other at the times of conjunctions—thus giving rise to the characteristic variation of light which has earned this type of binary system the name of *eclipsing variables*.

Eclipsing binaries represent, in fact, the oldest known close binary systems, recognised as such long before Herschel demonstrated the existence of visual binaries. The Italian astronomer Geminiano Montanari (1633–1687) appears to be the first man on record to have noted (in 1670) that the star Algol (*El Ghul* of the Arabs) in the constellation of Perseus did not always shine with constant light, but occasionally dropped below the normal; he became so impressed by this curious phenomenon as to prepare a special tract about it ('Sopra la spariziona d'alcuire stelle e altre novita celesti'), which appeared in the *Prose di Gignori Academici Gelati di Bologna* in 1671.† Twenty-five years

---

† The literal version of the relevant passage is, unfortunately, tantalising by its incompleteness. 'The brightest star that shines in it,' (i.e., in the head of Medusa) 'affected by frequent mutations, attains but occasionally its greatest magnitude. I had observed it for several years to be of the 3rd magnitude. It faded in 1667 to the 4th magnitude; in 1669 it recovered again its previous lustre up to the 2nd magnitude, but in 1670 it exceeded only by little the 4th magnitude ...'. This is all that Montanari

later after Montanari, the variability of Algol was fonfirmed by Giacomo Filipo Maraldi (1665–1729), nephew of Giovanni Domenico Cassini—then director of the Observatoire de Paris (founded in 1667), who brought his relative from Italy to Paris to assist him in his work; and Maraldi's observations are published in volume 2 of the *Mémoires de l'Académie Royale des Sciences* in 1773 (pp 139 and 223). But neither his nor Montanari's observations attracted much attention, and they gave rise to no contemporary comments; perhaps because neither Montanari nor Maraldi noted that Algol's light changes were *periodic*. This discovery was not made till nearly a hundred years later; and the real history of Algol (and of eclipsing variables in general) did not commence until that time.

The person destined to make this discovery was a young British amateur astronomer, John Goodricke (1764—1786) of York, an astronomical prodigy who magnificently overcame the handicaps of his life to become, in spite of his lamentably short life (he died not quite 22 years old), the founder of an important branch of double-star

---

had to say in his report, mentioning neither the month nor the day of his observations. Could any additional information about the times of the minima be extracted from the MSS of Montanari's observations?

A search in Bologna proved, however, fruitless; but Montanari left this city in 1678 for Padua in response to the invitation of the Venetian Republic to accept the chair once Galilei's at the Bo, and spent the rest of his unfortunately short life (died in 1687) in Padua. There one of his closest pupils became F Bianchini, subsequently canon at the Liberian Basilica in Verona, who observed many years with his teacher, and left after his own death (in 1729) a voluminous collection of manuscripts which were published by Manfredi under the title *F. Bianchinis Astronomicae ac Geographicae Observationes Selectae, etc* at Verona in 1737. The material left by Bianchini was, however, so extensive that Manfredi's edition included only a selection of what their editor considered to be most important at that time. This left out, unfortunately, all observations of variable stars—but he did mention them; and this offered the clue to later investigators.

The most recent research on the Bianchini–Montanari MSS which contains their observations of variable stars was undertaken by A Porro. In particular, one fascicle of 275 pages (no. 387 in the library of the Liberian Basilica at Verona) contains the magnitude differences of all stars which its author had found between their own observations and those of Bayer. And in this survey (on p. 240 of the Verona MS) Porro discovered a statement revealing that one minimum of Algol (of 4th magn.) was apparently observed by Montanari on 8 November 1670. This, then, is the date of the first recorded minimum of Algol's light; though its hour or minute is, unfortunately, unknown.

It may be added that the relevant bibliographical evidence was published by Porro in this edition of the MSS under the title *Observationes circa fixas* (Stabilimento Fratelli Pagano, Genova 1902). A brief account of it was published by Porro in 1891. It is not impossible—as was stressed by Porro himself—that a further bibliographical search in the Capitolar Library in Verona may reveal further evidence on the times of the early minima of Algol; but such evidence—if it is there—remains yet to be brought to light.

astronomy—the study of eclipsing variables, which also became the dominant interest of my own scientific life. Since Goodricke (as well as Michell) hailed from the north of England (a part of the country which became almost my own 'astronomical bailiwick' during my Manchester years) it was of great interest to me to find out from extant sources all I could about the roots of double-star astronomy in the north of England; and in what follows I shall share with the reader the results of the second part of this quest on the younger of these two remarkable men.

John Goodricke (so we read in most extant sources) was the descendant of an old family of English country squires, whose tree can be traced in Yorkshire to the Middle Ages, and who were related by marriage with most of the local gentry. One branch of this family (from which our astronomer descended) was raised to a baronetcy by the middle of the seventeenth century; and its members were occasionally called upon to fill minor diplomatic posts abroad. This was certainly true of the fifth baronet, Sir John Goodricke (1708–1789), who at one time served as British minister in Stockholm; and his son Henry (1741–1784), the father of the future astronomer, spent several years on some (unspecified) mission in Holland, where in January 1761 he married Levina Benjamina Sessler of Namur, at Woldthuzen in Friesland.† The family lived, however, most of the time at Groningen, where our astronomer was born on 17 September 1764, and baptised two days later in the local Anglican church.

Scanty records reveal but little of his early childhood, beyond a suggestion that he became deaf and dumb as a result of a severe illness in early infancy. We know, however, that at the age of eight he was sent back from Holland to Edinburgh, to be educated at a school for the deaf which Thomas Braidwood (1715–1806) was conducting there at that time. A lack of school records conceals the early development of young Goodricke; but his progress must have been satisfactory; for in 1778 he was able to enter Warrington Academy, then a well-known educational institution in the north of England, which made no special provision for handicapped pupils. There we are told by extant records, '... having in part conquered his disadvantage by the assistance of Mr. Braidwood, he attained a surprising proficiency becoming a very tolerable classicist and an excellent mathematician' (cf Clegg 1961). For the latter he had undoubtedly to thank Dr William Enfield, an

---

† The bride's first names suggest that she may have descended from an originally Jewish family; if so, this could partly account for the long stay abroad of Henry's family. But her marriage was celebrated in a Christian church; and, many years later, Levina Benjamina was buried at Hunsingore cemetery not far from the tomb of her husband and her son; having survived both by many years. This means that had she been originally Jewish, she must have been baptised before marriage.

## John Goodricke of York

outstanding teacher (author of several textbooks) and mathematician of some renown, who almost certainly awakened the boy's interest in astronomy and set him on his subsequent career.

Just when Goodricke left Warrington Academy we do not know; but certainly not later than in 1781; for his *Journal of Astronomical Observations* contained its first entry made on 16 November of that year at York (where his family returned from Holland in 1776); and it is to this source that we must turn for a description of Goodricke's discoveries.

On 12 November, 1782 (a few days before the first anniversary of the start of his diary) Goodricke recorded that . . .

> This night I looked at $\beta$ Persei, and was much amazed to find its brightness altered—it now appears to be of about 4th magnitude. I observed it diligently for about an hour—I hardly believed that it changed its brightness because I never heard of any star varying so quickly in its brightness. I thought it might perhaps be owing to an optical illusion, a defect in my eyes, or bad air, but the sequel will show that its change is true and that I was not mistaken...

He continued his observations until the end of the season when Algol could be seen above the horizon at York; and it was not until 12 May 1783 that he communicated (through the good offices of Rev Anthony Shepherd, then Plumian Professor of Astronomy at Cambridge) the results of his observations in the form of a letter read before the Royal Society of London on 15 May.

This communication, which promptly appeared in print (Goodricke 1783) at once created considerable interest in astronomical circles; and prompted the Society's Council to award to its youthful author one of the two Copley medals for 1783. Well did Goodricke deserve it; for not only had he discovered the first known short-period variable star, and established a remarkably accurate estimate of its period; but at the end of his communication we find the following sentence which truly made it prophetic:

> If it were perhaps not too early to hazard even a conjecture on the cause of its variation, I should imagine it could hardly be accounted for otherwise than . . . by the interposition of a large body revolving around Algol . . . (*op. cit.*)

Nature had denied much to young Goodricke, but certainly not the gift of a splendid imagination. For it seldom happens in the annals of science that the first conjecture of a discoverer strikes the nail on the head more accurately than this suggestion of a young man of eighteen, who within the short life vouchsafed to him found time to discover, besides Algol, also the variability of two other naked-eye stars, $\beta$ Lyrae and δ Cephei, which both became prototypes of other classes of variable stars. However, Goodricke's bold suggestion that Algol (and β Lyrae) was an *eclipsing variable*, as we now call them, was made too early to gain speedy acceptance by contemporary astronomers. Thus William Herschel (who at the request of Sir Joseph Banks apparently acted as a referee of Goodricke's paper) was again plainly noncommittal. 'The idea of a small Sun revolving around a large opake body has also been mentioned in the list of such conjectures' (Herschel 1783); but he thought he had reasons to doubt it; for (he said) prior to his report he had observed Algol repeatedly in the focus of his 7 foot telescope, and found it 'distinctly single'.

Twenty years later, it became Herschel's destiny to prove that several visual double stars observed by him indeed form physical binary systems. Whether or not the aging astronomer ever made up his mind about Algol we do not know; but it looks as if the boldness characteristic of his other speculations failed when confronted with this particular kind of stellar symbiosis. At any rate, Goodricke's brilliant suggestion of 1783 was destined to remain in the realm of hypothesis for a long time—until, in fact, the German astronomer Vogel discovered in 1889 that Algol was also a 'spectroscopic binary' whose conjunctions coincided with the light minima; it was not till then that the eclipsing nature of Algol and other similar variables was established beyond any doubt.

This was, of course, more than a hundred years later; but by 1783 Goodricke's short life was running out fast. The last observation recorded in his diary is dated 24 February 1786. In April of that year, the Royal Society of London elected Goodricke to fellowship; but, unfortunately, he no longer had a chance to accept it; for he died in York only two weeks later, on 20 April... 'in consequence of a cold from exposure to night air during astronomical observations' (cf Clegg 1961). The immediate cause of his death is unknown, and the event passed largely unnoticed; no stone over his tomb at Hunsingore (close to the former family seat at Ribston Hall in Yorkshire) suitably commemorates his final resting place to this day.

This was about all that one could learn about Goodricke from published literature in the early 1960s. Dissatisfied with the scarcity of these sources, and with what they conceal rather than disclose, my former student Dr Alan Batten (PhD Manchester, 1958; who has likewise made the study of close binary systems his life-long preoccupation) and I visited the ancestral lands of the Goodricke family in Yorkshire to search for John's tomb at the family cemetery at Hunsingore in May 1959. The search would have been fruitless had I not had Dr Batten with me; but Alan (who was born in a vicarage) knew his way about ecclesiastical establishments; and with the aid of the local

# John Goodricke of York

vicar we eventually located what we had come to see: a dilapidated tombstone overgrown with grass, on which only two barely-legible initials J. G. disclosed to a searching eye whose mortal remains are underneath, expecting resurrection. The realisation of this saddened us greatly; and on figure 6.2 you can see Alan ruminating on the transient nature of the glories of this world.

**Figure 6.2** The forgotten tomb of John Goodricke the astronomer, at the cemetery of Hunsingore, near Ribston Hall (seat of the Goodricke family) in Yorkshire. (Above) Dr Alan H Batten contemplating (in 1959) the vanity of this world. (Below) The neglected tomb itself is overgrown with grass. Only the initials J. G. on the tombstone allowed us to identify the last resting place of the youthful founder of the astronomy of close binary systems.

I recall that, while travelling back to Manchester, we discussed the ways in which this neglect could be remedied; and later that year I sent a 'Letter to the Editor' of the London *Times*, calling attention to the matter and soliciting unpublished information on the Goodricke family. This letter appeared in the *Sunday Times* on 20 December 1959, and elicited a remarkable response.

The primacy among it belonged to letters received from descendants of the Goodricke family still living at that time: one from Mr John H Goodricke of Parkstone, Dorset (the last surviving member of the family in England); the other from Mr John R D Goodricke of Natal in South Africa. These two branches of the family apparently knew of each other only by hearsay (they did know each other's address); but Mr Goodricke of Dorset very kindly lent me a copy of the privately-printed *History of the Goodricke Family* (London, 1885) in which the family tree can be traced to the beginning of the sixteenth century. On 14 August 1641, the family (which through Clare, wife of Richard Goodricke, who died in 1581, may have some royal blood in their veins going back to William I, and possibly to Alfred the Great) was raised to the rank of baronet by King Charles I. The fifth baronet, Sir John Goodricke (1708–1789) was grandfather of our astronomer, who would thus have succeeded to the title after the death of his father (in 1784) had grandfather not survived them both. Sir John's portrait, by Gustaf Lundeberg, painted in Stockholm in 1766 while Sir John (then 58 years of age) resided there as Minister of the Crown, has been preserved at Gilling Castle in Yorkshire; and is reproduced on Figure 6.3 by permission of the Rector of Ampleforth Benedictine College (now located at that castle). It shows the face of a rather formidable old man, who may not have been on too intimate terms with the younger members of his immediate family; and if so, this might perhaps explain why his son Henry spent so many years in Holland.

Moreover, according to tradition (cf Goodricke 1913) a portrait of Henry's son John (the astronomer) was also painted during his short lifetime by James Scouler in 1785; since 1913 this has been in the possession of the Royal Astronomical Society (to which the Goodricke family donated it at that time). This portrait can be seen in the Society's apartments in Burlington House to this day; it is shown in figure 6.4. However, one look at this picture is apt to give rise to some questions on whether or not the family tradition can really be trusted. Certainly it is hard to believe that it represents a man who (at the time when the painting is supposed to have been executed) was only 21 years of age! The background of the picture is obviously stylised; but how much licence went into the painting of the person, intended to make him conform to the traditional style of a country squire, rather than a star-gazer enchanted by the heavens, who spent precious clear nights of his short life watching the stars in their courses, we may never know for sure. But the possibility cannot be ruled out that, in Scouler's portrait allegedly of John Goodricke, we are not looking at the face of that young astronomer at all.

**Figure 6.3** A portrait by G Lundeberg of Sir John Goodricke (1708–1787), the fifth baronet, who was grandfather to the youthful astronomer who discovered the periodicity of Algol.

Many thoughts like these went through my mind as Alan and I stood at Goodricke's tomb at Hunsingore. First, why does he rest there forgotten by all his clan; why was he not buried with the rest of them in their family vault; and why are the remarks about him in the family chronicles so scanty? Moreover, what was his father doing for so many years in Holland; and was the fact that his son was born of a foreign mother (a Fleming of Namur, possibly of Jewish descent) the cause

410  *The binary stars*

**Figure 6.4** A portrait (apocryphal?) by J Scouler of John Goodricke (1764–1786), discoverer of the light changes of Algol (whose eclipsing nature he foresaw), and also of the variability of $\beta$ Lyrae and $\delta$ Cephei. The portrait, now in the possession of the Royal Astronomical Society, is reproduced by courtesy of the Ronan Picture Library.

why the family in England kept Henry at arm's length? One look at Henry Goodricke's father (figure 6.3) makes it unwise to dismiss the possibility that he might perhaps have disapproved of his son and his offspring; and the fact that his grandson John became deaf-mute, regarded at that time as a further blot on the family's escutcheon, may have added to their disapproval of his intellectual inclinations (which had indeed no precedent in the family). For as far as we can gather from extant sources, all Goodrickes were essentially country squires of the fox-hunting type, whose personal interests were confined to hunting and the turf. Significantly enough, the only exception to this rule seems to have been the young astronomer's father Henry, who was a man of at least some literary accomplishments; for in 1766 he published in Groningen a Latin work on jurisprudence.

Too much interest in the turf proved, in fact, the undoing of the Goodricke family; but during John Goodricke's life their fortunes still stood high; and although we know nothing about the attitude towards astronomy of his grandfather, interest in this subject was shared by John with his relative Edward Pigott (1753–1825) of the Fairfax family, with which the Goodrickes were connected by repeated ties of blood. In fact, Edward and John were not only cousins to each other, but also very good friends; and Edward's effigy in his young years can be recaptured from a contemporary painting preserved at Gilling Castle in Yorkshire (a former seat of the Viscounts of Fairfax); it is reproduced here (figure 6.5) by kind permission of the Rector of Ampleforth College.

**Figure 6.5** Edward Pigott (1753–1825), discoverer of the variability of $\eta$ Aquilae, cousin and intimate friend of John Goodricke, with whom he shared his astronomical inclinations.

There is no doubt that John Goodricke and Edward Pigott exerted a beneficial influence on each other (a partnership in which Edward seems to have been the organiser, while John had the more original mind of the two). The day of 10 September 1784 should be recorded with golden letters in the annals of our science; for that night Pigott discovered the variability of the naked-eye star $\eta$ Aquilae (a well-known cepheid) which turned out to possess a period of 7.176 days; while only hours later Goodricke detected the variability of $\beta$ Lyrae. There is more; within the same week Goodricke also discovered the variability of $\delta$ Cephei, another prototype of its class. This eminently fruitful partnership ended two years later with Goodricke's premature death.

Pigott himself, although older, still had almost forty years to live; and in the course of them he discovered also the variability of R Scuti and R Coronae borealis (prototypes of certain other types of irregular variables); and performed also other observations of considerable interest. However, none of these concerned variable stars, and are thus outside the scope of this chapter.

In the meantime, and following the death of their most famous member, the Goodricke family's place in history waned, and their fortunes were gradually drawing to a close. Nothing much can be said on Sir Henry Goodricke (1765–1802), John's younger brother and the sixth baronet; and what can be said of the latter's only son, Sir Harry James Goodricke (1797–1833), the seventh baronet, is not much to his credit. For while on his accession to the title he inherited an annual income estimated at £40 000—a real fortune in those days—during his short life he managed to squander pretty well all of it on the turf; so that the eighth (and last) baronet of his line (Sir Harry's first cousin twice removed, who succeeded to the title at the age of 71) was left to survive on a pittance of only £20 per annum. He died three years later; and with him the baronetcy became extinct. The family survived, of course, in collateral branches to our own days; but as both my correspondents in 1960 claimed to be the last of their respective lines (and their handwriting showed signs of advancing age), it is possible that now, 25 years later, the Goodricke family has become completely extinct.

In taking leave, in these reminiscences, of their most famous member, John Goodricke the astronomer (1764–1786), as well as of his Yorkshire compatriot John Michell (1724–1793), it may interest the reader (especially if he is a theoretician) to reflect on the historical fact that, while the first representatives of wide (i.e. visual) as well as close (i.e. eclipsing) binary systems were discovered by Riccioli (1650) and Montanari (1670), who both happened to observe in Bologna (the site of the oldest of all universities) under the clear Italian skies, the

## John Goodricke of York

binary nature of each group was recognised by pure reasoning by two British amateurs—Michell (1767) and Goodricke (1783)—long before observers like Herschel (1803) or Vogel (1889) provided the compelling observational proof. Today, both Michell and Goodricke have reached the stage at which their places in the history of science are fully recognised; and the University of York has named one of its halls of residence Goodricke Hall. Another building belonging to York University, located on the site from which Goodricke, according to tradition, carried out his observations of Algol (and, perhaps, contracted the pneumonia which carried him away) is now marked with a memorial tablet, reproduced in figure 6.6.

**Figure 6.6** Memorial tablet to John Goodricke on the wall of the Treasurer's House of the present University of York. One of the University's halls of residence has been named after that devoted early student of variable stars.

It would also appear that the legacy left behind by John Goodricke should have been taken up more quickly and developed by astronomers of subsequent generations; but this was not to be the case. In particular, the *interpretation* of the observations, and quantitative elaboration of Goodricke's brilliant suggestion as to the cause of Algol's variability trailed behind the observers for many decades. This was, to be sure, not the case with the 'visual' binary systems, for which an analysis of the observations was essentially a matter of geometry; this already received adequate attention at the time of Bessel. But while the prototypes of the eclipsing variables—Algol as well as $\beta$ Lyrae—were discovered in 1782–83, it was not until almost a hundred years later that Edward Charles Pickering (1846–1919) attempted to interpret their light changes in terms of a model originally proposed by Goodricke (cf Pickering 1880); and the line of research so initiated gathered momentum but very slowly, largely because of the very limited knowledge of the physics of the stars available at that time.

In order to place these in proper perspective, let us recall that whenever we wish to study the properties of *any* celestial body, be it a planet or a star, all information available for this purpose will reach us through two distinct channels: their gravitational attraction, and their light. Gravitational interaction between our Earth and its celestial neighbours is, however, measurable only at distances of the order of the dimensions of our solar system; and the *only* means of communication with the realm of the stars are their nimble-footed photons reaching us, with appropriate time-lag, across the intervening gaps of space.

As long as a star is single and emits constant light, it does not constitute a very revealing source of information. Spectrometry of its light can disclose, to be sure, the temperature (colour, or ionisation) of the star's semi-transparent outer layers, their chemical composition, and prevalent pressure (through Stark effect) or magnetic field (Zeeman effect); it can even disclose some information about the star's absolute luminosity or rate of spin. It cannot, however, tell us anything about what we should like to know most: the mass or size (i.e. density) of the respective configuration, its absolute dimensions, or its internal structure.

In order to disclose its mass, the star must be made to 'step on the scales' by entering into gravitational partnership with another star to form a 'binary system'. A certain amount of information on the masses of stars can be obtained from observations of nearby 'visual' binaries which are within measurable distances, and close enough to exhibit absolute motions about their common centre of gravity within 1–2 centuries of their observation. The number of such systems is, however, limited by their required proximity; and their available supply is not copious. Binary systems situated at greater distances can be discovered spectroscopically (from periodic variations of the Doppler shifts of spectral lines of their components) the more easily, the closer they happen to be; and although their spectroscopic observations can furnish absolute values for the lower bounds of their masses, they can say nothing about the (absolute or relative) dimensions of their components.

If, however, the system happens to be sufficiently close for the dimensions of its components to represent not too small a fraction of their separation, and if their orbit is not inclined too much to the line of sight, each component may eclipse (partly or wholly) its mate in the course of each orbit; and the system thus becomes an *eclipsing variable*, recognisable as such not only by fluctuating Doppler shifts in the spectra of each component, but also by the *variation of light* of the system within eclipses. The characteristic variations of light due to this cause represent an even more eloquent (and more easily observable) tell-tale feature of such systems.

How well can we measure the requisite data which constitute the basis of all subsequent reductions—namely, the brightness of the respective star at a given time? The timing of photoelectric observations can be made essentially error-free (in the sense that one second of real time, the typical example of instrumental errors, is likely to correspond to 1 part in $10^5$ of the orbital period, or less than one-hundredth of a degree in phase angle); but the same is not, unfortunately, true of the measurements of instantaneous light intensity during this time. A comparison of the errors of photometric and positional measurements of celestial bodies may be illuminating: for while ground-based positions can now be measured as a matter of course with a precision of the order of one ten-millionth part of the quadrant (a precision which may be increased further by a factor of 100 from space within the framework of Project *Hipparcos*) photometric measurements (or, rather, estimates of brightness which could be made by Goodricke and his contemporaries) were seldom more precise than to 10% of the measured intensity of light; and remained so till well into the second half of the nineteenth century. It is only the advent of photoelectric photometry in the twentieth century (and, in particular, of photoelectric photomultipliers since 1945) that has enabled us to increase the accuracy of photometric measures to 1 part in a thousand (i.e. 0.1% of the measured flux); and it is unlikely that we can expect this to increase by more than a further factor of 10 within our lifetime (the light curves of several eclipsing variables have now been observed from space, at altitudes where the measurement should be completely immune to atmospheric effects; but not, unfortunately, free from instrumental reasons inherent in the quantum structure of light).

In view of so great a disparity between the ultimate precision that can be attained by photometric and positional measurements in stellar astronomy, the need of a different approach to the determination of the elements of eclipsing binary systems from those of (say) asteroidal or cometary orbits will be immediately apparent. The difficulties of the eclipsing orbit work, based even on the best observations available at the present time, should be comparable with those confronting the computer of asteroidal orbits if the observed positions at the basis of his work were subject to errors amounting to 20 minutes of arc (i.e. almost the apparent diameter of the full Moon!). In both problems, asteroidal as well as eclipsing, three observations are sufficient in principle to furnish the desired answer. To obtain it on the basis of astrometric observations is indeed feasible (for all but 'definitive' orbits); but to obtain significant elements of an eclipsing binary from an analysis of its light changes the underlying photometric observations must obviously be very many to compensate for their low relative accuracy; and will require statistical, rather than individual, treatment.

But is it worth it? The answer is an unqualified 'yes', for a study of the phenomena exhibited by eclipsing binary systems occupies a very important position in contemporary stellar astronomy for several reasons. First, because of a truly prodigious abundance of the objects of its study. Surveys of stars in our neighbourhood disclose that at least 0.1% (probably more) of them form systems which happen to eclipse; and if stars with masses greater than that of the Sun were alone considered, the percentage would be much higher. If, moreover, the foregoing conservative estimate of their percentage were to apply to our whole galactic system, the total number of eclipsing variables within it should be of the order of $10^9$. Only a minute fraction (a few thousands) of these have been identified so far, and their periods determined; but their total number in our Galaxy is beyond the hope of individual discovery. Eclipsing variables are, therefore, manifestly not exceptional or uncommon phenomena!

The significance of eclipsing variables is further emphasised by the fact that they constitute the only class of double stars that can be discovered in more distant parts of the Universe. In the neighbourhood of the Sun, up to distances of the order of 100 parsecs, at least wide binaries can be recognised by their orbital motion (or, for very wide pairs, by common proper motions of their components). Spectroscopic binaries can be discovered as such up to distances of the order of one or two thousand parsecs (depending on their absolute brightness) with the aid of modern powerful reflectors capable of decomposing their light into spectra of sufficient dispersion. Beyond that limit close binaries can, however, be detected if, and only if, they happen to be eclipsing variables. Their characteristic light variations can be measured photometrically almost down to the limit attainable by our telescopes—not only in our Galaxy, but in any system resolvable by them into stars. At present dozens of them are known in external galaxies, down to distances exceeding one million parsecs.

But the significance of eclipsing variables in astronomy is not based only on their ubiquitous presence and enormous numbers; it rests also as much (or more) on the unique nature of the information which they (and they alone) can provide. We mentioned above that spectrographic observations alone can furnish only the minimum values of the masses of the components of close binary systems or of the dimensions of their orbits. The missing clue, necessary to convert these lower bounds into actual values, is the inclination of the orbital plane to the line of sight; and this can be obtained from an analysis of the light changes if the respective binary happens to be an eclipsing variable. Moreover, if so, an analysis of the observed light curves can specify not only the orbital inclination, but also fractional dimensions of the components—which, on combination with spectroscopic data, can furnish the masses, densities and absolute dimensions of the constituent stars.

Astrophysical data which can be deduced from the observed light changes of eclipsing variables transcend, moreover, information on the absolute dimensions of their components or the characteristics of their orbits; for even their internal constitution may not (under certain conditions) remain concealed from us. Even though the interiors of the stars are concealed from view by the enormous opacity of the overlying material, a gravitational field emanates from them which the overlying layers, opaque as they may be, cannot appreciably modify; and the distribution of brightness over the exposed surfaces of the components is governed by the energy flux originating in the deep interior.

With the stars, one is almost tempted to say, it is like with human beings. A solitary individual, watched at a distance, will seldom disclose to a distant observer more than some of his 'boundary conditions', insufficient in general to probe his real nature. If, however, two (or more) such individuals are brought together within hailing distance, their mutual interaction will bring out their 'internal structure' the more, the closer they come to each other. Similarly, as a star in the sky remains single, there is no way of gauging the detailed properties of its gravitational field, or of the distribution of brightness on its surface. Place, however, another star in its proximity to form a pair bound by mutual attraction: many properties of the combined gravitational field can be inferred at once from the observable characteristics of the components' motion, just as a distribution of surface brightness can be deduced from analysis of the observable light changes.

These and other possibilities opened up by the studies of eclipsing variables have long attracted due attention on the part of observers. Largely because of the recurrent nature of the phenomena which they exhibit, eclipsing variables have always been favourites for pioneers of accurate photoelectric measurements, and the total number of observations made in this field runs into several millions. Observations alone are, however, insufficient to disclose to inspection the wealth of information which they contain. To develop this information calls for the introduction of systematic methods, rooted in physically sound models of the phenomena we observe, to decipher what the observations have to say.

The problem at issue is one of astronomical cryptography: the messages these stars sent out on waves of light are encoded by the processes responsible for them. The task of the analyst is to decode the photometric evidence to yield the information which it contains. To do this requires some knowledge of the code; and to provide it is the task of the theoretician.

The decoding process constitutes an essentially mathematical problem; but the identification of the code is primarily one for the astrophysicist; and the extent to which this code can be set up determines the gain to be expected from its use. Eclipsing variables do not represent by any means the only sources of encoded information reaching us from the stars: indeed, virtually all stars to the right or left of the main sequence in the Hertzsprung–Russell diagram are known to be variable (though not necessarily periodic). The reason, however, why their light variations tell us (at best) only a part of their story is that, because of the physical complexities, we as yet lack the code to decipher some (or most) of it. In a physical sense, the quest for a code to unlock the information contained in the light changes of eclipsing variables will not confront us with unsurmountable problems. However, unless the theoretical basis of an analysis of their light changes is well understood, and due care is exercised in their interpretation, not only do we fail to do justice to the skill and perseverance of the observers but, worse still, we run the risk that a considerable part of stellar astronomy may rest on inaccurate empirical foundations.

It is, incidentally, astonishing to contemplate the range of inspiration which astronomy (and physical science in general) has received from the phenomena of eclipses. In the fourth century BC, the observed features of the terrestrial shadow cast on the Moon by the Earth enabled Aristotle (384–322 BC) to formulate the first scientific proof that the Earth was a sphere; and somewhat later, eclipses of the Sun provided Aristarchos (in the early part of the third century BC) and Hipparchos (second half of the third century) with the geometric means to gauge the distance which separates us from the Sun. In the seventeenth century AD, the eclipses of the satellites of Jupiter furnished Roemer (in 1676) with the first experimental proof of the fact that the velocity of light was finite. Total eclipses of the Sun have, in particular, proved a veritable godsend to students of many branches of science as well as the humanities—from the historians of ancient times (whom they enabled to straighten out many an obscure date of early chronology) to the geophysicists (who used old eclipse records to detect secular irregularities in the length of the day), or to the chemists who learned (in 1868) of the existence of at least one new element (helium) from the solar flash spectra before it was identified in the laboratory. Students of the motion of the Moon in the sky found the records of ancient eclipses invaluable for identification of its 'secular acceleration'; and disappearances ('occultations') of stars of known position behind the Moon's limb are valuable for accurate location of the instantaneous positions of our satellite; while high-speed photometry of such occultations led in several cases to a more precise determination of the angular diameters of such stars than would (at present) be possible by any other alternative optical method. In recent decades astrophysicists have also taken advantage of the fleeting minutes of total eclipses to measure the extent of light deflection in the gravitational field of the Sun, and thus to study the metric properties of space in the neighbourhood of large masses.

In the face of these facts, the reader may inquire whether or not a study of stellar eclipses might prove equally rewarding; and his expectations should indeed not be disappointed. The information which we should like to acquire is, of course, not being transmitted by them through space *en clair*, but encoded in a language peculiar to them, which we must break before the desired information can fall into our lap; and the code to this operation is rooted in the physical model of the system (about which, more later). Should the reader find this operation somewhat unusual (not to say exasperating), it is the novelty of approach rather than actual numerical work inherent in the process which may be the cause of such qualms. Above all, we should keep in mind that the phenomena of eclipses have never yet caused an astronomer to lose his head—at least not since the days of our legendary ancestors Hi and Ho in old China! Stellar eclipses will, to be sure, scarcely ever again have an opportunity to stop a battle as did (allegedly) the famous solar eclipse of 585 BC at the time of Thales. However, the results of their study may well exert a more profound and lasting effect on science than did that abortive skirmish between the Lydians and the Medes on the history of the human race.

But let us return from these centuries before our era to the time just before I was born. The analysis of the light changes of eclipsing variables, initiated by Pickering in 1880, was carried out for so restricted a model of such systems (assuming that both stars are spherical and exhibit uniformly bright discs; that the secondary (eclipsing) component is completely dark; and—the most restrictive assumption—that the eclipses causing the observed minima of light are grazing!) as not to be able to disclose very much about the nature of the respective systems; and in subsequent years these assumptions were removed one by one only with considerable difficulties. Thus the hypothesis of grazing eclipses was given up by J Harting (1889) in a paper in which the existence of partial eclipses was admitted; and E Hartwig (1900) allowed the eclipsing body to possess a finite luminosity of its own. Furthermore, G W Myers (1898) and A W Roberts (1903) allowed the shape of the components to depart from spherical form; and A Pannekoek (1902) with C Roediger (1902) first considered the effects which limb-darkening of the stars (so manifest on the apparent disc of our Sun) would exert on the shape of the light curves within eclipses.

However, all these investigators were concerned with specific eclipsing binaries—like Algol or $\beta$ Lyrae—and not with the general procedures which, by the turn of the century, were being discovered in increasing numbers. Such methods were not launched till 1912 by Henry Norris Russell, who in collaboration with Harlow Shapley first developed systematic procedures for an analysis of the light curves of eclipsing variables, designed to cope with the large numbers of observations of limited accuracy available by that time. This partnership, which inaugurated a new era in the study of close binary systems, was referred to in Chapter 4; figure 4.3 showed these veteran investigators as they looked in the fullness of years, some forty years later. However, to attain their aims with the technical means at their disposal (remember how the slide-rule remained Russell's favourite computing aid till the end of his life) they had to carry over from previous work several oversimplified assumptions of their predecessors, assumptions prompted by the need to render the problem tractable at all by their methods, rather than by their physical reasonableness. The most questionable of these was, of course, the assumption that the geometrical shapes of both components were spheres or similar ellipsoids. In addition, smooth curves drawn free-hand to follow the course of individual observations were substituted for the actual data; and several tables of artificial functions were constructed to make their solutions practicable.

The original version of the Russell–Shapley method, to which certain (non-essential) modifications were subsequently proposed by Vogt (1919), Fetlaar (1923), Stein (1924) or Sitterly (1930) remained without challenge in their field for the first third of this century; and excellent summaries of them could be found in the fourth part of the volume by Johann Stein in *Die Veränderlichen Sterne*, or in abridged form in Part 5 of K Schiller's *Einführung in das Studium der Veränderlichen Sterne* (Leipzig 1923); which at that time (when I was still ignorant of the English language) were my only sources of information. I met both these authors later in life—Father Stein at the Sixth IAU General Assembly in Stockholm in 1938, and Dr Schiller at the Eighth General Assembly in Rome in 1952, to be able to tell them how much their books meant to me in my student years. Although, by that time, the milestones of progress had already been carried forward far beyond the stage, both books—especially Stein's (containing also a comprehensive historical account of work in this field from Pickering to Russell)—continue to remain useful historical sources up to the present time. Both are still on the shelves of my room, where I write these lines; Schiller's inscribed to the late Professor R Prager (from whose estate I acquired it after Prager's death in 1945).

But let us return once more to the days of my innocence, when these books first came into my hands. The first step which a youth barely more than sixteen years of age could take after grasping the essentials of Russell's method with their aid was to apply it to practical cases. This I did and published the results of these applications in two notes (Kopal 1932a, b) which Professor Kobold accepted for publication in the *Astronomische Nachrichten*. In Chapter 2, I described the circumstances of the appearance of these first contributions of mine in an international journal; and while their contents are of course wholly obsolete today (and so are the light curves on which my work was based), they represent the first distant intimations of interest in a field which has become the fate of my life.

This fate was, however, not to unroll at too fast a pace. Some old papers still on hand from my early university years disclose that the first ideas that all was not well with the foundations of Russell's method, and that, perhaps, I could do better, began to germinate in my mind at that time. I recall that my teachers at that time—especially Professor Nechvíle (less Professor Freundlich, who kept an altogether more open mind on most subjects)—gave me no encouragement in such iconoclasm (so that responsibility for it should be solely my own); but several unrelated events took place later which inspired all that was to follow in subsequent years; and none was perhaps more pregnant with consequences than the acquaintance which I was privileged to make with Professor Shin-ichiro Takeda (1901–1939; see figure 6.7) of the University of Kyoto during my visit to Japan in 1936 (described in Chapter 3). Fate cut Takeda's life short soon thereafter, but vouchsafed him time to leave behind two papers (Takeda 1934, 1937) which became of historic importance for the further development of our subject. During the trip to our eclipse camp in northern Hokkaido in June 1936 we had ample opportunity to discuss our mutual interests; and Takeda gave me then a reprint of his 1934 paper to which, I confess, I did not react as promptly as I should have. By nature I have been a 'slow developer' from the days of my youth till the present time. The best work of my life did not come till I was past sixty; and with some of it I am engaged concurrently with this book, at a time when academic retirement gives me a better chance of concentration.

422  *The binary stars*

**Figure 6.7**  Professor Shin-ichiro Takeda (1901–1939) of Kyoto University (right), at the 1936 eclipse camp in Nakatombetsu, with Dr H Slouka of the Czechoslovak Solar Eclipse Expedition.

Two years after our return from Japan, in October 1938, I came to the United States and was able to work for several years in close proximity to Russell and Shapley, a team which had done so much in 1912 to advance the knowledge of our subject. As it happened, a reprint of Takeda's second (1937) paper reached Princeton around that time; and Russell passed it on to me 'to try to make some sense out of it' (which, truth be told, was not easy, if only because of a great many misprints which marred its text; and Takeda, who died in 1939, was no longer able to help). But I found my way through it unaided; and one of my early contributions of more permanent nature (Kopal 1942) grew out of this posthumous encounter with a man I had met in Japan a few years before his untimely death.

In late years, when it became my privilege to educate, or otherwise influence, many Japanese scholars of the next generation, I always had a feeling that I was only paying back the debt I owed to Takeda for inspiration derived from his work during the formative years of my own career, and for which I was vouchsafed greater longevity. To bring home in brief what Takeda had done to a more technical reader, he was the first in 1934 correctly to estimate the effects of gravity-darkening on the light changes of close eclipsing systems between minima; and, secondly, he set out to investigate the effects of distortion on light curves within eclipses as well.

But apart from such niceties (essential as they are for a proper study

## The need for a new method

of such systems), other considerations obtruded insistently on my mind in 1940 and later which cast ever increasing doubts on whether, in following Russell and Shapley in their 1912 work (and all its subsequent modifications), we were not altogether on the wrong track. The reasons go back to the way in which the observed data are used to furnish the desired results—or, more specifically, to the replacement of actual observations (which consist of sets of discrete measurements of the star's brightness at particular times) by continuous curves drawn freehand to follow the course of the individual observed points. There is no doubt that, at the time such a tactic was adopted, it constituted a shrewd move whose time-saving features might have outweighed its intrinsic disadvantages. However, with the advent of photoelectric photometry the time was soon reached when the initial assets turned into a liability; and whatever one might think of the tactics, the strategy was all wrong!

For, in replacing the actual observations by graphical inferences, we had 'thrown away the baby with the bath water' by giving up the possibility of ascertaining the *uncertainty* within which our results were defined by the data on hand, and of establishing the extent to which these results provided a *unique* representation of these data. For a mere agreement of the theoretical light curve (i.e., one computed from the elements resulting from such an analysis) with observations constitutes, to be sure, a necessary condition for the correctness of these results, but is *not sufficient* to establish their uniqueness. For the question must be asked: do other combinations of elements exist which could equally reproduce the same data within the limits of observational errors? This, unfortunately, turned out to be true, especially for systems whose components only partially eclipse each other; and the shallower the eclipses, the easier it becomes to represent the observed light changes by a 'right combination of wrong elements'. By confining itself to a mere representation of light curves drawn free-hand by *some* elements as the ultimate objective of the analysis, Russell's method renounced any possibility of serving as a basis for error analysis of the outcome. Moreover, by dropping enough degrees of freedom of our problem (inherent in the adoption of 'fixed points' on the light curves, the positions of which were supposed to be error-free), one always obtains *some* numbers for the results; but what they mean is another matter.

The principal reason for such an obviously unsatisfactory state of affairs is inherent in the mathematics of the eclipse phenomena. The relation of the observed brightness of the system to any phase is not algebraic, but transcendental; and a construction of its solution which does not arbitrarily suppress any degrees of freedom can, therefore, proceed only by *iteration*. This approach I began to develop in 1941. Although, truth be told, it was greeted by Russell with somewhat less than enthusiasm (cf Russell 1941), such iterative procedures were systematically developed in 1947 during a year of happy collaboration with Dr Stefan Piotrowski (then visiting Harvard Observatory), who brought to bear on the problem his mastery of least-squares methods acquired from Professor Banachiewicz. By that time Russell was, however, already seventy years old—still interested in the subject, but no longer in a position to contribute any original ideas to its further outgrowth.

In retrospect, one can perhaps wonder why it took so long for the iterative methods to be more widely accepted. The reasons were several, partly historical, and partly technical. By 1940 the Russell–Shapley methods (with roots going back to 1912) had attained a canonical age; and were reduced to a series of 'precepts for the computer' (human, not machine) enjoining him to 'do this, do that; and what you get is this'. To follow such precepts, backed by the reputation which their originators had earned since 1912 by work in other parts of our vineyard, called for only a limited amount of computation which could be carried out with a slide-rule and even less thinking. Who would wonder that, in these circumstances, the users of time-honoured precepts were reluctant to abandon them for alternative approaches to the solution of our problem, requiring (at least initially) hours of computation with desk-type machines—unless their reluctance could be overcome by the attraction of a basically different quality of the outcome; and unless its knowledge was appreciated enough to justify the increased effort to obtain it? At first not many answered the call; and yet how few were aware at that time that, in not many years to come, astronomers (as well as others) would come to possess automatic computers working 'faster than thought', and rendering practicable feats which would have staggered the imagination of our predecessors in the first half of this century?

We may add that while the new iterative methods were slowly taking root among the younger and more dedicated students of the subject, the earlier (1912) methods gained a new lease on life by the publication of new sets of auxiliary tables by Russell and Merrill in 1952. With their aid, the old methods were kept alive by a dwindling band of epigones, whose devotion to what they had learned when they were young helped to prolong the lifetime of obsolescent ideas well beyond their natural term. Merrill's tables certainly represented a beautiful piece of printing. However, whether or not their ready availability contributed positively to the further development of our subject is questionable, for several reasons.

First, their existence facilitated the easy production of second-rate results which have continued to dilute the professional literature up to the present time. Secondly (and worse) they helped to perpetuate a false impression that little or nothing of significance remained to be added to the methods already known—an impression which those willing to accept it must have found discouraging. They must have wondered why this particular part of double-star astronomy was destined not to be disturbed from its fifty-year-old slumber while everything else was burgeoning right and left, and soon turned their creative abilities to other fields of research; while the old practitioners of the art continued to plod down the beaten path merely because what was once convenient, became sacrosanct to those of a conformist state of mind. It is, of course, known also from other branches of science that 'vested interests' of conservative leanings are seldom prepared to welcome the advent of new ideas; and if 'ancestor-worship' of this type impeded, at least for a time, the natural advance of scientific progress, those who may have been its cause will bear a responsibility for it before posterity.

But, speaking of posterity, what assurance do we have that all the methods mentioned above have a good chance of survival? For different as they may be in the tactics by which they set out to accomplish their aim, they still possess one feature in common: they start with a plot of the observed light intensity (replaced, in Russell's method, by smooth curves drawn free-hand to follow the course of individual observations) against time (or, what is equivalent, the phase angle of the respective system). We can, in brief, describe such an approach as an effort to interpret the observed light changes in the *time-domain*. To do this appeared most natural to all investigators of preceding generations, because such an approach starts in the simplest way from the output of our measuring instruments. On the other hand, the principal drawback of such an approach is the fact that the relation between the positions of individual observations in the time-domain is very complicated: the equations relating these with the unknown elements we wish to determine are not algebraic, but transcendental. Such relations are difficult to formulate analytically with sufficient exactitude, and even more difficult to solve. It is, of course, not necessary to do so; for an approach to the solution of our problem via the time-domain constitutes only one alternative. Another approach, equally valid in principle, is to interpret the observed light changes in the *frequency-domain*: in other words, to analyse, not the light curves of eclipsing variables as a plot of their brightness versus time, but their *Fourier transform*.

What is a Fourier transform of the light changes of eclipsing variables (or, indeed, of any other function possessing a finite number of discontinuities)? It represents a *frequency-spectrum* of the respective function, from which this function can be recovered by an appropriate *synthesis*. Indeed, 'spectral analysis' and 'synthesis' of such functions constitute reciprocal operations containing the same amount of information; and the question arises only as to the simplicity with which the desired information (in our case, the elements of the eclipsing systems) can be developed from the observations at different stages of the Fourier transformation. Now while, in the time-domain, the relation between the elements of an eclipsing system and their observed characteristics are *transcendental* (and incapable of analytic inversion), in the frequency-domain these relations become *algebraic* and can be inverted.

To many readers the Fourier transform may sound an esoteric concept existing only in mathematical literature. Far from it! It is something we all keep performing in our heads without being aware of it, and could not live without it (nor could any ear-possessing vertebrate animals). When I address my students, by proper use of my throat muscles I generate a train of waves in the surrounding air, which transmit their message in space to reach the eardrums in the heads of my audience. However, my students do not 'hear' the pressure-waves in the surrounding air which acted as the transmitting medium, but rather the 'tones' corresponding to the frequencies to which the original message was Fourier-analysed by their inner ear, that wonderful analogue–minicomputer (which no engineers have come within miles of matching by their hardware) yet which will convert a sequence of pressure-waves reaching our outer ear into tones with the same ease as our optical nerve inverts the images formed on the retina of our eyes by the optical system focussed upon it. If I strike a chord on the keys of my piano, the Fourier spectrum of the string vibrations so produced will be 'discrete', corresponding to a limited number of tones; but if I enter a 'disco', the cacophony which will attack my eardrums will be identified by my inner ear as giving rise to a 'continuous spectrum' of vibrations.†

And what musicians can do with their ears, we can certainly do with our computers, equipped with programs which we can prepare for them. The 'music of the spheres' has been written for us by our eclipsing binary systems in the course of their evolution; and we 'listen' to it with our photometers. In more specific terms, the situation facing the observer of the phenomena exhibited by eclipsing variables can be compared with the function of a television camera which optically scans a three-dimensional image and transforms what it sees sequentially, at high speed, into a series of linear elements which are reassembled by

---

† Nor, indeed, is contact with the external world by means of eardrums really necessary for the Fourier analyser of the inner ear to play music for us; for the inner ear can also 'read' sheet music; and some of the greatest musicians of the past—like Beethoven or Smetana—never heard the masterpieces which they composed in the latter parts of their lives other than through visual input.

the receiver as a two-dimensional picture. An eclipsing variable represents, in principle, an analogous source of information; for as one component proceeds to eclipse its mate, a scanning takes place (by the eclipsing limb) which gives rise to characteristic changes of light that can be observed at a distance. Our eclipsing variable can, indeed, be regarded as an elementary one-channel television system, transmitting continuous light signals which our detector (i.e. photometer) records sequentially; and the aim of the receiver must be to reconstruct, from time-variations of the intensity of one picture-point, a time-independent two-dimensional picture of the system. Needless to say, the simplicity of the receiver is bound to add to the complexity of the interpretation of its messages: the task is more complicated, but it can be done.

Or, to use another parallel, in close binary systems, the photometric proximity effects between eclipses are caused by a superposition of partial tides ('diurnal', 'semidiurnal', etc) which sweep around in front of our view with discrete characteristic frequencies. A spectroscopist would say that the variations of light arising from this cause are 'multiplexed' by Nature, and reach us along the same (optical) channel (i.e. our photometer) with different frequencies. The task of the analyst is to separate ('unscramble') these frequencies by a suitable (mathematical) procedure, in order to reconstruct their message. The techniques of 'modulation' of the light changes for the removal of photometric proximity effects are familiar enough to us from the field of communications engineering.

Let us add that the observational errors, which are responsible for the dispersion of individual observations around the mean light curve in the time-domain, will manifest themselves in the frequency-domain by affecting the high-frequency tail of their (empirical) Fourier transforms in a way which a communications engineer would describe as 'white noise'. And while, in the time-domain, we minimise the cumulative effects of observational errors by least-squares solutions of the respective equations of condition in the frequency-domain, this noise will be filtered out by a disregard of the high-frequency end of the respective transform (to which a Fourier series terminating after a certain number of terms will provide a suitable approximation).

A sequential receipt and registration of all the information contained in periodically-varying light signals from the stars is forced upon us by the very low frequency at which these signals are emitted. Since the orbital periods of close binary systems are, on the average, of the order of $10^5$ seconds, periodic information from them on the waves of light will reach us at frequencies generally of the order of $10^{-5}$ hertz. The messages which these systems keep transmitting with such insistence are, therefore, played in a very low key—in fact, some 23–24 octaves below that of audible sound. Moreover, the photometric eclipse phenomena translated into the frequency-domain produce a dissonant cacophony (characterised as they are by continuous frequency spectra); while, in contrast, the 'signature tune' of the proximity effects consists of harmonic chords of individual tones whose wavelengths bear integral ratios to each other.

In all these instances, an analogy between astronomical and terrestrial phenomena encountered in other branches of science or engineering is manifest; and so also should be their treatment. As has happened so often in the history of science in the past, a dissolution of interdisciplinary barriers, and a transfer of methods in one branch to bear on problems of another, is almost invariably beneficial to both. And yet, in this particular instance, no-one seems to have thought, or drawn consequences from this analogy, before I did so in the late 1950s; and the work carried out in this field since that time I consider to be the most important contribution to double-star astronomy I made in my lifetime—and without the benefit of any predecessors. Its roots may indeed go back to the years spent in the 1940s at MIT in close association with many outstanding electrical engineers; and what may have stayed in the back of my mind was the fact that what we astronomers have been accustomed to call a 'light-curve' of an eclipsing variable, Norbert Wiener would have regarded as a 'stationary time-series'; to which all theorems and paraphernalia developed for their treatment should be directly applicable.

At any rate, I openly put forward the first delayed reaction to this situation at the Tenth General Assembly of the International Astronomical Union before my colleagues of IAU Commission 42 in Moscow in August 1958 (where, I remember, my young friend Dr Plavec, who had spent some time earlier that year with me at Manchester, commented in some perplexity, 'now you have confused us again'); and I re-iterated it at the Eleventh General Assembly in Berkeley in August 1961 (Kopal 1962). What a transfer of the problem from the time to the frequency-domain could mean in this connection was developed further in a paper presented at the 'Congress for the Commemoration of the 50th Anniversary of the Death of G V Schiaparelli' in Milan, and published in its proceedings (Kopal 1960a).

Unexpected preoccupations with problems of the solar system between 1961 and 1972 (cf Chapter 5) made me, however, postpone any further developments of this work for more than a decade; it was the gentle insistence and friendly persuasion of Drs Igor Jurkevich and Alan F Petty of the US Naval Research Laboratory in Washington that induced me to return to the problem; and soon after a re-start of the earlier work the individual pieces of the puzzle began to fall quickly into a more complete picture. For more than a decade which elapsed since that time the further development of this work has remained my principal preoccupation, assisted by a number of gifted students who elected to write their PhD theses in this field, and will probably remain one for the rest of my life. An algebraisation of the Fourier analysis of the light curves of eclipsing variables (obviating the need for any special tables) was virtually completed in one day (1 August 1973) during vacations at Kandersteg in the Swiss Alps. The next major advance, resulting in recognition of the functions representing the fractional loss of light during eclipses as a particular class of Hankel transforms (with all the consequences which this entailed), occurred four years later (Kopal 1978), and helped further to establish our subject as an integral part of physical optics; and a generalisation of this to mutual eclipses of distorted stars is being completed as this book goes to the press. A summary of all the work accomplished up to the end of 1978 has appeared already in a separate volume, under the title *Language of the Stars* (Kopal 1979); but by the time this book appears the contents of the 1979 volume will already be obsolescent, and should be replaced by a new edition in due course.

In the meantime, however, many computer-minded colleagues, impatient with the Fourier analysis of the observed data furnished by our photometers, have taken to a construction of 'synthetic' light curves from assumed values of the elements; in the belief that should they identify (by trial and error) sets of elements which match the observed light curves, they will have established the solution of their problem. This line of thought operates indeed under the protection of the second half ('synthesis') of the Fourier theorem; and if the curve we are endeavouring to match were known exactly, the results of such work should be formally equivalent to those resulting from Fourier 'spectral analysis'—except that while the latter leads to the elements of the system by a systematic process, the former must seek it by trial and error.

However, no results forthcoming from astronomical measurements are infinitely numerous or precise; instead, they are subject to observational errors which we may strive to lessen, but can never completely eliminate. And, in order to ascertain their effects on the elements we seek to determine, we must commence our work with an analysis of these observations, and not with a synthesis based on assumed values of what should be the outcome in our work. In approaching this outcome by 'spectral analysis' of the given data, we not only can arrive at the desired results by a logical process describable in algebraic terms, but we can ascertain also the uncertainty due to the limitations of the observational data underlying our analysis. On the other hand, *mere numerical 'modelling' of the light curves cannot guarantee the uniqueness of the outcome*. Such work may furnish particular sets of the elements which can do so, but will not disclose the range of other combinations of the elements by which the light changes can likewise be represented within the limits of observational errors.

In point of fact, the smaller the intrinsic determinacy of the case from available evidence (photometric, spectroscopic), the easier it should become to reproduce the observed evidence by a wide range of combinations of the elements; and this range may indeed encompass the trial values adopted to begin with; thus engendering spurious confidence in easy success. In cases bordering on indeterminacy, however, merely to establish that a certain set of the elements can reproduce the observed light changes within the limits of observational errors may still mean very little from the physical point of view; and the method of approach will give us no warning that this may be the case. So the less determinate the solution happens to be, the easier it becomes to match the observations by a 'right combination of wrong elements' — as more than one investigator has learned to his sorrow the hard way.

In more general terms, what really matters is the total *probability p* with which the respective elements are defined to lie within a given range. To summarise this in the language of formalism customary in the theory of probability, let $x$ stand for any one element (or combination thereof), deduced on the basis of a 'hypothesis' (or model) $y$. Let, moreover, $p(x)$ stand for the probability of the element in question, and $p_x(y)$ the probability of the hypothesis ('model') on which the analysis has been based. Then the total probability $p_y(x)$ of the element $x$ based upon the model $y$ is given by Bayes's equation of the form $p_y(x) = p(x)p_x(y)$.

Numerical construction of 'synthetic' light curves is, in this sense, equivalent to a 'pattern recognition' process, in which the probability $p_x(y)$ is arbitrarily set to 0 or 1 — a process which (in the latter case) is bound spuriously to increase the degree of confidence of the results. It is well to keep in mind that *any* process of reduction can only *lose* information contained in the observations (i.e., increase 'entropy'), but never generate new information. This entropy indeed becomes a minimum if we set (arbitrarily) $p_x(y) = 1$ (i.e. 'recognise' the system); but to what extent is such a procedure physically justifiable? Indeed, the obsession with a mere reproduction of the observations by *some* kind of elements has been with us like a 'hereditary sin' since almost the beginning of our century, turning a blind eye on the question of whether or not such an outcome can be regarded as unique — or, in blunter language, if it represents anything more than a 'computer-game'.

To make it mean something more (whether by Fourier 'synthesis' or 'spectral analysis') requires, of course, some prior knowledge of the general type of the celestial systems whose characteristics we are trying to decipher. To proceed blindfold from the beginning without some appeal to the physics of the task confronting us would be like attempting to decipher a message in an unknown language written in an unknown script—a task obviously impossible to accomplish on the basis of fragmentary data. In order to get on the right track to begin with, some knowledge of the general type of phenomena we are out to interpret is an obvious prerequisite.

The first element of the puzzle was already supplied in 1783 by Goodricke who rightly foresaw that Algol's light changes 'could hardly be accounted for otherwise than... by the interposition of a large body revolving around Algol...' (Goodricke 1783)—a hypothesis which had to await a hundred years for its independent (spectroscopic) confirmation. 'The wonderful train of consequences that can be drawn from such regular occultations (of Algol) should engage our utmost attention' wrote William Herschel in a preface to this branch of double-star astronomy more than 200 years ago (cf Herschel 1783); but this train of consequences too was rather slow to unroll. A meaningful pursuit of such a train required, of course, a connection to be established between the *geometry* of the proposed model and the underlying *physics* of close binary systems, relating geometry with their physical characteristics. This means, in principle, relying on information obtained through other channels than photoelectric photometers; and it is this information which will furnish the additional clues we need to solve our problem of celestial cryptography: namely, a knowledge of the way in which stars in the sky are likely to behave if placed in close proximity to each other, and which form stars of arbitrary internal structure should assume in their state of *hydrostatic equilibrium*.

It is of historical interest to note that an answer to this question is provided by a theory whose elements antedate the discovery of the eclipsing nature of Algol by at least 40 years; and the patron saint of this line of enquiry is Alexis Claude Clairaut (1713–1765), a remarkable personality and worthy *confrère* of other great mathematicians in that 'century of genius' (see figure 6.8). Born in Paris on 7 May 1713, he presented his first paper to the French Academy at the age of thirteen, and at eighteen he was elected to membership of that august body. His *Théorie de la Figure da la Terre tirée des Principes de l'Hydrostatique*, work which laid down the foundations of the theory of hydrostatic equilibrium of self-gravitating fluid bodies (and which, according to Ernst Mach, inaugurated the rise of mathematical hydrostatics as a scientific discipline) did not appear until 1743, when Clairaut reached the relatively advanced age of thirty; but he is known to have harboured the essential ideas of his method already by the time of his election to the Academy twelve years before.

Clairaut's death in 1765 brought to a premature end the career of one

432  *The binary stars*

of the truly remarkable mathematicians of the eighteenth century; and further development of the subject passed to the scholars of subsequent generations—among whom the contributions of Legendre and Laplace should in particular be mentioned. As is evident from the title of Clairaut's treatise, the applications he had in mind were primarily geophysical; and with the publication of the fifth volume of Laplace's *Mécanique Céleste* (Paris 1825) the development of the theory of the form of self-gravitating fluid bodies was developed to quantities of first order of their distortion; and it took a better part of the nineteenth century before Callandreau (1889) and Darwin (1900) set out to extend Clairaut's theory to terms of second order.

**Figure 6.8** Alexis Claude Clairaut (1713–1765) of the Académie des Sciences in Paris. The buildings of the Observatoire de Paris (figure 3.5) can be seen in the background of this woodcut.

The reason why progress in this field was so slow went back to the smallness of the second-order terms for the Earth, which is flattened at the poles by axial rotation by 22 km; but (hydrostatic) deformation of the rotational spheroid was found to amount to only a few metres—

i.e., to at most one part in $10^6$ of its dimensions. However, in close binary systems, in which the heights of tides raised by each component on its mate are comparable with the dimensions of these stars as a whole, it is quite a different matter; and first-approximation theory will no longer be sufficient.

It must have been a prescience of my future needs on the part of my old teacher Professor Nechvíle in Prague (who otherwise knew nothing of eclipsing variables) to introduce me to Clairaut's theory in his university lectures; and the rest I learned on my own in my undergraduate years. Some of my earlier papers communicated to the Royal Astronomical Society dealt with my first contributions to this theory (cf Kopal 1938), which the late Professor Milne (who presented them in my absence) described as 'breaking a new ground in getting away from polytropic distributions to allow for the effects of rotation in the stars' (cf Milne 1938); and today one can only blush to see how these have dated. But, at that time, I was encouraged by Milne's acknowledgment that Clairaut's theory was more general than those developed a decade earlier by von Zeipel (1924) and Milne himself (Milne 1923); and which nine years later were so successfully developed by Milne's pupil Chandrasekhar (1933).

The work of Milne and von Zeipel in 1923 and 1924, as well as the 1933 investigations by Chandrasekhar (which, a year later, were used by Takeda in his pioneer work on the photometric effects of gravity-darkening), was restricted to quantities of first order in surficial distortion; and referred, moreover, to polytropic gas spheres only. In the years which elapsed since that time I have returned, however, to Clairaut's theory to investigate with its aid the effects of rotational as well as tidal distortion of stars of arbitrary structure to terms of second order in my book on *Figures of Equilibrium of Celestial Bodies* (Kopal 1960b); and my pupil Dr Rahimi-Ardabili (1979) has extended our knowledge of the effects of tidal distortion to quantities of third order; while a third-order theory of rotational distortion (Kopal 1973) was subsequently extended by us at Manchester to terms of fourth order (cf Kopal and Kamala Mahanta 1974). And there the case rests at the present time; for further refinements of Clairaut's theory must be left to more courageous individuals of the next generation.

What was the impact of Clairaut's theory of the figures of equilibrium of self-gravitating fluids on the astronomy of close binary systems? For a long time almost none at all; for although this theory makes it evident that the components of a close binary system must assume the figures appropriate for the prevalent field of force (so that, therefore, the shapes of both stars should generally be expected to be dissimilar), little note was taken of this fact for a long time. Thus, for Pickering in 1880, both components of Algol still were spherical (we know today that this is far from being the case); and when in β Lyrae or other similar systems the photometric effects of distortion could not be ignored, its components were regarded by Russell in 1912 as similar ellipsoids—a case whose consequences could be treated by secondary-school geometry, and never mind the physics! Actually the first astronomer who pointed out that the components of Algol (and in Algol-like systems in general) had to be very dissimilar in form was the German astronomer Kurt Walter (1931); but neither he, nor anybody else at that time, could deduce the photometric consequences of this fact—not only between minima, but also within eclipses. To do so eventually became the task of the present writer—a work which commenced in 1940 (under the stimulus of Takeda's second paper) and has continued ever since.

But what about the future? In looking back at the development of our subject over the past fifty years, I am more than ever convinced that the future of the analysis (including error analysis) of the observed light changes of eclipsing variables lies in the frequency-domain; because in that domain the observed data and the model of the system are connected with each other more directly and in a mathematically more satisfying manner. However, the code of this cryptographic work will have to be continually improved and refined, not only to incorporate the effects of minor phenomena predictable from the theory, which may be observationally unimportant at the present time, but become significant as the quality of observations continues to improve, but, above all, to take account of possible departures of the components from the figures of equilibrium which have been postulated (on grounds of simplicity) too freely in the past, and on which our present codes are based.

For example, it has traditionally been assumed so far that the fractional dimensions of the components of close binary systems remain significantly constant over thousands, or tens of thousands of orbital cycles. But although such components may indeed remain in the state of equilibrium, they may *oscillate* about it with frequency depending on their internal structure; and if so, the ratio of their radii would not remain constant but oscillate with periods comparable with that of their orbit. Moreover, if—as is more than likely—the orbital period of the system is not commensurable with those of free oscillations of the components, photometric beat phenomena will arise which make the shape of the light curve not repetitive from cycle to cycle. Observations are already on hand, for such well-known totally-eclipsing systems as U Sagittae (cf Olson 1982), indicating that this may indeed be the case. A theory of the photometric consequences of such oscillations has also been developed to a certain point (cf Kopal 1982); but its application to practical cases still constitutes a task for the (albeit near) future.

# References

Chandrasekhar S 1933 *Mon. Not. R. Astron. Soc.* **93** 390—404, 449-461, 462-471, 539-574
Callandreau O 1889 *Ann. de l'Obs. de Paris* **19** E
Clairaut A C 1743 *Théorie de la Figure de la Terre, tirée des Principes de l'Hydrostatique* (Paris)
Clegg F E 1961 *The School Science Review* no. 147
Darwin G H 1900 *Mon. Not. R. Astron. Soc.* **60** 82—124
Fetlaar J 1923 *Utrecht Recherches* no. 9 (Dissertation, Utrecht)
Goodricke C A 1913 *Mon. Not. R. Astron. Soc.* **73** 3
Goodricke J 1783 *Phil. Trans. R. Soc. London* **73** 474-482
Harting J 1889 'Untersuchung über den Lichtwechsel des Sternes $\beta$ Persei' (Dissertation, Munich)
Hartwig E 1900 *Astron. Nachr.* **152** 309-310
Herschel W 1782 *Phil. Trans. R. Soc. London* **72** 82-111.
—— 1783 in an informal report 'Observations upon Algol', read before the Royal Society of London, but not printed until *The Scientific Papers of Sir William Herschel*, London 1912; vol. **I** p. cvii
—— 1802 *Phil. Trans. R. Soc. London* **92** 477-528
—— 1803 *Phil. Trans. R. Soc. London* **93** 339-382
Huyghens Chr 1698 *Cosmotheoros* (The Hague) (*Oeuvres*, vol. **21**)
Kopal Z 1932a *Astron. Nachr.* **245** 335-338
—— 1932b *Astron. Nachr.* **247** 117-120
—— 1938 *Mon. Not. R. Astron. Soc.* **98** 589-597
—— 1941 *Astrophys. J.* **94** 145-158
—— 1942 *Proc. Am. Phil. Soc.* **85** 399-431
—— 1959 *Close Binary Systems* (London: Chapman and Hall and New York: Wiley)
—— 1960a in *Atti del Convegno per le Celebrazion di G V Schiaparelli* ed F Zagar (Milano) pp 156-161
—— 1960b *Figures of Equilibrium of Celestial Bodies* (Madison: Univ. Wisconsin Press)
—— 1962 in *I.A.U. Transactions* ed D H Sadler (New York: Academic Press) p. 369
—— 1973 *Astrophys. Space Sci.* **24** 145-174
—— 1974 *Astrophys. Space Sci.* **27** 389-418
—— 1975 *Astrophys. Space Sci.* **34** 431-457
—— 1977 *Astrophys. Space Sci.* **50** 225-246; **51** 439-460
—— 1978 *Dynamics of Close Binary Systems* (Dordrecht: Reidel)
—— 1979 *Language of the Stars* (Dordrecht: Reidel)
—— 1982 *Astrophys. Space Sci.* **87** 149-191, **88** 313-329
Kopal Z and Kamala Mahanta M 1974 *Astrophys. Space Sci.* **30** 347-360

Laplace P S 1825 *Mécanique Céleste* (Paris) vol. 5
Michell J 1767 *Phil. Trans. R. Soc. London* **57** 234–264
—— 1784 *Phil. Trans. R. Soc. London* **74** 35–57
Milne E A 1923 *Mon. Not. R. Astron. Soc.* **83** 118–147
—— 1938 *Observatory* **61** 10
Myers G W 1898 *Astrophys. J.* **7** 1–22
Olson E C 1982 *Publ. Astron. Soc. Pacific* **94** 70–75
Pannekoek A 1902 'Untersuchungen über den Lichtwechsel Algols' (Dissertation, Leiden)
Pickering E C 1880 *Proc. Am. Acad. Sci.* **16** 1
Piotrowski S L 1947 *Astrophys. J.* **106** 472–480
Porro A 1891 *Astron. Nachr.* **127** 41–42
Rahimi-Ardabili M Y 1979 *Astrophys. Space Sci.* **66** 325–340
Roberts A W 1903 *Mon. Not. R. Astron. Soc.* **63** 527–549
Roediger C 1902 'Untersuchungen über das Doppelsternsystem Algol' (Dissertation, Koenigsberg)
Russell H N 1912 *Astrophys. J.* **35** 315–340, **36** 54–74
——1942 *Astrophys. J.* **95** 345–355
Russell H N and Merrill J E 1952 *Princeton Univ. Obs. Contr.* no. **26**
Russell H N and Shapley H 1912 *Astrophys. J.* **36** 239–254 and 385–408
Sitterly B W 1930 *Princeton Contr.* no **11** (PhD Thesis, Princeton)
Stein J 1924 *Die Veränderlichen Sterne* (Freiburg in Breisgau)
Struve F G W 1837 'Stellarum duplicium et multiplicium Mensurae Micrometricae' *Mem. de l'Acad. Impér. de Sci. de St. Pétersbourg*
Takeda S I 1934 *Mem. Coll. Sci. Kyoto Univ.* (Ser. A) **17** 197–217
—— 1937 *Mem. Coll. Sci. Kyoto Univ.* (Ser. A) **20** 47–86
Vogt H 1919 *Heidelberg Veröff.* no. **7** pp 183ff
Walter K 1931 *Königsberg Veröff.* no. 2
von Zeipel H 1924 *Mon. Not. R. Astron. Soc.* **84** 665–683, 684–701, 702–719

# Chapter 7

# *Astronomy of the Past and in the Future*

In the preceding chapters of these reminiscences, we have dwelt mainly with people and places known to me at different stages of my life; or with events which occurred within my lifetime (or not too long before for my contemporaries or myself to have lost touch with them). What we have witnessed, or can bear testimony to on behalf of the generations just retreating into the past, is but a very small fraction of the long history of astronomy, which has been the principal subject of this book, and whose origins in the distant past antedate the invention of writing. The night sky, and what it can disclose to the inquisitive naked eye, must have fascinated our ancestors from their cave-dwelling days with its portent; and when their knowledge began to emerge from the hieroglyphic or cuneiform records that have recently been deciphered (while others are still awaiting their turn) it disclosed a story which is an integral part of human culture. The present chapter of this book is not intended to outline this story in a coherent form; but rather to take a bird's-eye view of some of its essential parts, in so far as its extrapolation can help us to anticipate also its future.

To go back to the origins of astronomy, there is no doubt that these are to be located in those parts of our terrestrial abode where the favours of Nature (including climatic conditions) liberated at least some of the strata of human society from the incessant drudgery of the struggle for life, and permitted them to lift their eyes to the heavens. The racial origins of astronomical science—and it is essential to stress this for the sake of the historical record—are not Indo-European, but Semitic; and, geographically speaking, they are to be sought in the Near East and lands adjacent to the Mediterranean basin. Moreover, its tasks were initially descriptive—to divide the firmament of the sky into constellations; and (at a somewhat later stage) to establish the positions of their constituent stars with respect to some fixed frame of reference. The fact that a large number of stars in the sky still carry Arabic names (some of which, as attested by their occurrence in Homeric poems, must go back at least to the second millennium BC); or that a considerable fraction of astronomical terminology (in particular, as regards instrumentation) still in current use is of Arabic origin (remember such terms as zenith, nadir, azimuth; or alhidade, almucantar, etc) shows the extent of our indebtedness to Arabic culture in the Middle East. However, the greatest gift which we received from this culture is the institution which we call today the 'astronomical observatory'—a place where groups of astronomers work together, for more than one generation, advancing common goals. In this sense, the concept of an astronomical observatory represents an invention of the Arabs, and is one of their permanent contributions to human culture.

It may perhaps come as somewhat of a surprise to Western readers, as heirs to a culture rooted in Graeco-Roman civilisation, that the Greeks (or their other contemporaries) developed no such institutions for astronomical use. Yet it is a fact, based on all our knowledge of ancient astronomy, that these were completely unknown to the Mediterranean world before the birth of Christ. Not a trace of any indication has come down to us that the Academy or Lyceum of Athens, Syracuse, or Alexandria possessed any observatory. The Museion of Alexandria in Ptolemaic times represented perhaps the closest approach to scientific academies as they developed in Europe in the seventeenth century AD, and was connected with certain astronomical advances of the first magnitude (Aristarchos, Eratosthenes). But there is every indication that its philosophers worked as individuals, sharing no common platform or instrumentation; and no excavations in Alexandria (or elsewhere) have brought to light remnants of any structure indicative of astronomical use. The same was apparently true also of Hipparchos (second half of the second century BC) or Ptolemy (second half of the second century AD), down to the end of the Graeco-Roman era—until the advent of Islam.

With this event, accompanied by a release of new creative energy which Islam injected into the arena of the Near and Middle East in the seventh and eighth centuries, astronomical observatories began to make an appearance (henceforth never to vanish from the Earth) in ever increasing numbers. This was particularly true in Mesopotamia, then under the rule of enlightened caliphs of the Abbasid Dynasty. Astronomy is known to have been cultivated in the 'house of wisdom' of Harun al Rashid (763–809 AD), the fifth caliph, in Baghdad; and its study progressed greatly in the next century under Al Mamūn (813–833 AD), the seventh caliph.

In the seventh century, the same part of the world gave birth to one of the greatest astronomers of the Middle Ages: Al Battani (latinised to Albategnius), who was born before 858 AD in (or near) Harrar, flourished at Al Raqua on the Euphrates, and died in 929 AD near Samarra. Al Battani was, moreover, not the only outstanding Muslim astronomer of that age. To prove this one need only recall his close contemporary Al Sufi (in Shiraz) of the AD 964 *Uranometria* fame; and Al Biruni (†1048) or Ibn Sina (†1037) in the eleventh century (both flourishing in Mesopotamia); while Ibn Younis (†1009) practised the same art in Egypt. They all flourished at a time when Christian Europe had as yet nothing to contribute to our subject. But more was to come; for soon thereafter the world of Islam saw the foundation of the first true *astronomical observatory* in the modern sense of the word at Marâgha (south of Tabriz) in Azerbaijan.

Marâgha Observatory was founded by the famous Arab astronomer Nasir-el-Din (who died in Baghdad in 1274) under the patronage of Khan Hulaghu, grandson of the great conqueror Ghengiz khan, some time around 1261; and it lasted for not less than fifty years. It represented an endowed institution supporting a large staff of several dozen astronomers with their assistants; it possessed a working library and was (possibly) connected with a school of the subject. All this we know from its renown which reached from the shores of the Atlantic to the Pacific Ocean, as we gather from the testimony of contemporaries as far removed from each other as Roger Bacon in Oxford and the Chinese visitors in the second half of the thirteenth century who described Marâgha for posterity. It was also the first observatory which survived its founder; and after Nasir-el-Din's death in 1274 was directed for some time by his sons. It flourished under seven rulers; and although it ceased functioning effectively some time after 1316, its buildings were still extant when the place was visited by Ulugh Bek in the next century.

The second great observatory of the Islamic world was founded, a hundred years later, in Samarkand, by Muhammad Turgay Ulugh Bek (1394–1449), grandson of another great Mongol conqueror Timur Leng (1369–1405), who under the latinised name of Tamerlane struck horror and fear in Christian souls of eastern Europe in the second half of the fourteenth century. His grandson and successor once removed, Ulugh Bek, did not inherit Timur's ferocity and war-like spirit; instead, as recorded by his chroniclers in Latin, *'fuit Rex iustus, doctus, perfectus praesertim in mathematicis scientiarum et eiusdem cultores dilexit'* (see figure 7.1).

Ulugh Bek's interests in astronomy must have been awakened early in his life, while he was still prince and heir-apparent to his father Shah-rukh; at least work on the building of the observatory in

Samarkand was started some time between 1420 and 1427. The observatory itself functioned till about 1460 when it commenced to fall into oblivion, and its buildings were largely dismantled about fifty years after Ulugh's death. Its remnants (in particular the 50 m mural meridional quadrant) were, however, excavated at the beginning of our century by the Russian archaeologist Vjatkin and his school.

**Figure 7.1** Medresse of Ulugh Bek in Samarkand.

The principal achievement of Ulugh Bek and his astronomers (for he must have had a very large staff of assistants) was the *Catalogue of the Positions of 1018 Stars*, published in Arabic in 1437 during the reign of Shah-rukh, ten years before Ulugh himself ascended the throne. Such became the reputation of this truly monumental work that it was translated into Persian in 1498; its first English translation, by Thomas Hyde (Bodley Librarian) appeared in Oxford in 1665; the second Oxford edition (by G Sharpe) appeared in 1767; while the last critical edition of Ulugh's work was brought out by the Carnegie Institution in Washington as recently as 1917!

Ulugh Bek succeeded to the throne of his ancestors in 1447 as the ruler of a mighty empire. He was the second medieval astronomer (after Alfonso X of Castile) to become a head of state, but he did not enjoy this exalted position too long: for only two years later he was expelled (and murdered while in flight) by his own son! We should not, however, regard this ghastly turn of events only as the victimisation of Ulugh's intellectual inclinations, or a political change of regime. For the son, in his turn, proved to be an able ruler; and he continued to support the work of his father's observatory for at least another ten years. It seems, rather, that the origins of the feud between father and son were astrological (a superstition to which Muslims of that time were at least as prone as their Christian contemporaries): they may have misread in the stars that their co-existence was mutually incompatible—a belief which did not stop short of patricide. If so, it would have been neither the first, nor the last act of violence perpetrated in the name of the stars!

It was my good fortune, during our visit to Samarkand in 1958 (cf p. 283) to stand before the graves of Timur Leng, Shah Rukh and Ulugh Bek; and think of what they concealed. Grandfather Timur (whose probable effigy was reconstructed by Soviet scholars from the remnants of his skull) was a short and formidable-looking old man, who lived at the time of the Turkish onslaught against Europe. We were shown (copies of) the letters by which the Pope of Rome at that time implored the old conqueror (in elegant Latin; to which Timur replied in Tadjik) to relieve the Turkish pressure against western Christendom by attacking the Turks in the rear! But the wily old man would not be lured into the trap of pulling other people's chestnuts out of the fire, and left the Turks alone.

In his old age, Timur apparently tested (more than once) the loyalty of his entourage by pretending to be dead and listening to what was said about him (with fatal consequences for some), and so terrified his court that when he eventually did die (in 1405) nobody would believe it! In contrast, his grandson Ulugh Bek (whose effigy was similarly reconstructed; and is probably the oldest astronomer of whom we know what he looked like) was apparently a man more trusting of his family and friends; and that is why he did not live so long.

At any rate, the decline of his once-famous observatory in Samarkand, which in the late Middle Ages must have enjoyed a reputation (if only by the size of its instruments) equal to that of the Palomar Mountain Observatory in our own time, occurred in the 'Indian summer' of Islamic culture as far as science was concerned.

The last observatory of the Islamic world which we wish to mention here—the Istanbul Observatory—was founded in 1575 by Taqui-al-Din under the Sultan Murad III at the time of Islam's intellectual decline. It also did not survive very long; for in 1580 it was blasted to pieces by naval ordnance on the orders of the same sultan, on the grounds that the observatory's task of correcting the astronomical tables of Ulugh Bek had been accomplished. Once this was done, the observatory was thought to be needed no more.

This was indeed a sad epilogue to a once great astronomical culture nurtured by the Muslim world, in which (in the words of Mohammed) 'the ink of the scholar was more precious than the blood of its martyrs.' The world of Islam, especially at the peak of its intellectual creativity between the eighth and twelfth centuries, indeed made great contributions to observational astronomy (ending with the work of Ulugh Bek). However, truth compels us also to say that, in the interpretation of their observations, the Islamic scholars did not progress beyond their Greek heritage. In particular, their planetary theories remained strictly geocentric—which was fully true of the work of their last outstanding exponent, the Damascene astronomer Ibn-al-Shatir (1304–1376); his planetary theory (but recently brought to light) is only an elaboration of the Ptolemaic system. In no way was the Islamic astronomer ever touched by the heliocentric heresy of Aristarchos (third century BC); to revive this heresy became the historical task of Nicolaus Copernicus, a canon of the Christian Church of Rome!

But to return to our epilogue on Islamic astronomy, the ill-fated observatory at Istanbul was already a contemporary of the Uranienborg Observatory of Tycho Brahe (1546–1601), which between 1576 and 1597 flourished on the island of Hveen in Denmark—the last well-known astronomical institution before the advent of the telescope. It was founded under the direct royal patronage of the Danish King Frederick II, and did not survive much longer than the observatory of Ulugh Bek at Samarkand. By now both have become only the sites of archaeological excavations.

Like Ulugh Bek 150 years earlier, Tycho had also to make his own instruments (of very similar type, albeit different in size), carry out his observations, and print his own publications. However, while Ulugh Bek lived in the twilight of the Islamic world (although this was not apparent to him or his contemporaries), Tycho's life reached almost to the watershed dividing ancient astronomy from its modern times—the time of Galileo and Kepler.

How did the Muslim concept of astronomical observatories become eventually transplanted to the western world? At the time of Ulugh Bek, there was nothing as yet in Europe to match the grandeur of the observatory at Samarkand. With the exception of Alfonso X of Castile (1252–1284), no-one had as yet thought of setting up even a private establishment to observe the sky; and the European medieval universities taught the astronomy which their scholars had learned from the ancient Greeks (transmitted through Arabic channels). New interest in astronomy commenced at the time of the Renaissance; but the institution of the astronomical observatory was slow to take root in Europe. Neither Regiomontanus (1436–1476) nor Copernicus (1473–1534) founded any such observatories for themselves or their followers. The first European observatory in the true sense of the word was not founded until at Kassel, Germany, by Wilhelm IV (1532–1592), Landgrave von Hesse—the last crowned head of state to get actively involved in our subject. But it too did not survive the life-span of its patron; and became derelict soon after the Landgrave's death.

But before we cross the watershed marked by the discovery of the telescope, let us mention a few words as to the causes motivating astronomical observatories and their inhabitants, from their cradle in medieval Islam to their present state; for it is only by learning of their past that we may be able to foresee their future. Can we, in particular, discern the reasons why the Near-Eastern supremacy in astronomy, manifest throughout the Middle Ages, should have been so completely lost (nay, almost willingly given up) in the second half of our millennium, and inherited by the Atlantic community of western Christendom? As is often the case, no single reason may have been responsible for this 'changing of the guard' in the Temple of Urania; but at least some can be mentioned which may have contributed to this outcome.

First, there is no doubt that in most cultures of the world (including Muslim as well as Christian) original interest in astronomy was rooted in astrology; this was particularly true among the elite and rulers (witness the role of Stoics in antiquity!). Astronomy for its own sake emerged only gradually from the shadow of its more foolish sister, which was often called upon to provide sustenance up to the time of Kepler! And, secondly, its tender plant would have had but scant chance of development if it had not been for the interest exhibited in those days by the more enlightened section of the clergy of either religion. For, throughout most of the past, the clergy—Muslim as well as Christian—represented the only significant educated element of the population; and also one sufficiently endowed with worldly goods to be able to spend some time on disinterested pursuits.

An objection may be raised that the principal aim of the clergy (of any religion) is to preserve the purity of faith, and transmit it to posterity in unadulterated form. This is indeed the orthodox view held by most religions at different times of their development; and the backwardness of medieval Europe can be largely ascribed to this cause. However—and this is essential—history has shown that each clergy harbours at times in its ranks a creative minority, capable of giving thought also to other than orthodox views on matters concerned with this world. Moderate deviations from orthodoxy could be tolerated; though ideas too far ahead of their time were often subject to derision, and their champions persecuted.

Whatever their vicissitudes may be, it is, however, impossible completely to submerge new ideas. Now, it seems to me, at least one of the reasons for the current ascendancy of our Atlantic civilisation is the fact that, with the advent of the Renaissance, Christian society had developed a greater degree of tolerance towards unorthodox ideas precisely at the time when Muslim society gradually forgot the noble message of its Founder, or the deeds of some of his followers in the seventh and eight centuries, and relapsed into sterile orthodoxy. New ideas took root in Europe at that time which subsequent counter-reformations (Catholic as well as Protestant) were unable to extirpate; and so they survived (at least in clandestine form) to the time of Enlightenment and our own days. In particular, at the time of the Renaissance our ancestors rediscovered (and eventually brought to efflorescence) the *experimental method* of questioning Nature—a method contrasting starkly with the petrified dogma of 'It is written', which continued to hang like a millstone around the necks of the Muslims. In some parts of the world of Islam, this attitude has prevailed to our own times; and is largely responsible for most of the problems (including economic and military ones) which beset the Muslim world today.

In contrast, a Western student of science is inclined to regard the Holy Scripts of any religion only as a mirror of the times in which they were conceived (or codified); and does not shirk from textual criticism to interpret their meaning. For him the only Holy Writ is the open book of Nature, so far but incompletely understood—so incompletely that, to most questions raised by an inquisitive mind, our answer still remains: we do not know! We have learned to live with the fact that while solutions of many problems can be advanced in months or years, and others in decades, there are others that may require centuries or millennia for a fuller understanding. Who are we to presume that we should know everything today; or that we (or any of our ancestors) were the privileged generation to which everything had been revealed?

Modern students of science do not seek answers to their problems in past revelations or in the alleged wisdom of their ancestors. In fact, guided by experience of the past, they are highly sceptical of revelation (as distinct from intellectual inspiration) as a legitimate method of scientific enquiry; and look for the solutions of their problem to the future—not to the past! This state of mind (the roots of which go back to the European Renaissance) is very different from that in which men's minds were dominated by the heavy hand of the past. Modern man, who sets his life's goals in the future, exhibits indeed an attitude of mind completely alien to that of his ancestors. Ancient Egypt, Greece, Rome, or the vast Asiatic civilisations of India or China did not look ahead for the inspiration of their existence, but sought to find it in their prehistoric origins, in their ancient glories, their fabled heroes or pristine virtues. Unlike modern man who dreams of worlds to conquer, for his ancestors man's 'golden age' was in the past, not in the future! And, in this sense, one could say that the needle of the compass indicating the way to the future for Christian and Muslim societies swung in the opposite direction around the year 1500 AD; while that of Buddhist society and its offshoots has remained stagnant for almost 2500 years.

Nothing perhaps illustrates the extent of the dichotomy between Christian and Muslim society in more recent times better than the difference in their attitudes towards the concept of the 'laws of nature'. Since at least the time of the Renaissance, practising Christians accepted (tacitly, if not openly) the view that once God created the world, the laws by which the world is run remained fixed and immutable (at least the incidence of occasions when God is supposed to have suspended their validity to glorify his saints by miracles has been diminishing faster than exponentially with the passage of time). In contrast, up to quite recently, the leaders of the Muslim faith have accepted free changeability of the laws of nature according to God's instantaneous whims or intentions—with an outcome reminiscent of the effects produced by the whims of pagan gods in ancient times, working often at cross-purposes with each other, the solution of their problems being left to a *deus ex machina*. If this were the case, a study of the laws of nature could not disclose anything about God and his works except his immediate and possibly temporary intentions—a fact not overly conducive to the scholastic pursuit of science.

It can perhaps be objected that the Christian, Muslim or Buddhist cultures are basically different, resting as they do on different spiritual foundations. Different they are indeed; and their diversity should be cherished by all men of good will. However, *science is one*, representing as it does the only activity of the human mind which is truly *cumulative*, and one aspect of our existence *which all cultures should possess in common*. Needless to say, we are still at some distance from this goal; and how rapidly it will be approached only the future can tell.

Astronomy is indeed one of the oldest sciences which germinated in the human mind millennia ago; and a retrospective look at its age-long story reveals that its course has been like that of a meandering river, with long and quiescent periods of slow gestation interrupted now and then by rapids, reached as a rule when the advances of human technology provided astronomers with new tools of the trade with which to enlarge our horizons. One such dramatic period we have been privileged to experience in our lifetime with the advent of the Space Age; but the one which really marked the watershed between ancient and modern astronomy sprang from the discovery of the *telescope* in the first decade of the seventeenth century. This story deserves to be recalled by more than a few words.

The circumstances surrounding this discovery remain still (to some extent) shrouded in a mystery which does not lack its dramatic elements. It is, however, certain that, towards the end of the first decade of the seventeenth century, telescopes suddenly appeared in human hands in several places of western Europe—Middelsburgh in Holland, Paris, Venice; though whether their discoveries were independent, or due to rapidly spreading intelligence of the fact, is very difficult now to ascertain.

This is all the more true as, from the very beginning, the telescope was not regarded primarily as an instrument of scientific research, but one which lent itself also to other eminently practical uses. Listen to what Galileo Galilei (1564–1642), one of the principals of this episode in the history of science, wrote to the Doge of Venice in August 1609 in commending to him this novel invention:

> The power of my *cannocchiale* to show distant objects as clearly as if they were near should give us an inestimable advantage in any military action on land or sea. At sea, we shall be able to spot the enemy warships and their flags two hours before they can see us; and when we have established the number and type of the enemy's craft, we shall be in a position to estimate his strength and decide whether to pursue and engage him in battle, or take flight. Similarly, on land it should be possible from elevated position to observe the enemy camps and their fortifications; and even in open field we should be able to see all his movements or preparations and follow them in detail.

And less than a year later, when Galileo was seeking to exchange the academic life of Padua for Court service in Florence and was soliciting a position of the Principal Mathematician to the Grand Duke of Tuscany, this is what he wrote on 7 May 1610, to the Grand Duke's Prime Minister, Belisario Vinta:

> I have many and most admirable plans and devices; but they could only be put to work by Princes because it is they who are able to carry on Wars, build and defend Fortresses, and for their regal sport make most splendid expenditures.

We do not perhaps need to continue the quotations to gather that Galileo was a shrewd man, wise in worldly affairs, who would be quite capable of drafting suitable research proposals to Defence Departments of any one of the Great Powers which displaced Venice and its contemporary rivals from the seven seas of this planet in our own time. *Plus ça change, plus c'est la même chose.*

Was it the reported appearance of a telescope at the Rialto in the spring of 1609, or the news received by Galileo from Paris at about the same time, which drew his attention to the problem? Whatever happened, it is certain that soon thereafter he was able to construct superior tubes of his own making; and when he turned them to the sky—he gasped in awe! The autumn and winter months in Padua of that year, when the human eye beheld so many heavenly wonders for the first time, must indeed have been an enchanted period in the history of astronomy; and how we regret that Galileo was not endowed with the pen of a Kepler to recapture for us the excitement he must have felt when writing an account of his observations. 'Oh, much-knowing perspicil', exclaimed Johannes Kepler (1571–1630) enthusiastically on the pages of his *Dioptrice* (1610), 'more precious than any sceptre! He who holds thee in his right hand is a true king, a world ruler....'

Fuller details of these events have been amply told elsewhere; but a few words may be added on the optical properties of these early telescopes. Students of the history of science will recall the amount of incredulity and opposition which greeted Galileo's early telescopic discoveries. Was it due only to contemporary inertia of thought or professional prejudice? Unqualified answers in the affirmative often given by the historians overlook, however, the fact that the telescopes in the hands of Galileo's colleagues were probably—at least in the earliest days of telescopic astronomy—very inferior to those of his own. For, in spite of Galileo's assertion that he invented the telescope 'through deep study of the theory of perspective', there is little doubt that his process was essentially one of trial and error; and so were his methods of producing the lenses. Needless to stress, no methods for grinding and polishing optical surfaces had yet been developed by 1610. Galileo appears to have succeeded in this task better than his contemporaries, and scored his observational triumphs as much by the skill of his hands as by the fact that his mind was prepared to accept what he saw.

Galileo is reported to have made many telescopes; but knowing him as we do we shall probably not err in surmising that he kept the best ones for his own use. In a letter to Belisario Vinta of 19 March 1610, Galileo reported that out of one hundred and more glasses which he had ground 'at great fatigue and expense' only ten were able to show the newly discovered satellites of Jupiter; and his best telescope, which Galileo called affectionately the 'Old Discoverer' (still preserved at Arcetri near Florence) magnified only thirty times. May these facts not explain partly why his early telescopic discoveries—mountains on the Moon, phases of Venus, or the satellites of Jupiter which paved the way for acceptance of the Copernican system—were so slow to be confirmed by others?

The road of further development of astronomical telescopes was rather circuitous and, in retrospect, replete with misunderstandings. In the years following the generation of Galileo, the principal obstacle to increased telescopic power was seen in the chromatic aberration of the objective, which increases with the curvature of its surface. The easiest way to lessen it seemed to be to diminish this curvature and increase the focal length. As a result, telescopes grew inordinately in length at first, to usher in the first geological age of optical dinosaurs characterised by small heads on huge bodies. The apertures of their objectives seldom exceeded 6-8 inches; but their focal ratios became extremely large, leading to focal lengths in excess of those of most telescopes existing at the present time. Thus the telescope with which Hevelius (1611-1687) at Danzig carried out most of his observations of the lunar surface had a focal length of 158 ft!

Needless to say, telescopes of such great lengths could be but crudely mounted. Astronomers of that time had mostly to dispense with any kind of a tube (a series of diaphragms being a poor substitute for keeping away stray light), and the objective was often mounted at the end of a long pole, directed to different parts of the sky by means of ropes and pulleys. Sometimes, in desperation, the astronomer dispensed with the mounting altogether, and fixed his objective to the roof of a building, waiting on the ground for a transit of his celestial object literally with an eyepiece in his hand. No wonder that Hevelius, under such circumstances, preferred the unaided eye for the measurement of stellar positions to the end of his life. This was the truly heroic age of observational astronomy—the age of Hevelius, Huygens or the Cassinis—and their discoveries (rings and satellites of Saturn, motion and maps of the Moon, etc) are not seen in proper perspective until one considers the crude telescopic means at their disposal.

In the end, the long-necked telescopic dinosaurs of the second half of the seventeenth century vanished from the scene under their own weight as much as under the impact of new developments in astronomical optics which had taken place in the meantime; and one of them was the gradual introduction into practical use of the astronomical reflector. The idea of such an instrument was already known to Galileo, and described by him, through the mouth of Sagredo, in his *Dialogues on the Two Great World Systems* (1633); but it was not translated into practice until 1671 by Isaac Newton. Newton's instrument, of a type called after him (and still in the possession of the Royal Society), was too small for active celestial exploration; its principal mirror was only 37 mm in diameter, it was of 16 cm focal length and magnified 38 times (only a little more than Galileo's 'Old Discoverer'). Even though modest in size, however, it translated for the first time into practice the germ of an idea conceived half a century before in Florence, and which grew up to produce the giant reflectors erected in the second half of this century—instruments which have already started to continue their careers in space.

These developments were, however, slow to unroll; with many

detours on the way. Besides his positive contribution to astronomical optics in the form of the first reflecting telescope, Newton also left our science a negative legacy in the form of a mistaken assertion that it was impossible to achromatise a convex lens. The incorrectness of this assertion was, to be sure, proved in 1733 by Chester More Hall, and the first achromatic objective was actually produced by John Dollond around 1759. Such, however, was the weight of Newton's authority (as well as technical difficulties in producing achromatic objectives of larger size) throughout the eighteenth century that the pendulum continued to swing from dioptric to catoptric systems, and the stage was set for the first period of efflorescence of the astronomical reflector in the lifetime of William Herschel (1738–1822).

The story of this one-time musician and organist, who relatively late in life turned to astronomy to become one of the greatest observers of all time, has been told so often as hardly to call for a repetition in this place; but some remarks must be made at least on his achievements in astronomical optics. In the 1780s, at the time when he turned from music to astronomy, there were as yet no professional telescope-makers to whom one could turn with an order for any but the smallest telescopes; and practising astronomers were still very largely their own customers. Production of optical glass was still in its infancy; but the casting of metallic discs which could be polished into the shape of a mirror represented a much easier technical task, and one at which Herschel became singularly adept.

His *chef d'œuvre* was the famous 20 ft telescope of 18 inch free aperture and $f/13$ focal ratio; and it was mainly with the aid of this excellent instrument that Herschel at last '*coelorum perrupit claustra*' and opened to his contemporaries, concerned still essentially with solar-system astronomy, the vistas of a much vaster world. His subsequent *magnum opus* a 40 ft reflector (of 48 inch aperture; completed in 1789) was, to be sure, never a full success in the technical sense of the word; for the technological difficulties encountered in this work were still insuperable in Herschel's time (and remained so for another hundred years).

However, as in the case of Galileo 200 years before, the technical quality of Herschel's instruments was only part of the secret of his observational achievements; the person of the observer was another. To convince ourselves of it, it should be sufficient to cast a glance at the list of telescopes which Herschel supplied to others. It would perhaps be too much to expect any discoveries to be made with a 25 foot telescope furnished in 1810 to the King of Spain (the excellent optics of which is still in the possession of the Observatory of Madrid), or by other instruments of similar power ordered by such august customers as Empress Catherine of Russia, or Emperor Francis I of Austria. What was, however, accomplished with two costly mirrors supplied by

Herschel to Greenwich (and one to Oxford)? And who else but Herschel would have been able to keep a star in the field of view under a magnification of 1000–2000 (or even greater) with the mounting and means of control at his disposal?

Herschel's observational triumphs between 1780 and 1820 inaugurated the first great era of astronomical reflectors. But, alas, this era proved to be short-lived; and its duration limited with the decline of the aging astronomer, when it largely spent its momentum. It is true that the achievements of his successors, such as his son Sir John Herschel, W Lassell or Lord Rosse, who eventually produced reflectors exceeding Herschel's in size as well as performance, command our respect even today. But, at the same time, new developments in glass technology had opened up new possibilities for the development of refracting telescopes. Thus Guinand (1799) in Switzerland, Feil and others had at least mastered the art of producing flint glass of the necessary optical qualities; and these, in the hands of Fraunhofer and his successors, rapidly secured ascendancy for the astronomical refractor, which was to last till the end of the nineteenth century.

The Dorpat objective which founded Fraunhofer's fame in 1824 was, to be sure, only 10 inches in diameter; and still by 1865 the largest existing objectives (at Harvard and Pulkovo) possessed apertures no larger than 15 inches. Soon thereafter, however, refractors of ever-increasing size began to come from the hands of Alvan Clark and his successors, culminating in the 40 inch Yerkes refractor in 1897. Today, ninety years later, this refractor still remains the largest of its kind, as a witness to the fact that the evolution of its line once more became bogged down under its own weight—not because larger glass discs of the requisite optical quality could not be produced (they actually were), but rather because increasing absorption of light in several inches of glass threatened to defeat their light-gathering power.

As a result, in the opening years of this century the pendulum of progress swung once more from refractors to reflectors: and the twilight of refracting telescopes went hand in hand with a new triumph of astronomical reflectors which continue to reign supreme in astronomy today, and whose fortunes (in different form) have dominated the advancing front of observational astronomy ever since. The 100 inch (2.54 m) reflector of Mt Wilson Observatory (inaugurated in 1917 and retired from action in 1985, as a mirror telescope of almost the same size—of 2.4 m aperture—is to be launched into space in the near future) was followed in 1948 by a 200 inch (5 m) telescope at Mt Palomar; while in subsequent years a 6 m telescope was put into operation in the Caucasus in Soviet Russia. Even this may be exceeded before the end of this century by a reflector of 7.5 m free aperture planned by the Japanese at Hawaii, and a 10 m one by the Americans;

while a dozen or so telescopes with apertures close to 4 m (and located mainly in countries girdling the Pacific Ocean) are now regarded as instruments of only moderate size. Their cooperative efforts have (since the year 1900) provided our planet Earth with an unparalleled increase of celestial light-gathering power—several hundred times as large as that at the disposal of our astronomical ancestors of a hundred or so years ago. No wonder that, with its aid, we have been able to learn more about the sky in the past eighty years than our ancestors did throughout all the centuries of the past!

It has often been asserted that progress in any branch of science is largely controlled by the instrumental means which contemporary technology places at the disposal of its students; and the history of the telescope in the service of astronomy bears this out in a convincing manner. Thus the long line of astronomical refractors commenced with our ability to produce flint glass of the requisite properties and size; and the relatively slow progress of glass technology gave its first lease of life to astronomical reflectors towards the end of the eighteenth century. The reasons why this first efflorescence of the reflector proved to be relatively short-lived were again largely technological: namely, the inability to prevent rapid tarnishing of exposed optical surfaces (thus necessitating frequent polishing and refiguring of the mirrors), coupled with the inability to mount them in such a way that a relatively thin metallic mirror would not get disfigured by sagging under its own weight. The first of these drawbacks was, to be sure, lessened by the middle of the nineteenth century when Foucault introduced into astronomical practice glass mirrors with chemically silvered front surfaces—a practice which more recently was replaced by aluminisation to give the reflecting coat a longer life-span; but the second had to wait for more than a century following Herschel's time till advances in glass technology made it possible to produce large blanks of solid (or ribbed) glass or quartz, of low thermal expansion, which are sufficiently free from internal stresses.

In the meantime, techniques have also been perfected for grinding and polishing large optical surfaces to a high degree of precision (approximating the desired mathematical surface to within less than one-tenth of a wavelength); while parallel advances in servo-control mechanisms have made it possible to automate telescope control well within the requirements of the observer.

Most of the problems just mentioned arise, of course, mainly in connection with telescopes to be operated under the full gravity field obtaining on the surface of the Earth, and at the bottom of its atmospheric ocean. But since the late 1950s telescopes have also been launched into space, and operated in orbit under (essentially) gravity-free conditions; and thermostated to a degree unattainable on the ground. These new technological developments, now in full swing, are currently endowing our science with new dimensions which were unthinkable when I was young; and new knowledge obtained with their aid is bound profoundly to influence the further development of astronomy in the years and decades to come.

It is not only the increasing size of these telescopic light collectors which will materially influence such developments, but also new ways in which the light that reaches us from different parts of the Universe can be induced to disclose its secrets—i.e., the auxiliary instruments which can be made to scrutinise this light collected in the focus of the telescope. Until the middle of the last century, 'astronomy was indeed what you can see through a telescope'†; and the quality of telescopic observations depended as much on the eye of the observer as on the quality of the optics that formed the image in the focal plane. Since the second half of the nineteenth century the retina of the observer's eye was gradually replaced at the telescope by the photographic plate which, thanks to the photochemical effect of incident starlight, could, after long exposures, record the images of stars or nebulae much too faint to be seen with a telescope of a given aperture. Moreover, since the beginning of this century, photographic plates began to give way to photoelectric devices which make it possible to convert light into electrical currents susceptible of arbitrary amplification, and proportional to the luminosity of incident starlight much more accurately than anything based on the photochemical action of photographic plates. In the first half of this century, photoelectric photometry of starlight was very largely limited to single picture-points, such as the images of the individual stars. Since 1950, however, such techniques have been successfully extended to two dimensions. At present, TV image tubes, CCDs and other devices of this kind are on the verge of eliminating from astronomical practice photographic plates (which, in the long run, may have been no more than temporary makeshifts). Needless to say, the human eye as the prime instrument of observation has been hopelessly outclassed by these developments; and the remaining role of the observer has been relegated to the monitoring of the functions of auxiliary equipment mounted in the focus of the telescope, which have not yet been fully automated.

Indeed, the stage has now probably been reached at which it is the primary optics of the telescope that becomes an auxiliary of the instruments mounted in its focus. And, we may add, it is the potentialities of these developments which have caused astronomical observatories to abandon their time-honoured historical positions, and emigrate (largely

---

† A favourite saying of one of my elder colleagues at an ancient university in Britain until relatively recent times.

in our lifetime) to locations far removed from the centres of civilisation if a change in climate will suit them better for their primary missions. The first observatory at which telescopes were employed was apparently the one attached to the *Collegium Romanum* in Rome, then directed by the German Jesuit Fathers Kley (Clavius) and Grienberger, though its mission was mainly educational (and had no connection with the Specola Astronomica Vaticana, founded under Pope Leo XIII in the latter part of the nineteenth century). The next observatory in Europe (and still extant) was set up at Leiden (1632) and was followed by one at Copenhagen. The latter was established in 1637 under royal patronage, in this case of Christian V, King of Denmark; and the same was true of the observatories founded in France at Paris (1667) by Louis XIV, or at Greenwich in England (1675) under the somewhat less than bountiful patronage of Charles II.

Until the advent of the twentieth century, the scientific work of astronomical observatories was not influenced too much by their environment; and for some this remained true somewhat longer. However, with the increasing optical power of their telescopes (and increasing sensitivity of the light detectors employed), the overhead sky soon became the limit; and, in due course, this realisation led to a veritable exodus of astronomical observatories from thickly settled places to localities of better climate, where clear skies and (more important) telescopic images of superior quality could be expected to occur more often.

A quest for such sites in the past half-century led to a migration (of telescopes, which astronomers had to follow) which, astronomically speaking, has changed the map of our globe almost beyond recognition. Astronomical observatories (except those for purposes of teaching) have all but disappeared from the neighbourhood of large cities, and moved to sites of high altitudes (to gain in air transparency, and quality of the images); and they have often crossed national frontiers towards the equator (in quest of sites from which a greater part of the entire sky becomes observable).

A combination of these two requirements led to a virtual disappearance of large telescopes at high (northern or southern) geographic latitudes, in regions of poor climate (every observing hour at indifferent sites would cost so much more!). In this sense, by the end of this century Europe, the cradle where modern astronomy was born, will be almost depopulated of large astronomical instruments intended to operate in light at optical frequencies; as is already the case with the East Coast of the United States in favour of the Pacific South-West. 'Go west, young man', has been the rule of the day for young American astronomers since the middle of the century; and few Easterners elect to stay at home. If we look at the map of the world today, one fact stares us in the face: namely, that a majority of large astronomical telescopes are now situated in the mountain chains girdling the shores of the Pacific, in Australia, or on the Pacific islands (Hawaii) which already now offer hospitality to some of the largest telescopes of the world, and may attract even larger instruments in the future.

So there has been a dramatic shift in the centre of gravity of astronomical research within one's lifetime; though no more dramatic than the parallel shift in the human population on this planet—so different from the one we were taught at school when we were young. The countries bordering on the Pacific (and Indian) Ocean are now home to three-quarters of the population of our Earth; and the fact that, until quite recently, most of these were dominated (economically, if not militarily) by a relative handful of faraway Europeans is one of the paradoxes of history which the future world will comprehend only with difficulty—if at all! But *sic transit gloria mundi*—in the past as well as no doubt in the future; while astronomers observing the sky can only wonder.

As astronomical observatories and telescopes gradually evolved in the past four centuries (largely in response to advancing technology), so did the astronomers manning them and their aims. Both Galileo and Kepler, the founding fathers of astronomy of the telescopic era, held (like Tycho Brahe and others before) essentially court appointments (Kepler as 'Imperial Mathematician' to no less than three successive Holy Roman Emperors, and Galileo as 'Principal Mathematician and Philosopher' to two Grand Dukes of Tuscany) for major parts of their lives; and the case of the Observatoire Royal de Paris, the foremost of the astronomical institutions founded in Europe in the seventeenth century, illustrates the period of transition between royal observatories and those we know today. For the first hundred years of its existence, the Paris Observatory had no real directors appointed by the Crown. Each of its first four heads of the Cassini family was only a *primus inter pares*; all members of the Académie could work there at their discretion as they wished, on tasks (like measurements of longitudes, or triangulation of the dimensions of the solar system) sometimes assigned to them by the Government.

In such cases, each research worker (of their groups) had to raise subsidies from the royal purse necessary to pursue his tasks from time to time through his own efforts. The observatory was not assigned any permanent budget until the time of Napoleon; and it was only well into the nineteenth century that the Paris Observatory became fully integrated into the French Civil Service; the same was true of Greenwich in the time of G B Airy. This is the type of astronomical observatory which we have inherited in the twentieth century—for better or worse—from our ancestors. As time goes on, the heavy hand of central bureaucracies, lacking only too often the enlightened interests of at least some royal patrons of the past, continues to weigh more and more on the further pursuit of science; substituting their own aims for those of science.

We shall comment further on such developments later; but here we wish to return once more to the beginnings. If the first two centuries which followed the discovery of the telescope could well be described as the 'age of discovery'—in the sky as well as on the surface of our own planet—the nineteenth century became one of consolidation. The astronomical refractor (of which this was the last 'golden age') was an instrument primarily suited for measurements of the positions of celestial bodies in the sky (an art on which the twentieth century has improved but little) and a mastery in this field stimulated, in turn, parallel developments in applied mathematics needed to interpret these motions in terms of Newtonian mechanics. This was the time when the phrase 'astronomical precision' made its appearance, earning our ancestors a well-merited prestige; and one in which most 'special functions' of mathematical physics (Legendre, Bessel, etc) were born out of the problems arising in celestial mechanics. Astronomers of that age were, on the whole, a very sober lot (very different from what they have become since) who knew what they were doing; and their professional interests were largely centred on the last decimals of their results.

To give word on this subject to some of the high priests of that severe attitude to astronomical explorations—so different from that which motivated William Herschel only a generation before—let us quote the following statement by the great mathematician Carl Friedrich Gauss (1777–1855): 'Mere opinions end, and real knowledge begins, in astronomy only in such subjects which can be treated mathematically. This is the case with size and shape of celestial bodies, their distances, relative positions and their changes—or, in effect, their motions.' And his contemporary, Friedrich Wilhelm Bessel (1784–1846), generally regarded as the greatest astronomer of the nineteenth century, expressed himself on this subject in still more definite terms: 'The duty of astronomy to mankind is clear for all times: namely, to provide precepts [*Vorschriften*, in Bessel's native vernacular] which should enable us to compute the motions of the celestial bodies as we can observe them from the Earth. Everything else that we can learn about them—such as their appearance and surface properties—is not unworthy of our attention; but this does not really concern astronomy... but to investigate the motions of the celestial motion so completely that their positions can be evaluated for all times—this is, and will remain, the principal task for astronomical research' (Bessel 1841). Moreover, towards the end of his life Bessel stressed also that 'astronomy gains by new results only if they are unambiguously obtained. Not the premature guessing, but acquisition of fundamental data and knowledge must be the principal aim of our efforts' (Bessel 1844).

This simplistic approach to duty by our predecessors can be understood only if we remember that the astronomy of Bessel or Gauss was governed by one very simple physical law (that of gravitation, which represents the weakest force operating in nature). When, however, phenomena governed by atomic (or nuclear) physics of the twentieth century came to the forefront of observational interest, astronomy ceased to be simple; for the underlying laws of nature no longer provide a platform stable enough for accurate predictions; and when this becomes true, astronomy is apt to drift back to a semi-quantitative stage. This is perhaps why so few astrophysicists these days bother to learn enough mathematics to satisfy Bessel or Gauss. Should they have done so, many of their more unbridled speculations (for instance, about the evolution of close binary systems) would probably have never left their pens.

Needless to say, the spirit which pervaded astronomy throughout most of the nineteenth century was that which Bessel or Gauss would have approved of (and I was still brought up in it by my teachers); anything else would have been regarded as astrology. How much times have changed since! Gone are the days of a typical nineteenth-century observatory, whose most prestigious instrument was a meridian circle (or transit instrument), designed to measure the positions of the stars for the use of posterity; and followed in the contemporary telescopic hierarchy by a 'great equatorial' (refractor of large focal distance) used mainly for differential micrometric measurements, or plain visual observations. These two standard instruments were usually supplemented by smaller refractors clustering around the great equatorial, and used by different staff members of the observatory for different programmes of their own. Such programmes (in the pursuit of which the astronomer may often have spent a lifetime) were regarded very much as the individual's private enterprise; and 'butting in' by others would mostly have been regarded as uncalled for.

Today, less than a hundred years later, the most important instrument of any observatory (other than those set up for special purposes) is apt to be a large reflector, collecting much more light than all the refractors of an observatory of bygone days taken together; attended (usually) by a smaller wide-field short-focus reflector for survey purposes. The housing (and mounting!) of large reflectors is so much more expensive than those of the smaller refractors of old that few observatories can afford more than one; and the observing time with it is shared by the entire staff on mostly collaborative projects. Moreover, resident staff members seldom spend their entire life in one place; and like their auxiliary instruments, they are increasingly interchangeable among institutions all over the world.

And what is true of the instrumental equipment, is all the more true of the ways in which the results of their work are being made available to others by publication. The situation facing us today in this respect has already been commented upon in a preceding chapter. But how about their contents? Less than a century ago, the investigator publishing new results was expected to convince his contemporaries of their validity or significance; yet today almost the opposite seems to be the case! The scientific papers filling our journals these days are often replete with ideas which their authors do not bother to prove: instead, they challenge the reader to prove that they are wrong! What Gauss or Bessel would have thought of this inverted logic we can surmise from their opinions already quoted above; but who knows much about them today?

A hundred years ago, in the days when Newton's law still reigned supreme, the halo which canonised it never shone brighter than in the middle of the nineteenth century, when Neptune (the outermost planet of the solar system) was discovered in September 1846. The discovery was, to be sure, not unique; for only 65 years before William Herschel had discovered (in March 1781) the planet Uranus. But while Herschel came across his planet by sheer chance (in the course of his systematic sweeps of the sky with his 7 foot telescope), the refractor of the discoverer of Neptune (J G Galle of Berlin) was unerringly guided to the right place in the sky by the theoretical prediction of its position by the French mathematical astronomer U J J Leverrier (1811–1877), who deduced it from the observed perturbations of the motion of Uranus in the intervening years.† It was this feat, so spectacular in the eyes of contemporary laymen, which did more than anything else to establish in people's minds the ascendancy of Newtonian mechanics in the hierarchy of the Universe, before a decline of confidence in its validity as the ultimate law of Nature began to creep in, and experimental evidence (both celestial and terrestrial) began to pave the way for the advent of relativity.

Not all astronomers of that age were, however, equally impressed by that feat. For when the American astronomer Simon Newcomb (1835–1909)—himself one of the leading lights of his times—raised in his reminiscences (cf Newcomb 1903) the rhetorical question: who was the greatest astronomer of our age?, he answered: Hansen! Now Peter Andreas Hansen (1795–1874) was distinguished as a creator of outstanding tables of the motion of the Moon, which indeed remained unsurpassed until the work of E W Brown (1866–1938; see figure 4.4)

---

† Almost the same prediction of the existence of Neptune was made at about the same time by a young Cambridge mathematician, J C Adams (1819–1892). However, for the story of why it did not result in the discovery of that planet because of the red tape prevalent at Greenwich (and Mrs Challis's insistence on an untimely cup of tea!) see Smart (1947).

and the advent of modern automatic computers almost three-quarters of a century after Hansen's death.

> Modest as was the public position that Hansen held†, [went on Newcomb] he may now fairly be considered the greatest master of celestial mechanics since Laplace.... To many readers it will seem singular to place any name ahead of that of the master who pointed out the position of Neptune before a human eye had ever recognized it. But this achievement, great as it was, was more remarkable for its boldness and brilliancy than for its inherent difficulty. If the work had to be done over again today, there are a number of young men who would be as successful as Leverrier; but there is none who would attempt to reinvent the methods of Hansen, let alone even to improve upon them (*op. cit.*, p. 315).

By the time he wrote these words, Newcomb no doubt spoke not only for himself, but also for his peers; and, for this reason, they deserve to be remembered even at this time. But ask any PhD candidate of our subject at this time who Hansen was; and in 99 out of 100 cases you will get no coherent answer. And if you ask their examiners for the reasons for Hansen's fame in the nineteenth century, you will fare no better. Today, we are only a hundred years younger; are we to surmise that the leading lights of contemporary astronomy will be similarly forgotten by posterity equally soon? A chilling thought!

More than likely, I should say; for the motivation of Newcomb's verdict discloses also a great different in emphasis for deserving the accolade. Note the reason why, in his opinion, Hansen deserved to be placed ahead of Leverrier on the honour roll of history. What Leverrier did was, in Newcomb's words, mainly a 'stunt' which others could have done equally well had they only thought of it (Adams, as we know, did!); while a life's work which outlasted a century belonged to a different category of achievement. How differently do we judge the merit of work today!

We must, of course, remember that Newcomb wrote his verdict around the year 1900—a time which posterity may view as a watershed between two epochs almost as distinct as pre- and post-telescopic astronomy around 1610. For the advent of the new century heralded also the advent of the electrons and quanta; the 'new physics' arriving in their wake would compel the stellar spectra (viewed in the second half of the nineteenth century as curiosities worthy of only descriptive or statistical treatment) to deliver their message; optical telescopes (in

---

† For almost fifty years of his life, Hansen (a Dane by birth) served as director of the (no longer existing) private observatory of the Duke of Mecklenburg at Seeberg, near Gotha in Germany.

## Marriage of astronomy and physics

the second half of the twentieth century) would be supplemented by those collecting radio-waves, x-rays or $\gamma$ rays; and optical astronomy from space would be joined by 'particle astronomy'—with light waves (in the full domain of the spectrum) no longer our sole link with distant celestial bodies. In the course of our age, classical astronomy has gradually been transformed into astrophysics—a leading branch and pathfinder of physics today—and astrophysics continues to provide clues to new laws of nature, the evidence for which could not have come from terrestrial laboratories.

But it was not only the mixing of two scientific disciplines, astronomy and physics, which proved so extraordinarily fruitful in our time; but also the interaction of their students, coming from very different backgrounds. The late Walter S Adams (1876-1956), the second Director of Mt Wilson Observatory (figure 4.1) who was at that Observatory since its foundation, has preserved for us some delightful stories in reminiscences of his own on the early days at Mt Wilson (cf Adams 1947) which are too good not to be recalled in this place.

The first goes back to the first decade of this century, when the observatory was visited by Simon Newcomb, the epitome of a classical astronomer, who considered it a part of his duty to report on the new observatory in California to Washington.

> The only thing which interested Newcomb at the Observatory to any extent [wrote Adams] were the Riefler clocks in the laboratory, and the Snow telescope building.... The coelostat and concave mirror combination he dismissed briefly as not forming a real telescope; and the spectrographs and spectroheliograph were not merely a closed book to his mind, but one which ought to remain closed.† In fact, Newcomb's attitude towards the spectrum resembled closely that of Burnham [S W Burnham, 1838-1921; outstanding observer of double stars at Yerkes Observatory] whose famous remark to Hough: 'Gale [a physicist, and one of the first co-editors (with G E Hale and E B Frost) of the *Astrophysical Journal*] showed me some soda lines at the Ryerson Laboratory [of the University of Chicago] but I did not think much of them' has become a classic in astronomical literature.

The second story is some twenty years younger; and goes back to the decade between 1920 and 1930, when Mt Wilson Observatory attached

---

† Even outside astronomical observatories, Newcomb had the reputation of an austere and rather forbidding character, as is illustrated by the following story passed on to us by A A Michelson (cf Strutt 1968, p. 415). At a dinner party one day Newcomb was seated next to a young lady, with whom he did not exchange a single word during the whole occasion; and when remonstrated with by the host for his lack of civility, Newcomb merely noted that 'I had nothing of particular importance to communicate to her.'

to its staff A A Michelson (1852–1931), a great experimental physicist, as an associate and active investigator. As is well known, Michelson came to Mt Wilson to apply interference methods to the measurement of the diameters of the stars; and his first successful determination of the apparent diameter of the star Betelgeuse ($\alpha$ Orionis) created a sensation in the scientific world at that time.

> Michelson [wrote Adams (*op. cit.*)] was, however, a physicist and not an astronomer. This is well illustrated by the story told of him by Moulton, [F R Moulton, 1872–1952; see Chapter 4] then a colleague of his at the University of Chicago. One clear night in the winter following Michelson's measurements on Betelgeuse at Mt Wilson, the two men were walking across the University campus when Moulton stopped and pointed to a bright red star well up in the sky. 'Michelson, what star is that?' asked Moulton. 'How the devil should I know?' replied Michelson testily. To him a star in the sky was a point of light to feed his instrument, and should be treated as such.

I may add that, during my years at Manchester (and elsewhere), I met only too many physicists who shared Michelson's limitations as far as the knowledge of the night sky is concerned (though without possessing perhaps quite the same experimental skill).

## Will Our Civilisation Survive?

So much for a brief sketch of the history of astronomy which antedated our times. How does it, however, fit into the more general framework of the history of science, of which astronomy is only a part? Is it on the up-swing (as it appears to have been since the discovery of the telescope) or are there signs on the horizon that the future may hold something else in store for our descendants? The aim of this concluding section of the book is to share with the reader some thoughts on this subject which are not without misgivings, and in the hope that my premonitions may turn out to be wrong. Yet the history of science (and, in particular, of the exact sciences) in the past few centuries indicates that *the incidence of talent*—the number of the carriers of our sacred fire—*fails to keep pace with the general rise of human population* on our planet; and astronomy is no exception.

In order to demonstrate this, let us, for the sake of brevity, single out for our statistical investigation the incidence of men of the very first rank, whose names and work shine like beacons in the world of darkness to guide us in our quest. To single them out, we shall require that their contributions to human civilisation meet the following stringent criteria.

First, their knowledge should depend but little on what they could have learned from their predecessors—what they knew, they discovered (or could have discovered) all by themselves. Secondly, the problems which they failed to solve in their lifetimes remain (mostly) still unsolved to this day. Third, whatever they pronounced *ex cathedra* was never faulted by subsequent generations; and, last but not least, they possessed the power to discern which problems are solvable by the means at their command and which are not—and never mixed up the two!

You doubtless guess by now whom I mean; for only three names of the past three centuries of the history of exact sciences would seem to pass muster: Isaac Newton, Carl Friedrich Gauss, and Henri Jules Poincaré. Although all three are perhaps best known by contributions they made to the science of mathematics, their spirits were truly universal; and their interests ranged far and wide over different fields of human endeavour, both pure and applied, all of which they enriched by valuable contributions.

Those of Isaac Newton are too well known to call for a special recount in this place; but it may be less well known that Gauss was almost as accomplished a linguist as he was a mathematician (and in his adolescence found it hard to decide which of the two fields to study for a career). He mastered many languages (including Chinese) as effortlessly as he did the theory of numbers. Although his greatest achievements were in the domain of pure mathematics, throughout most of his life he was a professional astronomer, as Director of the University Observatory at Göttingen. Although (truth be told) Gauss did not enrich our science with many new observations, he developed the 'calculus of observations' for their reductions (think only of the 'method of least squares' or the law of 'gaussian distribution of errors') to meet all the needs of the next hundred years. And—as if this were not enough—in the latter part of his life Gauss also became one of the pioneers of the electrical telegraph (the unit of magnetic field strength does not carry the name 'gauss' without good reason).

Henri Poincaré, the last of these universalists, was perhaps less experimentally inclined than Newton or Gauss; and least endowed with gifts of imposing physique. This grand master of celestial mechanics possessed eyesight too poor to enable him to observe the celestial bodies himself; and, one of the creators of topology, he could not draw a diagram that anybody could recognise. But what a magic instrument a pen became in his unskilled hand! Henri Poincaré was probably the most prolific contributor to mathematics since Leonhard Euler—and an intuitionist *par excellence*—whose interests ranged far and wide over all branches of mathematical physics (including relativity); astronomy; and even biology or the psychology of mathematical invention were within the domain of his professional competence. Poincaré was also the only scientist in modern times to be elected as one of the 'forty immortals' of the Académie Française on the literary merits of his work alone—as probably the greatest expositor of science (and of its role in human life) for the general public who has ever lived. Incidentally, Poincaré was the only one of our trio who wrote all his works, no matter where they were published, in his native vernacular; while Newton as well as Gauss still wrote only in Latin.

The basic facts on the lives of these three men are simple. Newton was born in 1642 (the year Galileo died) and attained a ripe old age of 85 years. For fifty years after his death his intellectual throne remained vacant, until Gauss came in 1777 to claim it; and when he died early in 1855 at the age of 77 years, Henri Poincaré was already ten months old. But since he passed away in 1912 at the relatively early age of 58 years, still at the height of his intellectual powers, the world has been waiting for his successor—so far in vain! And as we contemplate this fact, certain disturbing questions come to mind which may bear on our future and which I find very troublesome: namely, why have we not seen more minds of this calibre emerge from our midst in more recent times?

In order to appreciate the full significance of this question, let us consider the reservoir of talent from which these three men emerged in the past 300 years. Demographers tell us that by the time Newton was born the human population of the world was close to 300 million; and by the birth of Gauss, it had risen only to some 400 million. When Gauss died and Poincaré was born, it already reached the level of approximately one thousand million; and was close to 1500 million when Poincaré left us in 1912. However, in the subsequent 70 years which separate us from that time, the world's population has almost trebled, becoming ten times as large as it was at the time of Gauss; but no genius of comparable stature has emerged among us since: why?

In any attempt to find an answer to this question, we should keep in mind that talent—of any magnitude—is created by 'natural selection' on the level of genes—a biological mechanism which represents an essentially *random* process. If we can learn anything from the history of science, it is the fact that genius is *not* hereditary; nor does it seem to recur in a finite number of generations. Newton left no blood descendants behind him to enable us to test this conclusion in his case; and as regards Gauss or Poincaré, the time-span which has elapsed since they died is still too short when measured in terms of the human lifetime. Consider, however, the case of Johannes Kepler, who was born more than 400 years ago and left numerous offspring. German scholars with a diligence characteristic of them traced the Kepler family tree up to the present time. There are hundreds of his blood descendants still living among us today, but none of them since the seventeenth century has shown any sign of intellectual distinction; and neither (as we learned from Kepler himself) did any of his ancestors.

But—to return to our main theme of inquiry—if the natural production of men of genius occurs by a process which operates essentially at random, with the increasing reservoir of human beings now populating our planet their numbers should increase in proportion; if so, ten new Gausses should have been born to us since 1800; and four more Poincarés! And yet, how well we know that this is not the case!

It could perhaps be objected that these figures are too small to provide a sufficient basis for meaningful statistics. Yet if we set out to increase the size of our sample by descending below the level of our giants to lower steps on the intellectual ladder, we continue to encounter essentially the same situation. To make myself clear, let me stress the obvious fact that the spectrum of talent generated by random heredity is a continuous one, bordered at each end by talents whom we could describe (for lack of a more suitable term) as 'crystallisers' and 'innovators'. Newton, Gauss or Poincaré were innovators *par excellence*; for many of the ideas they left behind were so far ahead of their time that humanity has not caught up with them yet. On the other hand, men like Galileo, Darwin or Einstein (of his special-relativity days) were crystallisers of the ideas—the idea of the heliocentric solar system at the beginning of the seventeenth century, of the evolution of species in the mid-nineteenth century, or of special relativity in the first decade of the twentieth century—by bringing together ideas for which their time was ripe; their historical role was to crystallise them to a form in which their contemporaries could no longer ignore them. To be sure, Einstein's crowning achievement, the general theory of relativity, was certainly on a par with the best work of Newton, Gauss or Poincaré; and if we did not include him in this trinity, it was because Einstein had only one such achievement to his credit, and thus failed to measure up to our requirement of universality (the same could, perhaps, be said also of Maxwell).

In this sense, the 'crystallisers' of thought are as a rule only a few years ahead of their time (indeed, they cannot be more if they are to fulfil their historical role); and this is also reflected in the extent of attention or recognition they receive from their contemporaries. The 'crystallisers' usually cash in on their accomplishments during their lifetimes (both tangible rewards, and sometimes also—like Galileo or Darwin—the stick, if the crystallisation occurs too early and its results run counter to the prejudices of the majority). In contrast, the true innovators are usually immune to such vicissitudes; for their contemporaries are as a rule incapable of assessing their real greatness, which may emerge in proper perspective only after the lapse of whole centuries (see the role of Aristarchos in recognising the true structure of our solar system, or of Archimedes as a forerunner of the infinitesimal calculus, almost two millennia before the time of Copernicus, Kepler or Newton). But whatever else we may think of the contributions of both 'crystallisers' and 'innovators' to the advance of human civilisation, a glance at its more recent past leads to a conclusion which becomes almost inescapable: *the occurrence of talent in the past century or so continues to lag behind the general trend in human population on this planet to an ever-increasing extent.*†

Why should this be so? *What are we doing to ourselves to inhibit* not the birth but *the development of talent in human society*; and what kind of 'brain pollution' is responsible for this disturbing phenomenon? The causes must obviously be *social* rather than biological; and, in certain respects, some of them should be expected actually to work the other way. For instance, the sharing of information through the twentieth century has certainly been much more effective than it was in the past for technical reasons alone; and the general schooling system now under development in all countries should reduce wastage of talent by neglect. And yet the net outcome points in the opposite direction! What may, then, be the adverse influences which can more than counterbalance the beneficial effects just mentioned?

When Isaac Newton was once asked how he managed to make so many discoveries in his lifetime, he was reported to have replied 'by thinking unto them.' Is this the point on which we may have been falling short of requirements for some time; but if so, why? Is it merely because we have gradually less and less opportunity (or incentive) to do so? It is certainly true that throughout most of our lives now, almost from cradle to grave, we are being constantly bombarded by information which, to a scientist, is largely irrelevant for the main purpose of his life. The sources of this 'noise' are multifarious—from the 'yellow press' in the days when people used to read more, to radio and television in the more recent past. All these 'media' see to it that we have less and less time to think for ourselves. 'Don't think, scheme!' seems to be the slogan of the day. It is true that scheming constitutes a mode of ratiocination offering better and more immediate rewards than pure thought in the rat-race which has penetrated even the sacred walls of academe—the more effectively so, the more developed the country; and life within these walls is increasingly alienated from its original purpose; with a tendency to conformism—a

---

† The same seems, moreover, to be true not only of science, but equally of the world of arts (such as music). How many Mozarts or Beethovens have been born since the beginning of the nineteenth century (or, in the performing arts, Liszts or Paganinis) out of a global population now almost ten times as large?

curse of any large social unit—replacing a quest for truth (for this minimises risks).

The question can, of course, be asked: is it conceivable that a Newton, Gauss, or Poincaré could have been alienated from his manifest destiny by exposure to such a social climate as began to pervade our life at all stages since the advent of this century? That a young Gauss (a prodigy if there ever was one) could have been lured away from thinking arithmetic by watching TV seems unlikely. However, the frightening fact is the impossibility for the historian to give an unequivocal answer in the negative. Newton certainly did his best in his fifties to get away from Cambridge and join the Civil Service in London; now he could no doubt do it sooner and with less effort. What was regarded with approbation in Gauss at the time of enlightenment (the age of prodigies) might now be viewed as oddity or queerness—especially if his mother and the benevolent Duke of Brunswick were not on his side to tip the scales in his favour. Would the academic bureaucrats of today agree to admit Poincaré to the Ecole Polytechnique after he failed (because of unskilled hands) in his entrance examination in drawing—a subject then compulsory for future military engineers? And how many American colleges (including Harvard) would hesitate today to confer a degree on students who failed to pass (say) their swimming test (or how many students are currently being 'pushed through' the institutions of higher learning for being good at ball games or other forms of athletics; and why)?

But, as my final witness to this disquieting state of affairs I should like to call on someone who, though no scientist, has shown penetrating prescience of the shape of things to come. His name may indeed be familiar to many, since the year of 1984 is now already behind us. On pages 271–272 of his famous novel (Orwell 1949) its author engaged two of his characters, the 'organisation man' (or *apparatchik*) O'Brien and his luckless victim Winston Smith, in a conversation on astronomy, which should continue to be of more than passing interest to us as well.

> 'What are the stars?' said O'Brien indifferently. 'They are bits of fire a few kilometres away. We could reach them if we wanted to. Or we could blot them out. The earth is the centre of the universe.... For certain purposes, of course, that is not true. When we navigate the ocean, or when we predict an eclipse, we often find it convenient to assume that the earth goes round the sun and that the stars are millions upon millions of kilometres away. But what of it? Do you suppose it is beyond us to produce a dual system of astronomy? The stars can be near or distant, according as we need them. Do you suppose our mathematicians are unequal to that? Have you forgotten double think?'

Of course, it is clear from this quotation that George Orwell was no astronomer; for otherwise he could have engaged his characters in a discussion on subjects more topical than eclipses or navigation. Of course, neither O'Brien nor Winston Smith knew anything about (say) quasars—a fashionable topic of today; could *these* be placed 'near or far, according as we need them'? A casual perusal of contemporary professional literature discloses that only too many astronomers may indeed think so. And how about if the discussion drifted into the realm of cosmology beyond the limits of the solar system: was O'Brien so wrong in surmising that our mathematicians are quite capable of double-think? At least some of our current journals appear not to shy away from 'newspeak' (with the full blessing of their referees); and we engage in it, as a matter of course, whenever the facts on hand are not sufficient to admit of a unique interpretation, and when we deliberately camouflage this by double-talk intended to conceal the true situation. In the last century this would still have been regarded as professional misdemeanour, but does this happen only rarely in our contemporary scientific literature?

And, last but not least, the 'newspeak' in the Orwellian sense has already established a durable place for itself in another branch of literary effort, of ever-increasing practical importance: in contract proposals for government support, or reports on the results obtained. Perhaps the bureaucrats at the receiving end get only what they deserve. However, the sad part of the story is the fact that this new style is seeping into the veins of the upcoming generation so irresistibly that, given a little more time, it will overspill into literature so far considered legitimate; and soon no-one may be able to note the difference.

Your opinions on these trends may, or may not, be the same as mine; but what Poincaré would have thought of them I have no doubt; and maybe this is why we have not seen one such as him since the first decade of this century. But *how long can we get along without such leaders?* The history of science projected into the future may again be pointing to the outcome.

Our immediate predecessor in the hierarchy of civilisations, one which flourished around the shores of the Mediterranean two to three thousand years ago, is known to have run out of original inspiration by the second century BC; what followed was a gradual decline (at least in Europe) into the darkness of the Middle Ages which lasted for more than a thousand years. To be sure, our forefathers emerged eventually from their stupor to regain, in the past few centuries, the intellectual leadership of mankind. However, is it not impossible that with the advent of the twentieth century of our era we once more have reached the stage at which the genetic game has started to produce diminishing returns; and if so, is there a risk that soon we may find ourselves drifting without pilots of sufficient calibre to steer us through the dangerous seas of the future?

In the past 5000 years—ever since man invented writing to communicate with the future—several distinct civilisations (in Egypt, Mesopotamia and around the shores of the Mediterranean) came and went which, different as they were in many aspects, possessed also certain features in common. Thus at the commencement of each we encounter an efflorescence of three-dimensional arts (architecture, sculpture), followed by literature (led by epic poetry), drama, history, and eventually science, whose ascendancy began in each case with a decline of the arts; and whose autumnal flowers were eventually destroyed by the frosty chill of barbarian invasions. So it was in Egypt; and such was the case of the Mediterranean civilisation, from archaic Mycenean architecture and Homeric poems, through the golden age of Periclean Athens to the Hellenistic civilisation which spread around the shores of the Mediterranean, eventually to perish under the onslaught of Roman military might.

When we turn to the next and last cycle of European cultural history—the one commencing with the Renaissance, and best known to us because we are its direct descendants—it is as if a magic wand wakened a culture that had slept for some 1600 years; and brought it to life again in almost the same sequence: the efflorescence of architecture and visual arts in the time of Leonardo, Michelangelo and Raphael was followed by that of music and literature culminating in the eighteenth and nineteenth centuries (our 'romantic' era); while science with its accompanying technology, manifested in the release of nuclear energy on a macroscopic scale, in the exploration of our Moon or of the planets by spacecraft, or in current advances of molecular biology, remains our chief title to fame in the twentieth century—not necessarily because the spirit of our science is superior to that of our predecessors, but because it may represent our remaining creative contribution to the last act of the present cycle.

Is this inevitable? Perhaps; for like all large and complicated systems, the course of human events may possess its own system of built-in checks and balances, whose logic continues to escape us, but which may come into play if any of its components has stepped out of line to out-distance the others unduly. So it may have been with architecture and the visual arts at the time of the Renaissance; or with literature and music during the epoch which historians classify as romantic. Does the current ascendancy of science in our time, like in the Hellenistic period of the Mediterranean civilisation which preceded the current Atlantic one, presage the end of another era in the tides of man; and is the increasing scarcity of its great torch-bearers, noted earlier in my remarks, a mere harbinger of this fact?

The late Arnold Toynbee, who devoted his lifetime to a study of the rhythms in human history, demonstrated convincingly that civilisation on our planet seems to evolve, not at a steady pace, but by a 'rout-and-rally' effect, as if the human spirit were incapable of sustaining a long haul for more than a few centuries at a time. Have we, in the twentieth century of our era, reached once more the stage at which our creative efforts on a broad front are running out of breath? The question can be asked; but the future will provide the answer.

Its gravity is underlined by the fact that the parallel growth of human population on the Earth (with all its consequences) should make us expect something else than actual observations seem to be indicating. When the present century was young, around the time I was born, the total population of our planet stood at about 1500 million people. At present, their number has more than trebled to almost 5000 million: as I write these lines, several people are being born each second; and, at this rate (if unchecked by any calamity of global dimensions) the population should reach a staggering total of 20–30 000 million individuals some time by the middle of the next century—i.e., during the lifetime of our grandchildren! Accordingly, of all the human beings who have ever lived on our planet from the time *Homo erectus* descended from the trees in the last few million years, about one-quarter are our contemporaries, living with us today! Certainly no other species has ever attained such absolute mastery of its environment since our planet was formed; and its avowed enemies (long efficient in regulating our numbers) have now been suppressed to microscopic and sub-microscopic worlds, where they are successfully held in check by modern medicine.

But can present numbers—let alone their further growth in the proportions anticipated—be long sustained? Indeed, the current overgrazing of resources essential for our support has become painfully evident and has been responsible for a major part of the political history and social strife of this century. In the next, it will not be merely an exhaustion of precious metals (which caused the breakdown of Rome, and of other ancient empires of the past); or of technologically important ones (no less than seven of which—bismuth, chromium, lead, mercury, nickel, silver and zinc—are likely to have their proven supplies exhausted by the end of this century; with copper and tin—not to speak of oil!—soon thereafter); but of the soil to plant our food in, of fresh water (three-quarters of which constitutes our bodies), or of clean air to breathe; to preserve these will be the primary technological problem of the future!

How are these problems likely to be managed—and will they indeed prove manageable? Whatever happens, there is no doubt that the parts of the world which have been the traditional cradles of higher civilisations will find themselves increasingly at a disadvantage. The Mediterranean basin and countries adjacent to it have already become museums, rather than crucibles, of our civilisation; and historical processes are already at work preparing the same fate for the Atlantic nations, which have been carriers of human civilisation since the time of the Renaissance. Present demographic trends disclose that, since the beginning of this century, the centre of gravity of the human population on our planet has been systematically shifting from the Atlantic to the Pacific area. More than three-quarters of it already live now in countries bordering on the Pacific and Indian Oceans. Indeed, one-quarter of the Earth's population are now the Chinese; and China with India accounts for more than two-fifths of the total!

That the explosion of population which we now witness is bound to devalue man as an individual is inevitable. But if, as we surmised earlier in this chapter, talent (including scientific talent) represents a statistical phenomenon of proportionally equal incidence in human population (where conditions for it are ripe), there is but little room for doubt that the torch of science (including astronomy) will soon pass from Caucasian hands to other races of our planet, inhabiting another hemisphere from that which produced Newton, Gauss, or Poincaré in the past, and is sadly running out of them.

This fact, which we have already deplored, may be only another symptom of the general state of a society in decline. In the first centuries of the post-Renaissance era, the principal economic source of sustenance of society was agriculture, to which since the nineteenth century we must adjoin industry. Up to the first half of the present century the backbone of this society (on both sides of the Atlantic) was heavy industries, bent on the production of durable goods (as is still the case, say, in the USSR or China, which emerged from their agrarian past not too long before).

In eastern lands bordering on the Atlantic this is, however, no longer the case. Their agrarian past has been relegated to oblivion by imports from former colonies. Moreover, as the twentieth century is approaching its end, heavy industries themselves are in retreat from their once predominant role as the prime employers of urban population (being gradually replaced in this role by electronic industries); but even the nature of the end product of industrial effort has undergone a profound change. If previous efforts were directed towards the production of goods which were (or were supposed to be) useful, the new needs place emphasis on the development of what can be called the *entertainment industry*, if under this term we consider theatre (both legitimate and illegitimate), film, television, pop music, the yellow press with its comic strips, professional sports, etc—in brief, those activities whose aim is not to provide us with anything of durable value, but to *dispel the boredom* of a population which enjoys (if this is the right term to use) the benefits of an increasing degree of leisure.

In more specific terms, why is kicking a ball in the playground, or hitting it with a bat of some kind, watching two men beat each other with their fists, or driving mechanical contrivances at unreasonable speed from nowhere to nowhere in particular, why are these things so much more attractive for an increasing fraction of our present human population than to read great literature, to listen to great music—or to push the frontier of the unknown further ahead in any field of science? In the Olympic Games of ancient Greece, the programme also incorporated artistic contests by poets, dramatists or musicians; and they were preceded by a period of generally respected peace—humanitarian features sadly missing in the modern counterparts of such occasions resurrected in more recent times, and devoted largely to fostering 'tribal conceit' rather than the peaceful co-existence which could benefit humanity as a whole. What we see instead in the western world more often are throngs of sedentary 'supporters' watching professional games on their television sets at home—an activity which benefits primarily the performing gladiators (or their sponsors), earning astronomical sums for the entertainment which can release (albeit temporarily) the spectators from chronic boredom without any need on their part to exert their muscles or brains.

History discloses, however, that civilisations which fall prey to this way of life are unable to stand up long against those which are not bored, or content to rest on past glories, but eager to gain new laurels. There is a rhythm in the life of societies as well as of the individuals constituting them; and the decline of old age goes side by side with the exuberance of youth—they only unroll on different timescales. If what happened in the past can offer any clue to the future, the decline and fall of the greatest civilisations of the past, once started, requires only half a dozen lifetimes for its consummation (witness the three centuries which ancient Rome managed to survive off the legacy of the Hellenistic world) and the faster rhythm of the contemporary world may shorten this time in the future.

Those of us living today have indeed been born to the time of decline of what we can call western (or Atlantic) civilisation; the two great wars of the twentieth century were not the cause, but the consequence and external manifestation of a malaise whose roots antedated them. The spiritual atmosphere which preceded the wars created situations whose feedback can only accelerate the process. And, in the meantime, the intensity (not to say desperation) with which an increasing fraction of the population in 'developed' societies pursues this quest for 'entertainment', and pushes up the 'gross national product' only to finance their escapism, is manifest by the inordinate offers of tangible rewards to those who can cooperate to this end.

We know, of course, that after an exhausting time in the office, or on the factory production line, the mind may be in need of vacuous entertainment rather than of meaningful recreation. However, the worst aspect of this mental asphyxia of our society is the fact that our contemporary struggle to keep boredom at bay can also atrophy interest in the future; or in any remedial action to relieve boredom before it is too late. Fortunately, however, the history of mankind on our planet discloses that all its population has never been stricken by the same affliction at the same time; and as the Sun is setting over one part of the world, it may be rising elsewhere. As the population curve of our Earth soars to its empyrean heights, can it happen that, out of the ever-increasing number of human beings, Nature may produce by a genetic lottery 'super-geniuses', intellectually much greater than Newton or Gauss (who, after all, emerged from population samples 10–100 times smaller)?

This is indeed not only possible, but likely; at least there seems nothing to prevent this from happening in individual cases (wherever they may come from). And so, let us hope that the day may not be far in the future when human brains may triumph again not only over the adversities of their cosmic habitat, but also over the evil angels of their nature, to navigate us safely through the maelstrom of the impending explosion of the population and across the great divide to enable our more distant descendants to reach, if not Paradise, at least a better world.

Yet for those of us who are not afraid to look the facts in the face, optimism is far from being the order of the day for the long-range prospects of life on this planet. It is true that, since the nineteenth century, we have defeated all bacterial (and most virus) diseases to prolong human life by a factor of more than two over what it was only a few centuries ago. However, while this is being accomplished we continue filling the hospitals of the 'developed' countries with mental patients, incapable of looking life in the face and responding to its challenge. Can one acquire immunity against such kinds of diseases in our armchairs, watching surrogate life on our television sets (or sleeping in front of them?).

But a still very much worse fate may be in store for our society before this century draws to its close. As the stresses to which we are currently exposed in daily life keep mounting (in the form of, say, boredom, or chronic unemployment, which only war has so far been found to cure), it is disturbing to observe the rising fraction of the population (mostly young, and precisely in the developed countries) who, unable to face the facts, are 'opting out' by resort to chemical drugs—the traffic in which is rapidly becoming one of the world's principal businesses†.

Legal (or police) action can, at best, provide only a limited and temporary remedy against the deluge which threatens to overwhelm our entire society in its wake; human greed (i.e. the cost of the drugs) constitutes a somewhat more effective deterrent; but none of these will be of any avail if and when chemists synthesise the dangerous drugs in their laboratories from cheap materials, and thus obviate the need for expensive imports of raw materials provided by Nature. Indeed, the only really effective panacea against this deluge would seem to be to *remove the causes* which lead people to reach for drugs to begin with—i.e., to dissolve the stresses from which frail human nature vainly seeks release. They will not disappear by mere denial; nor can we opt out of them at will.

But if the way is not found in time to free our society from the bondage of these evil angels of human nature, the possibility—nay, probability—cannot be ruled out that, not long after the commencement of the next century, the average span of human life in countries we now call 'developed' will drop from the present 70–75 years again to less than a half of that, back to where it was in the fourteenth century at the time of the 'black death'; and with a corresponding drop of the entire human population. Nuclear explosives in military hands, coupled with a pandemic of mind-destroying chemical drugs, represent the greatest threats facing humanity (and more than humanity—life!) on our planet by the end of this century; and the possibility that they may yet engulf us in the destructive maelstrom cannot, unfortunately, be ruled out. Shall we find navigators capable of taking us across while there is still time? If so, they cannot come too soon.

† According to the latest figures, the money turnover in unlawful drug traffic in the United States actually exceeds the country's entire annual budget; with western European countries not far behind!

## References

Adams W S 1947 *Publ. Astron. Soc. Pacific* **59** 285
Bessel F W 1841 *Astron. Abhandlungen* **I** 32
—— 1844 *Astron. Nachr.* **22** 145
Newcomb S 1903 *The Reminiscences of an Astronomer* (London and New York: Harper Bros)
Orwell G 1949 *Nineteen Eighty-Four* (London: Secker)
Smart W M 1947 in *Occasional Notes R. Astron. Soc.* **2** (no.11)
Strutt R J 1968 *Life of Rayleigh* (Madison: Univ. Wisconsin Press)

# *Index*

Abell, N H, 380
Abetti, G, 65
Abhyankar, K D, 332*f*
Adams, J Couch, 85, 457
Adams, John, 284
Adams, W S, 151*f*, 459*f*, 472
Aga Khan, 327
Agnew, Spiro, 297
Agricola, Gaius Iulius, 229
Airy, G B, 454
Aitken, R G, 219
Akcyali, M (Mrs E Hamzaoglu), 260
Al Battani (Albategnius), 439
Al Biruni, 439
Aldrin, E E ('Buzz'), 293, 295
Alfonso X (King of Castile), 420, 442
Alfvén, H, 153, 235, 300, 302*f*, 320, 378
Alfvén, K (Mrs H Alfvén), 153, 300, 320
Alladin, S H, 333
Allen, Harvey, 206
Aller, L H, 172*f*, 389
Al Mamûn (7th Caliph), 438
Al Naimiy, H M K, 328
Al Sabti, A, 328
Al Sufi, 439
Al Zelatny, 331
Ambartsumian, V A, 321, 387, 392
Amundsen, Roald, 135

Anděl, K, 269
Ando, H, 352*f*
Andoyer, H, 93*f*
Anne-Marie, Princess (later Queen) of the Hellenes, 291
Anti, G, 253
Antonakopoulos, G A, 319
Antonov, G N, 232
Archimedes, 372*f*, 464
Argelander, F W A, 91
Aristarchos, of Samos, 319–321, 418
Aristotle, of Stageira, 418
Armstrong, N A, 293
Arnošt (Arnestus) of Pardubice (Archbishop of Prague, and first Chancellor of Charles University), 4, 7, 15, 33, 82
Arp, H C, 216
Arthur, D W G, 287
Asaad, A, 344
Attlee, Clement, 241
Auwers, A F G, 91
Awadalla, N, 323, 352

Baade, W, 163, 246, 392
Babcock, Harold, 142
Babier, M, 291
Bacon, Roger, 439
Baker, J G, 162
Baker, M H, 193, 200
Baker, R H, 193

Balbín, B, 10, 12
Baldwin, S, 187
Balucinska, M, 323
Banachiewicz, Th, 210f, 315, 424
Banér, Jan (Gen.), 7
Banks, J, 406
Banos, C G, 319
Banos, G G, 319
Bappu, M K V, 216, 246, 337, 341–343
Barrande, Joachim, 53
Barbier, D, 390
Barlow, B, 339
Bartlett, M S, 242
Batten, A H, 98, 406f
Bauschinger, J, 193
Beals, C S, 296
Beattie, R, 232
Beaudoin, D, 175
Bečvář, A, 214f
Beer, A, 109, 251, 392
Beethoven, L van, 426, 464
Bellarmine, R (Cardinal), 381
Beneš, E (President), 81, 138
Beneš of Vartenberk (Lord), 11
Beran, J (Cardinal), 250
Bessel, F W, 91, 375, 414, 455–457, 472
Betz, C, 165
Betz, M (Mrs Harlow Shapley), 158, 163, 194, 224
Bevan, Aneurin, 184
Bianchini, F, 404
Biermann, L, 285
Billing, H, 260
Blaauw, A, 383
Blackett, C E, 241
Blackett, P M S, 160, 226f, 237–242, 247, 300, 392
Blegen, C, 318
Blunt, Anthony, 175
Boas, R P, 146
Bok, B J, 159, 167, 172, 174, 346
Bonev, N, 286
Bonifazi, A, 323
Borman, F, 309
Bouguer, P, 400
Brachtl, A, 47

Bradbury, Norris, 199
Bradford, W R, 259
Bragg, W L, 233, 237
Brahe, Tycho, 134, 140, 246, 442, 454
Braidwood, Th, 404
Bramble, C C, 195, 199
Brandeis, L, 264
von Braun, W, 293f, 302
Breindl, V, 51, 53
Brenton, V K, 193
Brodowski, W, 232
Brouwer, D, 210
Brown, E W, 163, 457
Brown, Harrison, 267
Brown, R Hanbury, 238, 245–247, 343, 393
Brück, H A, 147f
Brückner, B, 133
von Brunn, A, 97
Brušák, Karel, 72, 227
Buchar, Emil, 306, 309–311
Budding, E, 131, 323, 346–348, 356
Budding, P, 323
Bumba, V, 92
Burbidge, E M, 242
Burchi, R, 323
Burnham, S W, 459
Burša, Milan, 92
Bus, J S, 345
Bush, Vannevar, 217
Butler, C C, 239
Bydžovský, B, 92, 215f

Caesar, Gaius Iulius, 58
Calame, O, 287
Callandreau, O, 432, 435
Cambresier, Y (Mrs L Rosenfeld), 242
Cameron, A G W, 175, 308, 378, 382, 392
Camichel, H, 272
Camm, G L, 236
Campbell, K ('K2'), 220
Campbell, L, 63, 172
Campen, Ch, 271f
Cannon, A J, 176

*Index*

Čapek, Ema, 27
Carder, R W, 272, 276f, 301, 324, 394
Carling (Finlay), E B, 139, 244, 260, 323, 327f, 339, 355, 394
Carlyle, Thomas, 360
Carmichael, J R B, 193f
Carnegie, A, 164, 170
Carpenter, J, 395
Carrus, P A, 221, 224
Carson, D, 289, 392f
Cassini, G D, 403, 448, 454
Cather, Willa, 249
Catherine (Empress of Russia), 449
Cauchy, A, 380f
Cavendish, Henry, 398
Čech, E, 80
Cerruti-Sola, M, 323
Challis, Mrs J, 102, 457
Chalonge, D, 254
Chaloupecký, V, 5
Chamberlain, J W, 295, 299
Chamberlain, Neville, 148, 187, 202f
Chandrasekhar, S, 137, 149, 182, 184f, 211f, 224, 228, 231, 256, 387, 392, 433, 435
Chapman, S, 146, 234–236, 244, 364
Chappas, W H (Col), 287
Charles II (King of England), 453
Charles IV (Emperor), 4, 33, 78
Charles V (Emperor), 336
de Chéseaux, L, 400
Chevallier, S, 132
Chiappe, M, 107
Christian V (King of Denmark), 453
Chubb, S, 167
Churchill, W S, 201–203
Clairaut, A C, 93, 137, 393, 431–433, 435
Clark, Alvan, 63, 305, 450
Clausius, R, 381
Clavius (Kley), Ch, 453
Clegg, J A, 238, 245
Clement VI (Pope), 33
Collins, G W, 186, 224

Colombo, G, 287
Colvin, B H, 292
Compton, K T, 22, 201, 221, 223
Constantine (King of the Hellenes), 291
Cooper, W M, 228, 277
Copernicus, N, 243f, 374, 442, 464
Corben, C, 144
da Costa, A A, 260
Coulomb, C A, 398
Cowling, T G, 236f
Craig, H, 267
Crawford, J A, 258, 393
Cromwell, Oliver, 133
Cronkite, W, 294
Curran, S C, 148

Dalton, John, 229
Daniels, Price (Governor of Texas), 281
Danjon, A, 209, 224, 283
Dante, Alighieri, 33
Darwin, Ch R, 85, 463
Darwin, G H, 265, 393, 432, 435
Davidson, M E, 286f, 392f
Davies, J G, 245, 286
Davies, J K W, 259
Davies, R D, 245
Dawes, W R, 63, 305
Dee, P I, 148
Demircan, O, 323
Denis, Ernest, 141
Denisse, F, 383
Desai, M R, 335
Deslandres, H, 94, 352
Devons, S, 245
Dichter, E, 387
Dimbleby, R, 294
Dimitroff, G Z, 174
Dolejšek, V, 80f, 83
Dollfuss, A, 272, 286
Dollond, J, 449
Dopita, M, 260
Doppler, Ch, 28
Dornberger, W (Gen.), 294
Downie, C S (Col), 287
Doyle, A Conan, 148
Drtina, F, 21

Dubček, Alexander, 250
Duerbeck, H W, 347f
Dufay, J, 110
Dugan, R S, 181f
Duke, Ch M, 297
Dvořák, Antonín, 36, 70
Dvořák, Max, 140
Dvořáková, Anna, 140
Dvorník, F, 2
Dyson, J E, 244, 260

Eccles, W, 232
Eckhardt, D H, 287
Eddington, A S, 80, 85f, 96, 109, 137f, 141–146, 149, 182, 185, 190, 224, 230, 232, 236, 364, 370, 377, 393
Eddington, H, 148
Edward VIII (King of England), 196, 201
Eginitis, D, 319
Einstein, A, 80, 96, 167, 206f, 319, 363, 403
Eisenhower, D D (President), 223, 369
Elliott, E H, 239
Elliott, K H, 260
Ellison, M E, 391
Elphick, M J, 259
El-Shaarawy, M B, 260
Emden, R, 231f, 376
Emiliani, C, 267
Emler, Jan, 140
Emler, Josef, 140
Enfield, W, 404
Engels, F, 230
Eratosthenes, of Cyrene, 331, 438
Erdélyi, A, 264
Erro, L E, 174
Esclangon, E, 107
Etzel, P, 323
Euler, L, 461
Evans, E J, 232
Eve, A S, 144, 149, 233, 393

Fahim, M, 344
von Falkenhayn, E G A (Gen.), 38
Falle, S A E G, 260

Faraday, M, 85, 379f, 393
Farnsworth, A H, 193
Feil, F, 450
Ferdinand I (Emperor), 336
Fetlaar, J, 420, 435
Field, G B, 171
Fielder, G, 259
Figulus, G V, 45
Filip, J, 3
Finlay, E B, 244, 259f, 272, 277, 285, 287, 301, 339, 394
Finlay-Reid, J, 287
Firkušný, R, 72
Flammarion, C, 76
Flammarion, Mme C, 109
Flaubert, G, 28
Fleischman, A, 12
Fletcher, A, 77, 149
Florja, N, 67
Foch, F (Marshal), 221
Fontenay, P, 397
Foster, Stephen, 177
Foucault, J B L, 451
Fowler, R H, 228
Fox, P A, 221f, 224
France, Anatole, 13, 28
Frank, Ph, 206f, 216, 222, 363, 393
Franklin, Benjamin, 318
Franz I (Emperor), 449
Franz Ferdinand (Archduke), 249
Franz Josef I (Emperor), 37f, 112, 132, 363f
Fraunhofer, J, 450
Frebonia of Pernštejn, 35f
Freundlich, E F, 93, 95–99, 112, 149, 153, 177, 206f, 226f, 421
Freundlich, H, 96
Frič, Jan, 285, 305
Frič, J J, 90, 305
Fricke, W, 341, 393
Frost, E B, 459
Fukumi, N, 125

Gale, H G, 459
Galileo Galilei, 253, 374, 381, 442, 446–448, 454, 463
Galle, J G, 457

*Index*  477

Galois, E, 380
Gaposchkin, S I, 173–175
Garibaldi, G, 241
de Gaulle, Charles (Gen.), 385
Gauss, C F, 375, 380f, 455–457, 461–463, 465, 469, 471
Geake, J E, 259
Gebauer, Jan, 10, 20f
Geiger, H, 232f
Gelon II (King of Syracuse), 373
de Gentili, G, 272
Getting, I A, 189
Ghengiz Khan, 439
Giacconi, R, 291
Gimenez, A, 323, 336
Gingerich, O, 182, 224
Giscard d'Estaing, V (President), 321
Godard, O, 174
Goddard, R H, 293
Gold, T, 286, 294
Goldberg, L, 171
Goldstein, S, 228
Goodricke, Henry (father), 404, 411
Goodricke, J H, 408
Goodricke, J R D, 408
Goodricke, John, 403–415, 431, 435
Goodricke, Levina B (*née* Sessler; mother), 404
Goodricke, Sir Harry James, 412
Goodricke, Sir Henry, 412
Goodricke, Sir John (grandfather), 404, 408f
Gorman, M, 260
Gottwald, K, 28
Goudas, C L, 285, 287, 301, 315–318, 322, 343, 392–394
Graff, K, 112
Green, H E, 103, 178
Gregory, D F, 260
Grienberger, J, 453
Grotrian, W, 97
Grouiller, H, 66
Gruss, G, 269
Grzedzielski, S, 211, 312
Güdür, N, 323

Guillemin, E A, 217
Guinand, P L, 450
Güssow, M, 100, 149
Gustav Adolph (King of Sweden), 7
Guta (Queen of Bohemia), 10f
Guth, V, 112, 310
Guthnick, P, 100, 309

Haas, F, 221, 224
Hacar, B, 63
Hack, M, 323
Hadjidemetriou, J, 318
Hagihara, Y, 353
Hailsham, Lord (Quintin Hogg), 277f
Hale, G E, 159, 161, 164, 182, 217f, 365, 393, 459
Hall, A C, 251
Hall, Ch M, 449
Hallows, J, 286
Hamdy, M, 287
Hamed, A, 329
Hammurabi (Emperor), 329f
Hamzaoglu, E, 260, 323
Hanč, J, 189
Hansen, P A, 457f
Hanson, J, 277
Haramundanis, K, 184, 224
Harting, J, 420, 435
Hartl, F, 14
Hartree, D R, 227f
Hartwig, E, 90, 420, 435
Harun al Rashid (5th Caliph), 438
Harvey, M, 177
Harwood, W A, 232
Hasan, A, 331
Hashimoto, M, 119, 125
Haškovec, P M, 23, 26
Hattink, J, 355
Hattori, A, 325
Haymes, W, 260
Hazard, C, 238, 246, 393
Hazen, H L, 217
Hazlehurst, J, 259f, 340
Hearn, L, 28
Hearnshaw, J B, 355
Heinrich, W W, 91f

Hejný, R, 115, 135
Helin, E, 344f
Helwecke (Hevelius), J, 448
Herczeg, T, 347f
Heřman of Ralsko, 3
Herschel, C L, 165
Herschel, J, 450
Herschel, W F, 91, 159, 397, 399, 401, 406, 413, 431, 435, 449–451, 457
Hertzsprung, E, 185
Heydrich, R, 26
Heyrovský, J, 80, 85
Hibbs, A R, 284, 286
Hidayat, B, 394
Hieron (King of Syracuse), 373
Hieronymus (Jeronym) of Prague, 78, 103, 105
Hill, G W, 361
Hipparchos, 418, 438
Hirayama, K, 125
Hirayama, S, 125
Hirschfelder, E (Mrs J), 264
Hirschfelder, J, 264
Hitler, Adolf, 100f, 249
Hlavatý, V, 80, 94
Hodgden, L, 177
Hodgkinson, M, 259
Hoffleit, D E, 177, 193
Hoffmeister, C, 66
Holden, F, 259
Hollingsworth, G, 292
Holmes, Sherlock, 144
Hooke, R, 397
Hoover, H (President), 164
Hopgood, G, 333f
Hopmann, J, 287
Hori, G, 353
Howe, L McH, 193
Hoyle, F, 242
Hromádka, J L, 189
Hrozný, A, 140
Hrozný, B, 86
Hruška, A, 92
Hubble, E P, 163, 184
Hübnerová, M, 16
Huffer, C M, 186, 224, 263f
Huffer, E (Mrs C M), 263f

Hulaghu Khan, 439
van de Hulst, H C, 383
von Humboldt, A, 49
Hunt, M S, 272, 287
Hurban, V S (Col), 188
Hurban-Vajanský, S, 188
Huruhata, M, 174
Hus, Mistr Jan, 34, 78, 105, 132
Husák, G, 30
Hussein (King of Jordan), 297
Huygens, Chr, 397, 400, 435, 448
Hyde, Thomas, 440

Ibanoglu, C, 323
Ibn-Al-Shâtir, 442
Ibn Sina, 439
Ibn Younis, 439
Iijima, T, 323
Inamura, K, 121
Ingraham, M H, 263
Ivanowska, W, 153, 202
Izera, J, 59
Izera, V, 56, 58f
Izerová, L (Mrs Ridley), 59

Jacchia, L G, 67, 174f, 193f
Jacoby, K, 377
de Jager, C, 382, 386
James, J, 231
Jamsheddi, Naushir, 333
Jan of Středa (Bishop of Litomyšl), 33
Jan IV of Litomyšl, Bishop (and later Cardinal), 34
Janko, J U, 20
Jansky, K G, 238, 393
Jarník, J U, 20
Jaruzelski, W (Gen.), 315
Jaschek, W, 112, 125
Javorek, S, 132
Jayewardene, J R (President of Sri Lanka), 344
Jeans, J H, 76, 113, 137f, 145, 150, 364
Jeffreys, H, 370
Jelínek, G, 72f, 174, 227
Jennison, R, 245
Jirásek, A, 44

Jiří (George) of Poděbrady (King of Bohemia), 11, 38
John XXIII (Pope), 250
Jones, H S, 147, 300
Jones, M, 273
Jones, R (Mrs B G Karpov), 177
Jones, T F, 220*f*
Joule, J P, 229, 381
Joy, A H, 211
Julian, R, 287
Jurkevich, I O, 186, 224, 323, 428

Kabátník, M, 35
Kadavý, F, 64, 67, 73*f*, 307
Kadouri, T, 328
Kahn, F D, 244, 246, 259*f*, 336*f*, 340, 393
Kaiser, T, 238, 245
Kamala Mahanta, M, 260, 433, 435
van de Kamp, P, 336
Kapteyn, J C, 159
Karamanlis, K, 321
Karandikar, R V, 287, 332*f*, 335
Katayama, S, 125
Kavanagh, K E ('K1'; now Mrs Hanson), 193, 220
Kay, W, 232
Kearns, A G (Mjr), 272
Kennedy, J F (President), 187*f*, 224, 264, 271
Kennedy, J P, 187
Kepler, Johannes, 5*f*, 11, 31, 88, 374, 400, 442*f*, 447, 454, 462–464
Khomeini, Ayatollah, 330
Kiasatpoor, A, 329
Kibe, S, 117–119, 353
Killian, J R, 223
Kim, T, 353
Kimura, H, 124
Kinoshita, S, 232
Kipling, R, 183
Kippenhahn, R, 336
Kirchhoff, G R, 230
Kitamura, M, 122, 131, 261, 351–353, 382
Kitamura, M (Mrs M), 261, 352

Kizilirmak, A, 323
Klages, L, 287
Kleczek, J, 92, 305, 346*f*
Klein, F, 96
Klepešta, J, 73*f*, 301, 394
Klik, J, 51
Klír, L, 51*f*
von Klüber, H, 97
Knight, C, 260
Knowles, V, 226
Knox, John, 133
Kobold, H, 68*f*, 75, 421
Koestler, A, 303, 371, 393
Komenský (Comenius), J A, 34, 44*f*
König, R, 61*f*
Kopal, A (*née* Müldner, wife), 272
Kopal, Eva M (daughter), 272, 309
Kopal, Francis (grandfather), 9, 43, 49
Kopal, Francis (uncle), 9
Kopal, Georgiana L (Mrs W E Rudge; daughter), 201, 250, 261, 294, 312
Kopal, Jan (great-great-grandfather), 8
Kopal, Jan (great-grandfather), 8
Kopal, Josef (father), 9, 11–13, 20–25, 27, 81, 306, 364
Kopal, Josef (cousin), 3, 7
Kopal, Miloš (brother), 19, 24, 27, 38, 41, 56, 59
Kopal, Miloš Jr (nephew), 56
Kopal, Vladimír JUDr, 8
Kopal, Zdenka A (Mrs D F Smith; daughter), 49, 153, 222, 249*f*, 252, 272, 294, 308
Kopalová, Anna (*née* Pažoutová; grandmother), 3, 8*f*
Kopalová, Ludmila (*née* Lelková; mother), 13, 18, 24, 27, 310
Kopecký, M, 92
Kosmas, (Dean), 31
Kothari, D S, 332
Kotsakis, D, 319
Kozák, J, 72
Koziel, K, 210*f*, 272, 286*f*, 312–315

Kratochvíl, J, 139
Kraus, O, 20
Kresák, L, 215
Kříž, S, 92
Kron, G E, 52, 211
Krushchev, N, 282
Kruszewski, A, 211
Krzeminski, W, 211
Kuiper, G P, 286, 294, 394
Kukarkin, B V, 67
Kukula, O, 18
Kurth, R, 259
Kurutac, M, 323
Küstner, K F, 91
Kvíz, Z, 92

La Follette, P (Governor), 264
La Follette, R (Senator), 264
La Follette, R M (Senator), 264
Lamb, H, 230, 234
Lamb, P, 260
Lambert, D L, 251
Langer, R E, 263, 293
Langer, W, 263
Laplace, P S, 91, 362, 432, 436
Lassell, W, 450
Lauisberry, W C, 232
Lazarus, D, 387, 395
Leavitt, H S, 175
Lebrun, A (President), 106, 108
Legendre, A M, 432
Lejay, P, 132
Lelek, Francis Josef (great-grandfather), 14
Lelek, Josef (great-great-grandfather), 14
Lelek, Josef (grandfather), 13$f$, 18$f$, 33, 38–41, 43, 49, 99, 112
Lelková, Marie (née Neumannová; grandmother), 16, 18$f$, 42
Lemaître, G, 248$f$
Leo XIII (Pope), 453
Leonardo da Vinci, 467
Leuschner, A, 151
Levanevsky, S, 135
Leverrier, U J J, 107, 370, 457$f$
Lighthill, M J, 206, 222, 242, 244
Lin, C C, 221, 225

Lincoln, Abraham (President), 47, 200
Linfoot, E H, 98
Link, F, 112, 310
Linnell, A P, 186, 225
Liszt, Franz, 464
Livaniou, H (Mrs P Rovithis), 260, 323
Longo, G, 323
Lorell, J, 287
Lorenzi, F, 323
Loreta, E, 66
Louis XIV (King of France), 104, 107$f$, 453
Louis XVI (King of France), 330
Lovell, A C B, 238$f$, 245, 247, 340, 391, 395
Lovell, J A, 295
Lowe, Ch, 271
Lowell, A B, 162, 183
Ludmila (Saint), 2, 32
Luheshi, M, 260
Lundeberg, G, 408$f$
Luther, Martin, 34
Luyten, W J, 357
Lyot, B, 169, 270
Lyttleton, R A, 144

McCarthy, Eugene (Senator), 164
McCarthy, Joe (Senator), 164, 175
McDowell, C S (Capt.), 218
MacDuffie, C C, 263
Mach, Ernst, 28, 431
Mack, J, 264
McKinley, W R, 238
McMath, R R, 270, 395
MacRae, D B, 174
MacRae (Mjr), 319
Mahdy, S, 323
Mahler, K, 242
de Maintenon, Mme, 107
Makower, W, 232
Malinovsky, R (Marshal), 282
Mammano, A, 323
Manfredi, E, 403
Mannion, C, 260
Mantegazza, L, 323

Mao-tse Tung, 350*f*
Maraldi, G F, 403
Marcus Aurelius (Emperor), 1
Mareček, J, 132
Marinatos, S, 318, 322
Markvart of Březno, 3, 11
Marsden, E, 232
Marshak, R E, 178
Martis, N, 321
Martyn, W Ross, 284
Marx, Karl, 230
Masaryk, Alice, 22
Masaryk, Herbert, 22*f*
Masaryk, Jan, 22, 140
Masaryk, Olga (Mrs Revilliod), 23
Masaryk, Thomas G (President), 18, 20-23, 30, 51, 85, 140, 197, 364
Mašek, B, 150
Maslowski, J, 284, 314
Mason, A C, 286
Mason, D, 260
Massey, H S W, 285
Matoušek, L, 64
Matoušek, O, 100
Mattingly, T K, 297
Maury, A C, 176
Max, F, 132
Maximillian II (Emperor), 15
Maxwell, A, 384
Maxwell, J Clerk, 381, 463
Mazzini, G, 241
Meaburn, J, 244, 255, 260, 334, 337, 340, 395
Mendis, A D, 344
Menzel, D H, 168-171
Mercouri, Melina, 322
Merrill, J E, 186*f*, 208, 211, 424, 436
Mersenne, Marin, 374
Methodios (Saint), 32
Meyer, D, 287
Michelangelo Buonarotti, 467
Michell, J, 397-401, 404, 412*f*, 436
Michelson, A A, 459*f*
Mietelski, J, 287, 314
Mikhailov, A A, 97, 285*f*

Mikhailov, Z K (Mrs A A), 286, 301, 394
Mikuláš of Litomyšl, 34
Milano, L, 323
Miller, S, 267
Millns, P Y, 253*f*
Mills, G A, 287
Milne, E A, 234, 364, 433, 436
Milne-Thomson, L M, 264
Minnaert, M G J, 283, 287, 386
Miyamoto, Sh, 324*f*
Mohr, Jenka, 171
Mohr, J M, 310
Montanari, G, 402*f*, 412
Moore, Ch E (Mrs W B Sitterly), 186, 225
Mossadegh, M, 330
Motamedi, G, 329
Moulton, F R, 195, 460
Mourek, J, 20
Moutsoulas, M D, 287, 300, 317-324, 340, 378
Mozart, W A, 464
Mussolini, Benito, 247
Myers, G W, 420, 436
Mylonas, G E, 318

Nagaoka, H, 125
Nagayo, H, 125
Nagy, B, 267
Napoleon I (Emperor), 107, 454
Nasir-el-Din, 439
Nasmyth, J, 395
Navrátilová, M, 95
Nechvíle, V, 83, 93-95, 104, 112*f*, 153, 306, 310*f*, 421, 433
Nechvílová, Z, 95
Nehru, Jawaharlal, 161, 331
Nehru, Mme Pandit, 161
Nejedlý, Z, 306*f*
Nelson, B, 323
Němcová, Božena, 18
Němec, B, 80
Němec, Z, 54, 69-71
Neumann, I, 15
von Neumann, J, 196, 198, 206
Neumannová, A, 15

Newcomb, S, 57, 180, 457–459, 472
Newell, H E, 268, 395
Newman, M, 242
Newton, Isaac, 361, 372, 448*f*, 461–465, 469, 471
Niarchos, P, 323
Nielsen, A, 67, 74
Nijland, A A, 68, 110
Nixon, R M, 264
Nováková, B, 112
Nowicki, A, 287
Nušl, F, 59, 64, 80, 83, 89, 90, 92, 137, 139*f*
Nuttall, J M, 233

Oberth, H, 293*f*
O'Connell, D J K, 248*f*
Odložilík, O, 189
Okazaki, A, 352
Olczak, T, 125
Olle, T W, 259
Olson, E C, 434, 436
Oort, J H, 246, 383, 395
Oppenheimer, J R, 199
Orszag, A, 287
Orwell, G, 254, 465*f*, 472
Ostrowski, A M, 264
Otmianowska, K (Mrs Mazur), 323
Owen, J W, 259

Paargeeter, J, 289
Pacholczyk, O, 211
Paczynski, B, 211
Paganini, Niccolo, 464
von den Pahlen, E, 97
Palacký, František, 9*f*, 51, 86
Páleníček, J, 71*f*
Palmer, H, 245
Panagiotopoulou, H (Mrs F Hörz), 260
Pannekoek, A, 420, 436
Papadopoulos, G (Col), 317
Papandreou, A, 322, 340
Parenago, P P, 67, 74
Parr, Geoffrey, 391
Parsons, W J (Vice-Admiral), 194
Pastori, L, 323

Patakos, S (Gen.), 320, 322
Patterson, F S, 174
Payne, P ('Master English'), 78
Payne-Gaposchkin, C H, 173–175, 211*f*
Pažout, V, 9
Pearce, J A, 211
Pecker, J C, 383–386
Pierce, B O, 365
Pekař, J, 9*f*, 25, 51, 86
Peraiah, A, 333
Perek, L, 92, 307
Perrault, C, 104
Pétain, H P (Marshal), 107
Petrarca, Francesco, 33
Petrie, R M, 211*f*, 395
Petržílka, V, 148*f*
Petty, A F, 323, 428
Pickering, E C, 161*f*, 179, 182, 225, 243, 365, 397, 414, 419, 421, 433, 436
Pierce, N L, 211
Pigott, E, 411*f*
Piotrowski, S L, 168, 210*f*, 225, 311*f*, 424, 436
Pişmiş, P, 174
Pius XII (Pope), 249
Planck, Max, 207
Plassmann, J, 67
Plavec, M, 92, 259, 305*f*, 311*f*, 382, 428
Poincaré, H J, 91, 315, 461–463, 465, 469
Polanyi, M, 303
Pols-van der Heijden, N M, 337*f*, 355
Porro, A, 403, 436
Pottasch, S R, 382–386
Pračka, L, 59, 63, 90
Prager, A, 205
Prager, K (Mrs R), 205
Prager, R, 65, 67, 74, 100, 174–176, 193*f*, 205, 421
Pražák, A, 211
Přemysl Otakar II (King of Bohemia), 36
Prey, A, 96
Pring, J N, 232

Proctor, D D, 186, 225
Prokeš, V, 51
Ptolemy, 397, 438
Purkyně (Purkinje), J E, 36

Rackham, T W, 272f, 276, 286f, 301, 326f, 394
Rahe, J, 394
Rahimi-Ardabili, M Y, 433, 436
Rais, K V, 14
Rajchl, R, 63
Rallis, G, 320, 340
Ramadhan, T Y, 341
Ramsey, W H, 244
Rao, P V, 323, 333
Raphael, Santi, 467
Rayleigh, J W Strutt (3rd Lord), 85, 230, 380f
Raymond, A, 124f
de Récamier, Mme, 18
Redman, R O, 124, 147, 296
Regiomontanus (Müller, J), 442
Reidel, A, 355, 378, 382f, 385f
Reidel, D, 378
Reissner, E, 377
Riccardi, N, 381
Riccioli, G B, 397, 412
Richaud, P, 397
Richmond, M, 340
Riefenstahl, E ('Leni'), 100
Rifaat, A (Mrs Molokhia), 287
Ring, J, 253f, 259, 272, 286, 395
Ritchey, G W, 94
Roberts, A W, 420, 436
Roberts, L, 291
Roberts, W O, 170
Roche, E A, 255f, 395
Rochester, G R, 239
Roediger, C, 420, 436
Roemer, Chr, 419
Rolland, Mme R, 30
Rolland, Romain, 28
Romeo, G, 323
Ronca, L B, 293
Roosevelt, Eleanor, 201
Roosevelt, Elliott, 200
Roosevelt, F D (President), 193, 199–201

Rösch, J, 272f, 286f, 300
Roscoe, H, 229f
Rosenfeld, L, 242
Rossi, R, 232
Roustan, M, 106
Rovithis, P, 323
Royds, T, 233
Rucinski, S, 211
Rudge, Julia K, 294
Rudge, W E IV, 261
Rudge, W E V ('Robin'), 294
Rudolf (Crown Prince of Austro-Hungary), 38
Rudolf of Habsburg (Emperor), 10
Ruffini, R, 336
Russ, S, 232
Russell, A, 180
Russell, H N, 40, 108f, 153, 157, 166f, 178–186, 206–213, 225, 255, 365, 388, 420–425, 434, 436
Rutherford, E (Lord), 85, 144, 230–234, 240, 366f

Saari, J, 293
Sadik, A, 328
Şadler, D H, 365, 383, 395, 435
Šafařík, P J, 63
Šafařík, V, 63, 305
Šafaříková, P, 305
Şagan, C, 378
Salda, F X, 21, 25f, 28f
Salisbury, J W, 272
Samaha, A H, 287, 325
Samii, A H, 329
Sand, George, 18
Sargent, W L W, 259f
Sarton, G L, 222
Sasser, J, 287
Sato, T, 117, 122f
Saward, D, 395
Sawyer, H E, 395
Sawyer-Hogg, H B, 177
Schatzman, E, 383–385, 389
Scheglov, V P, 283
Schiaparelli, G V, 428
Schiller, Friedrich, 361
Schiller, K, 420f

## 484  Index

von Schleicher, K (Gen.), 101, 106
Schlesinger, F, 90, 106, 150, 183
Schmidt, J F J, 319
Schmitt, H H, 296f
von Schrutka-Rechtenstamm, G, 272
Schüller, F, 63
Schumacher, H C, 375
Schuster, A, 230–233, 243, 395
Schwarzschild, K, 91, 231f, 395
Schwarzschild, M, 172, 231
Scott, Walter, 45
Scouler, J, 408, 410
Seares, F H, 157
Sedmak, G, 323
Seitter, W C (Mrs H W Duerbeck), 347f
Sekanina, Z, 92
Sekiguchi Mrs N, 352
Sekiguchi, N, 352
Sekiguchi, R, 125
Semeniuk, I, 211
Serkowski, K, 211
Shah-rukh, 439, 441
Shakespeare, William, 53, 105
Shapiro, I, 171
Shapley, Alan, 167
Shapley, Carl, 167
Shapley, H, 108, 148, 151, 153, 155–165, 170f, 177–179, 182f, 188–192, 208–213, 216, 221, 223, 225, 248, 307, 314, 388, 420, 424, 436
Shapley, Lloyd, 167
Shapley, M B (Mrs H), 153, 166f, 193f, 205, 216
Shapley, Mildred (Mrs Matthews), 167
Shapley, Willis, 167
Sharma, R C, 333
Sharonov, V V, 286
Sharpe, G, 440
Shepherd, A, 405
Shibata, N, 125
Shoemaker, E M, 286, 294
Shorthill, R W, 293
Sidky Mikhail, J, 325, 327
Siegbahn, M, 81

Sienkiewicz, H, 45
Silva, G, 258
Simpson, Wallis, 102, 196
de Sitter, W, 265, 395
Sitterly, B W, 420, 436
Skála (Rocher), K, 23, 77f
Slade, R L, 232
Slouka, H, 91, 105, 112, 125, 422
Smak, J, 211
Smart, W M, 102, 150, 472
Smetana, Bedřich, 36, 426
Smith, D F, 308, 356
Smith, F G, 383
Snow, C P, 389
Sobouti, J, 329
Šorm, F, 308f
Spitzer, L, 388
Stalin, J V, 27, 160, 201, 203
Stanovský, O (Mgr), 249f
Stansfield, H, 232
Stassen, H, 202
Šťastný, M, 43
Stebbins, J, 263
Štefánik, M R (Gen.), 61, 356
Stein, J W, 248, 420, 436
Steinberg, J L, 383
Stelčovský, M, 56
Šternberk, B, 91, 305–307, 309, 311
Steward, Balfour, 230
Stiefel, E L, 264
Stoddard, L G, 286, 289, 392
Stodolkiewicz, J, 211, 312
Stone, A H, 205
Stoner, G, 291
Stopford, Sir (later Lord) John S B, 243
Štorek, J, 113, 137
Stoy, R H, 287
Strand, K Aa, 357
Stratton, F J M, 124f, 146f, 244
Strawinski, I, 167
Strömgren, B, 258
Strutt, R J (4th Baron Rayleigh), 395, 472
Struve, F G W, 400, 436
Struve, H, 96
Struve, O, 211f, 218, 258f, 332, 388, 390, 395

Sudbury, B H (Mrs P V), 287
Sudbury, P V, 273f, 287
Suess, H, 267
Sullivan, W T, 395
Suthers, J M, 339
Švestka, Z, 92, 305, 310
Swaminathan, R, 333
Swings, P, 242, 258, 285, 377f, 383
Swope, H H, 176

Tacitus, Publius Cornelius, 1, 229
Tagaris, A (Gen.), 317f
Tai, Wen-san, 144
Takahashi, M, 122
Takeda, Shin-ichiro, 116, 389, 395, 421f, 434, 436
de Talleyrand, Ch, 362
Taqui-al-Din, 441
Tebutt, N, 260
Teleki, G, 341, 393
Teller, E, 281
Temple, B, 273
Tenorio-Tagle, G, 260
Teplý, F, 7
Thackeray, A D, 147
Thales, 419
Theodorakis, M, 324
Thompson, W (Lord Kelvin), 85, 370
Thomson, J J, 230
Thurn, Henry Matthias (Count), 11
Tille, V, 19
Timur Leng (Tamerlane), 439, 441
Tomášek, V J, 70
Tomkin, J, 251, 395
Torstenson, Leonard (Field-Marshal), 7
Trakowski, J W (Lt Col), 271
Trčka, Adam Erdman (Count), 11
Tsatsos, Mme K, 319
Tsesevich, V P, 387, 395
Tsiolkovsky, K E, 293
Tsouroplis, A G, 323
Tuomikoshi, T, 232
Tuominen, J, 174
Turing, A, 242
Twiss, R W, 247

Tyson, M, 227

Ulugh Bek, M T, 439–442
Underhill, A, 386
Urey, F (Mrs H C), 299
Urey, H C, 265–270, 280, 284, 286, 299f, 301, 316f, 378
Ursell, F, 222

Václav (Wenceslas) (Saint), 2, 78
Václav II (King of Bohemia), 10
Václav IV (King of Bohemia), 11
Vand, V, 67f, 74, 76f, 150
Vanýsek, V, 92
Vashakidze, M A, 246
Verne, Jules, 31, 47f
Vibert Douglas, A, 150
Victor Emmanuel III (King of Italy), 249
Viktorin, V, 149
de la Villemarqué, E, 132
Vinogradov, A P, 284, 296
Vinta, Belisario, 446f
Vjatkin, J L, 440
Vlček, J, 20, 112
Vogel, H C, 402, 406, 413
Vogt, H, 420, 436
Vojan, E, 16
Vojtěch (Adalbert) (Saint), 2
Vorontsov-Velyaminov, B A, 67
Vratislav of Pernštejn (Lord), 35
Vrchlický, J, 21, 28f
Všetečka, J, 18

Waland, R L, 98
Wallenstein, Albrecht (Duke of Friedland), 5f, 9–11
Walraven, Th, 246, 395
Walter, K, 137, 150, 258, 323, 395, 434, 436
Washington, G (President), 232
Wasserburg, G, 267
Waterston, J J, 381
Wavell, A P (Gen.), 160
Wayman, P, 343
Weigner, K, 138f
Weimer, Th, 272, 287
Weiss, E, 90

Whaley, F W, 232
Wheeler, F, 260
White, M, 232
Whittaker, E T, 142
Wick, G C, 231, 387, 395
Wickramasinghe, Ch, 344
Wickson, G, 185
Wiener, L, 197
Wiener, N, 197–199, 428
Wilcock, W L, 252
Wilhelm IV (Landgrave von Hesse), 442f
Wilkes, M V, 227
Williams, Marjorie, 193
Wilson, A G, 332, 377
Wilson, Ch (Lord Moran), 201
Wilson, H, 241
Wilson, R, 260
Wilson, Thomas Woodrow (President), 157, 180
Wilson, W, 232
Windischgrätz (Princess Ludwig), 38
Windsor, Duke of, 196f, 201
Winthrop, J, 45
Wirtz, C, 91
Wolf, A, 51
Wolf, F, 141, 146
Wolfendale, A, 239

Woolf, N J, 259
Woolley, R van der R, 146f, 365
Wright, F W, 177
Wurm, K, 97
Wycliffe, J, 78

Xanthakis, J, 318f

Yamamoto, N (Mrs Y), 120
Yamamoto, Y, 116, 324, 356
Yamasaki, A, 352
Yamasaki, S (Mrs A), 352
Yerkes, Ch T, 367
Yoshimura, H, 353
Young, J W, 297f

von Zach, F X, 374
Zafiropoulos, F, 321, 323
Zagar, F, 435
Žakavec, F, 23
Záviška, F, 80f, 83, 112, 137
Zdík, J (Bishop), 33
von Zeipel, H, 433, 436
Zinnecker, E, 347f
Žižka, Jan, 133
Zurhellen, W, 96
Zverev, M S, 286
Zwicky, F, 184, 225